ANTS

STANDARD METHODS
FOR MEASURING
AND MONITORING
BIODIVERSITY

Biological Diversity Handbook Series

Series Editor: Don E. Wilson

This series of manuals details standard field methods for qualitative and quantitative sampling of biological diversity. Volumes focus on different groups of organisms, both plants and animals. The goal of the series is to identify or, where necessary, develop these methods and promote their adoption worldwide, so that biodiversity information will be comparable across study sites, geographic areas, and organisms, and at the same site, through time.

ANTS

STANDARD METHODS FOR MEASURING AND MONITORING BIODIVERSITY

EDITED BY DONAT AGOSTI, JONATHAN D. MAJER,
LEEANNE E. ALONSO, AND TED R. SCHULTZ

SMITHSONIAN INSTITUTION PRESS
WASHINGTON AND LONDON

Copy editor and typesetter:
    Princeton Editorial Associates, Inc.
Production editor: Duke Johns
Designer: Amber Frid-Jimenez

Library of Congress Cataloging-in-Publication Data

Ants : standard methods for measuring and monitoring
    biodiversity / edited by Donat Agosti . . . [et al.].
        p.   cm.—(Biological diversity handbook
        series)
    Includes bibliographical references (p.   ).
    ISBN 1-56098-858-4 (cloth : alk. paper)—
    ISBN 1-56098-885-1 (pbk. : alk. paper)
    1. Ants—Speciation—Research.   2. Biological
diversity—Measurement.   I. Agosti, Donat.
II. Series.
QL568.F7 A575   2000
595.79′6—dc21   00-021953

British Library Cataloguing-in-Publication Data available

Manufactured in the United States of America
07 06 05 04 03 02 01 00   5 4 3 2 1

∞ The paper used in this publication meets the
minimum requirements of the American National
Standard for Information Sciences—Permanence
of Paper for Printed Library Materials
ANSI Z39.48-1984.

This book is dedicated to the memory of William L. Brown Jr.,
with affection, respect, and gratitude. For the inspiration you provided,
for the firm foundation you built for ant systematics, and especially
for your generous soul and irreverent good humor, we will never forget you, Bill.

Contents

List of Figures   ix
List of Tables   xiii
Foreword   xv
Edward O. Wilson

Preface   xvii

Chapter 1
Biodiversity Studies, Monitoring, and Ants: An Overview   1
Leeanne E. Alonso and Donat Agosti

Chapter 2
A Primer on Ant Ecology   9
Michael Kaspari

Chapter 3
A Global Ecology of Rainforest Ants: Functional Groups in Relation to Environmental Stress and Disturbance   25
Alan N. Andersen

Chapter 4
The Interactions of Ants with Other Organisms   35
Ted R. Schultz and Terrence P. McGlynn

Chapter 5
Diversity of Ants   45
William L. Brown Jr.

Chapter 6
Ants as Indicators of Diversity 80
Leeanne E. Alonso

Chapter 7
Using Ants to Monitor Environmental Change 89
Michael Kaspari and Jonathan D. Majer

Chapter 8
Broad-Scale Patterns of Diversity in Leaf Litter Ant Communities 99
Philip S. Ward

Chapter 9
Field Techniques for the Study of Ground-Dwelling Ants: An Overview, Description, and Evaluation 122
Brandon T. Bestelmeyer, Donat Agosti, Leeanne E. Alonso, C. Roberto F. Brandão, William L. Brown Jr., Jacques H. C. Delabie, and Rogerio Silvestre

Chapter 10
Sampling Effort and Choice of Methods 145
Jacques H. C. Delabie, Brian L. Fisher, Jonathan D. Majer, and Ian W. Wright

Chapter 11
Specimen Processing: Building and Curating an Ant Collection 155
John E. Lattke

Chapter 12
Major Regional and Type Collections of Ants (Formicidae) of the World and Sources for the Identification of Ant Species 172
C. Roberto F. Brandão

Chapter 13
What to Do with the Data 186
John T. Longino

Chapter 14
The ALL Protocol: A Standard Protocol for the Collection of Ground-Dwelling Ants 204
Donat Agosti and Leeanne E. Alonso

Chapter 15
Applying the ALL Protocol: Selected Case Studies 207
Brian L. Fisher, Annette K. F. Malsch, Raghavendra Gadagkar, Jacques H. C. Delabie, Heraldo L. Vasconcelos, and Jonathan D. Majer

Appendix 1
List and Sources of Materials for Ant Sampling Methods 215

Appendix 2
Ant Survey Data Sheet 219

Appendix 3
List of Materials for Ant Specimen Processing 221

Glossary 223
Ted R. Schultz and Leeanne E. Alonso

Literature Cited 231

Contributors 271

Index 275

Figures

Figure 1.1. Composition of total animal biomass and species composition of insect fauna near Manaus, Brazil   4

Figure 3.1. Classification of communities in relation to stress and disturbance   27

Figure 3.2. Functional group model of ant community organization in relation to environmental stress and disturbance   29

Figure 3.3. Behavior of ants at tuna baits at desert, woodland, and forest sites in southeastern Arizona, illustrating high, moderate, and low levels of behavioral dominance, respectively   30

Figure 3.4. Effects of vegetation on functional group composition in the monsoonal tropics of northern Australia and in cool-temperate southern Australia   32

Figure 3.5. Effects of disturbance on functional group composition of rainforest ants in the humid tropics of Queensland   33

Figure 5.1. Body parts of an ant   70

Figure 7.1. Abundance of 11 species of ant guard on *Calathea ovandensis* in a Neotropical rainforest   91

Figure 7.2. Abundance of two harvester ants, *Pogonomyrmex occidentalis* and *P. salinus,* over 15 and 9 years, respectively, in North America   92

Figure 7.3. Changes in two Chihuahuan desert ant assemblages   93

Figure 7.4. Model illustrating the response of an ecosystem to perturbation   93

Figure 7.5. Relationship between the number of ant species in 3-year-old rehabilitated mines and annual rainfall for a range of sites throughout Australia   94

Figure 7.6. Pattern of recolonization of ants in rehabilitated sand-mined areas on North Stradbroke Island, Queensland   95

Figure 8.1. Locations of 110 leaf litter collection sites   106

Figure 8.2. Species richness of ant leaf litter samples as a function of latitude   110

Figure 8.3. Species richness of ant leaf litter samples as a function of altitude   110

Figure 8.4. Proportion of ant species in a sample belonging to the subfamily Ponerinae as a function of latitude   120

Figure 9.1. An aspirator   125

Figure 9.2. A tuna bait monopolized by *Solenopsis xyloni* in a desert grassland in New Mexico   127

Figure 9.3. A pitfall trap placed in desert soil, a polypropylene sample container used as a pitfall trap, and a pitfall-trap scoop   129

Figure 9.4. Construction of the litter sifter, external dimensions of the "mini-Winkler" sack, and construction of the "mini-Winkler" sack   134

Figure 9.5. Leaf litter extraction using the Winkler extractor   135

Figure 9.6. The pattern used to create a Berlese funnel and the appearance of the assembled funnel   137

Figure 10.1. Assessment of each of 17 leaf litter ant sampling methods in Brazil   150

Figure 10.2. Assessment of Winkler sampling methods in Brazil   152

Figure 11.1. An ant mounted on a point and pinned with labels   159

Figure 11.2. Sample labels for mounted and alcohol specimens   161

Figure 11.3. Mounted ants arranged in unit trays in a collection drawer   169

Figure 13.1. Raw versus smoothed species-accumulation curves from 16 Berlese samples of litter-soil cores   190

Figure 13.2. Cost in samples of adding an additional species to an inventory   191

Figure 13.3. Within-habitat versus combined species-accumulation curves for the Berlese data   191

Figure 13.4. Comparing methods that differ in cost   192

Figure 13.5. Rank abundance plot from Berlese data   193

Figure 13.6. Hypothetical communities with contrasting species-accumulation curves   195

Figure 13.7. Lognormal distribution of species abundance   196

Figure 13.8. Contrasting relative abundance distributions for the Berlese data   197

Figure 13.9. Michaelis-Menten estimates of species richness based on the Berlese data   199

Figure 13.10. Chao2 estimates of species richness based on the Berlese data   200

Figure 15.1. Assessment of leaf litter ant sampling technique for the most species-rich site, that at 825 m on the Masoala Peninsula, Madagascar   209

Figure 15.2. Frequency of capture for species in nine 9-m^2, 16-m^2, and 25-m^2 plots in Malaysia   210

Figure 15.3. Species-accumulation curves for the three plot sizes of 9 m^2, 16 m^2, and 25 m^2 in Malaysia   210

Figure 15.4. Dendrograms comparing different sampling methods by ant species trapped in India   211

Figure 15.5. Mean number of ant species sampled by pitfall traps, Winkler sacks, and both methods combined along ten transects extending from the field into the rainforest in Bahia, Brazil   212

Tables

Table 1.1. The importance of ants   3

Table 3.1. Ant functional groups in relation to stress and disturbance, with major representatives in Australia and the New World   28

Table 3.2. Generalized myrmicines as subdominant ants to Dominant dolichoderines   31

Table 4.1. Major exotic tramp and invasive ant species   43

Table 5.1. Distribution, biology, and ecology of the ant genera of the world   46

Table 6.1. Correlations between ants and other taxa in changes in species richness across plots along a disturbance gradient   82

Table 6.2. Comparisons between the species richness of ants and the species richness of other taxa in Australia   83

Table 8.1. List of Winkler leaf litter collection sites   101

Table 8.2. Summary of Winkler leaf litter samples: taxonomic content   107

Table 8.3. Forty ant genera most frequently encountered in the survey   108

Table 8.4. Multiple regression analysis of sample species richness   109

Table 8.5. Data on Winkler samples from different biogeographic regions   111

Table 8.6. Analysis of covariance of sample species richness   112

Table 8.7. Analysis of covariance of sample genus richness   112

Table 8.8. Percentage of Winkler samples in which one or more species of a given subfamily were present   113

Table 8.9. Mean number of species per subfamily per Winkler sample   113

Table 8.10. Mean proportion of species per subfamily per Winkler sample   113

Table 8.11. Most frequent genera in each biogeographic region   114

Table 8.12. Distribution and prevalence of leaf litter ant genera in different biogeographic regions   116

Table 9.1. Relative efficacies of field techniques used to study ants   143

Table 9.2. Percentage of species recorded uniquely by different techniques in single communities   143

Table 10.1. Actual number of ant species sampled by 17 methods   149

Table 10.2. Observed number of ant species evaluated at different sample sizes for each of the 17 sampling methods   151

Table 10.3. Observed number of ant species evaluated at different sample sizes for the extended Winkler sampling experiment   153

Table 10.4. Combinations of two and three sampling methods that obtained the maximum number of ant species in the sampling methods experiment   153

Table 11.1. Ant genera with at least one species in which the worker caste is divided into physical subcastes   163

Table 11.2. Rules for submitting specimens for identification   166

Table 12.1. Major Formicidae type and regional collections of the world   176

Table 13.1. Example data set from Berlese samples   189

Table 13.2. Calculating a species-accumulation curve   189

Table 13.3. Diversity indexes from Berlese samples   194

Table 13.4. Complementarity of paired Berlese samples   201

Table 15.1. Observed number of ant species evaluated at different sample sizes for Winkler sacks, pitfall traps, and both methods for each 800-m zone site in Madagascar   208

Table 15.2. Number of ant species per square meter for 9- and 25-m^2 plot samples in Malaysia   210

Table 15.3. Number of ant species collected using three different sampling methods in forest fragments near Manaus, Brazil   213

Foreword

Edward O. Wilson

The development of rapid, consistent techniques for the study of the systematics and biogeography of ants is important—if for no other reason than because these insects are ecologically dominant in almost every terrestrial environment around the world. Ant species, although they constitute only 1.5% of the known global insect fauna, make up as much as 10% or more of the total animal biomass in tropical forests, grasslands, and probably other major habitats. They are among the principal predators of arthropods across virtually the entire size range down to and including oribatid mites, and they are the prevailing scavengers, by a wide margin, of small terrestrial animals of all kinds. Worldwide they turn and enrich more soil than earthworms.

For these reasons alone this volume contributes significantly to studies of the organization of terrestrial ecosystems. But there is a second, even more compelling, reason for its publication. The natural ecosystems in which terrestrial plant and animal species live are vanishing before our eyes. Tropical rainforests in particular, which harbor more than half the species diversity of plants and animals, are being destroyed at the rate of 1–2% annually, while the remaining forests are being fragmented and degraded at an even faster rate. In desperate response, systematists and biogeographers must lead the way in monitoring this destruction: theirs is an indispensable role in global conservation.

A key part of the global conservation effort is the exact mapping of biodiversity in order to pinpoint the hot spots, which are the most seriously threatened ecosystems as well as those with the largest number of unique species. To date the bulk of the data for this purpose has come from vertebrates, flowering plants, and a few especially well-studied invertebrate groups, such as butterflies and mollusks. Comparative studies have shown that no one such "focal group" will serve as an adequate proxy for all the rest. Therefore the sampling net must be thrown more widely in future surveys.

Ants are a logical focal group to add to the field biologists' repertoire. As the contributors to this volume repeatedly document, they are among the most abundant and easily collected of all animals. They are, moreover, diverse—but not so diverse as to be taxonomically confounding. The number of species in a square kilometer of lowland Neotropical forest, for example, ranges from as low as 20 or 30 (in the Lesser Antilles) to over 500 (in the Peruvian Amazon). Localness of distribution and endemicity are also relatively high. Being highly social, having colonies forming "superorganisms" with unique patterns of resource control, ants will add a distinctive new life form and possibly new patterns of diversity to the roster of focal groups. Last but far from least, they are perennial, and their foragers and nests are relatively easy to find even in the driest or wettest of seasons.

This book is an important contribution to defining and standardizing the methodology of ant collection techniques. It represents an essential step toward realization of the great potential of ants in biodiversity surveys around the world. It will serve as a stimulus and set the standard, not only for myrmecologists but also for ecologists and conservation biologists engaged in biodiversity surveys.

Preface

The Biodiversity Challenge

Biological diversity is the term used to describe the variety of life forms on earth. It encompasses at least three levels of diversity: those of the individual, the species, and the ecosystem (Convention on Biological Diversity 1992). This book focuses on the species level. The study of the distribution and abundance of species forms the core of the biological sciences, encompassing systematics, biogeography, ecology, and evolutionary theory. The study of species distributions and abundances also provides a wealth of information about the state of particular environments.

The increasing pressure exerted by the human population on the environment has resulted in an accelerating decline in global biodiversity. This decline has led not only to the Convention on Biological Diversity, a legally binding in-

strument for conserving biodiversity, but it has also resulted in an urgent need for mechanisms for surveying our existing biological resources. Such surveys are essential for establishing the legislation and other protocols for conserving biological resources for the future.

The accurate assessment of the biological resources present at a given time and place forms the basis of most conservation decisions. However, owing to the lack of standardization of survey protocols, the data produced in such assessments rarely contribute to the continued monitoring of biological resources over time or to the comparison of biological resources across different locations (United Nations Environment Programme 1995). The most widely used distribution models are based on extrapolations from a minimal number of field observations and thus suffer from the problem of biased field surveys. As global remote-sensing (satellite) land-cover data sets become available, it is important to have complementary biodiversity information. Such remote-sensing data sets, combined with field surveys, provide powerful tools for biological resource assessments.

The measurement of biodiversity is a complex matter. Whereas no single group of organisms can indicate the full range of biodiversity at a particular site, it is also impossible to survey all the organisms at any location. One solution to this problem is to choose to survey those groups of organisms that are ecologically important, relatively easily collected in a standardized way, reasonably diverse at the site, identifiable, and for which a critical minimum amount of scientific information is available, both in the form of publications and professional expertise.

Ants, particularly ground-dwelling ants, are the perfect candidates for such an approach. Though ants do not have the popular appeal of birds, large mammals, butterflies, or flowering plants, we predict that they will play an increasingly important role in conservation planning throughout the coming century.

Investing now in the assessment of ant diversity will prove a wise strategy for the long-term future of conservation efforts. As ants are increasingly used in biodiversity studies, the backlog of undescribed species will rapidly diminish. Stabilized species concepts and the use of a standardized collecting protocol will permit the comparison of local surveys across broad geographic distances, and such comparisons will provide an important tool for understanding such global phenomena as climate change. The information superhighway will permit the development of a leaf litter sample bank, similar to GenBank, through which researchers across the globe currently share gene sequences. The relative ease with which ants can be surveyed makes them an ideal taxon for the assessment of areas for which virtually no biodiversity data exist, such as large parts of the Amazon and Congo basins.

Origin of This Book

The use of ants in conservation and biodiversity research was seriously discussed for the first time at the International Conference of the International Union for the Study of Social Insects (IUSSI) in Paris in 1995. The study of ant diversity was subsequently proposed as a new approach to conservation from within the Social Insects Specialists Group of the Species Survival Commission of the World Conservation Union (SSC/IUCN). The strong positive responses from these organizations led to the Ants of the Leaf Litter (ALL) conference on the use of ants in biodiversity studies, upon which this manual is based.

In August 1996, 24 scientists from around the world met in the idyllic surroundings of Ilhéus in the Atlantic rainforest of Bahia, Brazil. Each was invited to elaborate on a specific topic related to the overall question of using ants as a focal taxon in biodiversity and conservation research. After a week of lively discussions, we all agreed that ants could serve well in this role, and a standard protocol was born.

That protocol, the ALL Protocol—which is the main focus of this manual—is much narrower than those elaborated in previous books in this series. The consensus of the ALL conference participants was to produce a manual that encouraged a standard protocol for collecting ants rather than a manual that described all possible ant-collecting methods, thereby leading to the production of many data sets that could be readily compared.

Thanks to Jacques Delabie and his crew in Bahia, a large data set was available for testing various combinations of collecting techniques (Chapter 10). The result is a simple but powerful tool: the 1-m^2 leaf litter sample, the minimum common denominator. This is the minimum area from which a number of ant species large enough to be meaningful in statistical analyses can be collected. The number of samples needed to obtain a reliable estimate of the expected total number of species for a given surveyed locality was also determined from this data set. This square-meter sampling approach allows extrapolations of the number of nests per area and of the total biomass. The modular protocol can be supplemented with other collecting techniques, such as pitfall traps or baiting, to address any specific objectives of a particular biodiversity study.

Our focus on ground-dwelling ants is based on the unique features of this fauna and the specific, tested methods by which it can be surveyed. Adequate methods for the standardized sampling of ants in the canopy and on vegetation have not yet been developed. We feel that a wider focus that includes these ant faunas is inappropriate for this manual, because it would dilute the essential strength of the ALL protocol: the ability to collect rigorously comparable biodiversity data.

The vitality of the standard protocol proposed in this manual is demonstrated by its increasing application worldwide. The protocol was largely developed in tropical and subtropical regions and it is in these regions that it is most effective. The decreasing diversity of ants with increasing latitude obviously sets some limitations on its application in colder regions. Increased aridity sets another limitation. Preliminary results, however, show that sifting leaf litter in savanna ecosystems is entirely possible, and yields interesting results. Other collecting methods, detailed in the manual, can add more information to surveys in such areas.

ACKNOWLEDGMENTS

The ALL conference was supported by, and would not have been possible without, the generous help of the U.S. National Science Foundation Biological Survey and Inventory Program; the Center for Biodiversity and Conservation and the Office of the Provost of Science at the American Museum of Natural History in New York; Edward O. Wilson; the Centro de Pesquisas do Cacau, Comissão Executiva do Plano de Lavoura Cacaueira (CEPEC/ CEPLAC), Ilhéus, Brazil; the State University of Santa Cruz, Bahia; the International Union for the Study of Social Insects (IUSSI); the Hotel Jardim Atlantico in Ilhéus; and Susi and Ruedi Röösli, Ilhéus, Brazil.

Many people contributed to the preparation of this book. We thank the Smithsonian Institution Press for taking an interest in ants and for including this book in their Biological Diversity Handbook Series. In particular, we thank Peter F. Cannell, Vincent Burke, and an anonymous reviewer for valuable comments that greatly improved the content and format. Rebecca Wilson painstakingly revised and formatted all the figures and illustrations, Beth Norden provided helpful research support, and Debbie Gowensmith prepared the index. The Smithsonian Institution Press, Edward O. Wilson, and the North American Section of IUSSI provided financial support for the preparation of the final publication.

ANTS

STANDARD METHODS
FOR MEASURING
AND MONITORING
BIODIVERSITY

Chapter 1

Biodiversity Studies, Monitoring, and Ants: An Overview

Leeanne E. Alonso and Donat Agosti

The goal of this book is to encourage and enable anyone involved in conducting biodiversity inventories, monitoring programs, or both to include ants among their focal organisms. The information provided here should be sufficient to guide principal investigators, station directors, natural resource managers, technicians, and graduate and undergraduate students in the study of ant diversity patterns.

Although biodiversity inventories should seek to sample as many taxa as possible, logistical constraints dictate that only a subset of all organisms can be sampled. Effective biodiversity studies may do best to focus on organisms that constitute a diverse group, make up a large proportion of the biomass in the area, and perform important or diverse ecological functions in the ecosystem. Ants

meet all these criteria and therefore should be given serious consideration for inclusion in biodiversity studies.

Nevertheless ants and other invertebrates are usually not included in such studies, despite their high diversity, numerical dominance, and ecological importance (Wilson 1987). Some of the reasons for this include

1. The high diversity of most invertebrate groups (particularly in tropical areas).
2. The poor status of taxonomic studies of many invertebrate groups (New 1987).
3. The lack of understanding and recognition of the ecological importance of invertebrates in ecosystem functioning.
4. The small size of invertebrates, which causes them to be overlooked.

Such negative perceptions make the incorporation of invertebrates into biodiversity programs more difficult, but not impossible. As this book demonstrates, with proper training and background information, it is possible for any biodiversity program to include invertebrates, and specifically ants. This book is an attempt to make the process as smooth and straightforward as possible, with the hope that data on ant diversity will be included in a network of biodiversity programs throughout the world. Such a network would build a substantial database and enable analysis of global patterns of ant diversity.

To facilitate the inclusion of ants in biodiversity studies, this chapter provides a general overview of the process of sampling and monitoring ants, with reference to the appropriate chapters of this book for more information.

Reasons for Including Ants in Biodiversity Programs

Ants have numerous attributes that make them ideal for biodiversity studies (Table 1.1). These attributes include high diversity, numerical and biomass dominance in almost every habitat throughout the world (e.g., Fittkau and Klinge 1973; Agosti et al. 1994; Fig. 1.1), a fairly good taxonomic knowledge base (Chapters 5 and 12), ease of collection, stationary nesting habits that allow them to be resampled over time, sensitivity to environmental change (Chapters 3 and 7), and important functions in ecosystems (Chapter 2), including interactions with other organisms at every trophic level (Chapter 4).

Ground-dwelling ants, in particular, are the focus of this book because they represent a subset of ants that can be fairly completely sampled using only a few target methods. Ants living in vegetation and tree canopies are more difficult to sample effectively. The ground-dwelling ant community is a good candidate for use in biodiversity inventory and monitoring programs owing to its relative stability, moderate diversity, and sensitivity to microclimate.

Using Ants in Investigative and Management Programs

Before any biodiversity inventory or monitoring program is started, the questions and goals of the study must be clearly defined. For example, for a biodiversity inventory: What is the purpose and scope of the inventory? Will a record of the number of ant species in the area or a list of morphospecies be enough information? Or is the goal to compare the ant fauna of one site with those of other sites, requiring that species names be compiled?

Likewise, monitoring programs must state clear objectives. These objectives in turn will dictate which organisms will provide the best answers to monitoring questions. Are ants the most sensitive indicators of the environmental factor under study? What types of data are needed to provide answers to the monitoring goals? How long must the study run before patterns are revealed?

Knowledge of the diversity of ants in an area can provide a great deal of useful information for conservation planning. First of all, an inventory of the species of ants in an area will provide data on their distribution and will document the presence of any rare, threatened, or ecologically important species, such as introduced species or those found only in particular habitat types. The number and composition of ant species in an area can indicate the health of an ecosystem (Chapter 7) and provide insight into the presence of other organisms, since many ant species have obligate interactions with plants and other animals (Chapter 4).

Data on the species richness and composition of ants provide the baseline needed for using ants to monitor environmental change or recovery. Although many ant species are capable of living in a wide range of nesting sites, many

Table 1.1 The Importance of Ants

Biomass
- Ants constitute up to 15% of the total animal biomass in a Central Amazonian rainforest (Fittkau and Klinge 1973).
- Of the more than 750,000 described species of insects, some 9500 are ants (Arnett 1985).
- Of all insect specimens collected in the celebrated forest canopy fogging samples in Peru, 69% are ants (Erwin 1989).
- Some 5300 individual ants were enumerated in 1 m^2 of tropical lowland forest soil near Manaus, Brazil (Adis et al. 1987).

Diversity
- In 20 m^2 of leaf litter and rotting logs in Malaysia, 104 ant species representing 41 ant genera were collected (Agosti et al. 1994).
- A single tree in Peruvian tropical lowland forest yielded 26 genera and 43 species of ants (Wilson 1987).
- In 250 m^2 on a cocoa farm in Ghana, 128 species and 48 genera of ants were reported (Room 1971).
- In approximately 5 ha of Peruvian tropical lowland forest, 365 species from 68 genera of ants were found (Tobin 1994).
- In 18 km^2 of semiarid South Australia, 248 species from 32 genera of ants were documented (Andersen and Clay 1996).
- In 5.6 km^2 in temperate Michigan, 87 species from 23 genera of ants were observed (Talbot 1975).

Biology
- All ants are social. Their nests are perennial and thus can be collected all year round.
- There is little variation in ant abundance between rainy and dry seasons (Adis et al. 1987).
- Fragmentation affects ground-dwelling ants (see Chapter 15).
- Together they turn more of the soil than do earthworms in New England (Lyford 1963).
- The density of leaf cutter ant (*Atta sexdens*) nests is up to 20 times greater in secondary forest than in primary forest (Nepstad et al. 1995).
- Leaf cutter ants are the dominant herbivores in tropical forests: the ground volume occupied by a single 6-year-old nest of *A. sexdens* weighed approximately 40,000 kg, and this young colony was estimated to have gathered 5892 kg of leaves (Wilson 1971).
- The seeds of 35% of all herbaceous plants are estimated to be dispersed by ants (Beattie 1985).
- Ants rank among the principal granivores in the southeastern United States (Davidson et al. 1980).

Systematics
- A catalogue of all described ant taxa exists and lists 9538 species (Bolton 1995b).
- An illustrated key to the ant genera of the world is available (Bolton 1994).
- The taxonomy of ants is based on the ubiquitous worker cast.

Leaf litter ant surveys are cost efficient
- A statistically representative sample of the ant diversity of a given area can be completed in one week.
- In comparison, other taxonomic groups require:
  Sampling and identification of tree species in 1 ha in the Atlantic Forest, Brazil: 4 person-years (Thomas, pers. comm.). The number of new tree species (DBH > 10 cm) in an Amazonian rainforest still readily increases after 4 ha have been sampled (Ferreira and Prance, 1998).
  Representative sample of snakes in the Brazilian Amazon: >1000 km walked (Zimmerman and Rodriguez 1990).
  Representative sample of frogs near Manaus, Brazil: >350 person-hours (Zimmerman and Rodriguez 1990).
  Representative sample of birds in Western Amazonia: >800 catches; 1.2–8 catches per day using mist nets (Robinson and Terborgh 1990).
  Representative sample of butterflies in Ecuadorian rainforest: >1000 catches (specimens) over one year (De Vries et al. 1997).
  Representative sample of ithomiine butterflies in Cartago, Costa Rica: 4 days (Beccaloni and Gaston 1995)

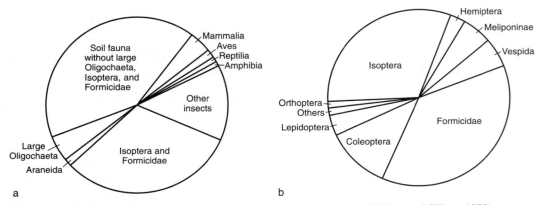

Figure 1.1. (a) Composition of total animal biomass near Manaus, Brazil (Fittkau and Klinge 1973). (b) Species composition of insect fauna near Manaus, Brazil (Fittkau and Klinge 1973).

others have specific requirements and can therefore be used as indicators of habitat change or restoration success (Chapter 7). On the flip side, there are several ant species in all parts of the world that are well adapted to living in disturbed areas and are the first to colonize these areas. The presence of these ant species is a reliable indicator of habitat disturbance. Since most ant species live in stationary colonies and do not move readily between habitats, they are ideal for monitoring because they can be resampled repeatedly over time using the same methods, providing information about how vegetation structure, prey abundance, soil quality, or predator density may be changing over time (Chapter 7).

The First Steps: How to Begin Incorporating Ants into a Study

In order to get the most useful information from an inventory, it is best to begin with a good knowledge base about the organisms under study. Therefore, the first step, for those who are not particularly familiar with ants, is to read the chapters in this book that cover basic ant biology and ecology (Chapters 2–5 and 8). The most comprehensive and readable overview of ants can be found in *The Ants* by Hölldobler and

Wilson (1990). One should also read the chapters on how ants can be used in monitoring programs, as indicators of diversity or environmental change (Chapters 6 and 7), to evaluate whether ants are appropriate for the goals of the biodiversity project in question. Ants are sensitive to many types of environmental disturbance, but other organisms might be better suited for particular challenges.

The next step is to learn some basic ant taxonomy (see Chapters 5, 11, and 12 for basic taxonomic information and additional references).

What Sampling Involves: Resources, Time, and Helpful Hints

Once the necessary background in ant biology has been acquired, the standardized sampling protocol for ants, the ALL (Ants of the Leaf Litter) Protocol (Chapter 14), can be implemented with ease. As with all research projects, the first step in conducting field work is to gather all equipment needed (Appendix 1). Second, the sampling site must be chosen. The placement of the sampling transect should be determined based on the research objectives. For example, a transect may be placed randomly if an objective overview of ant diversity in the habitat is desired, or it may be positioned so that it traverses several microhabitats within the sampling area, thus collecting ants from a variety of habitat types. Alternatively, the transect may be placed in the same areas where mammal or reptile surveys have been carried out, in order to draw some comparisons between taxa. Sampling need not be limited to only one transect per site; several transects can be utilized at each site.

The ALL Protocol, described in Chapter 14, is very simple to implement. The entire process can be carried out in less than 3 days if all the necessary equipment has been obtained. The standard protocol relies on two principal sampling methods: (1) leaf litter samples, which are extracted through mini-Winkler sacks for 48 hours, and (2) pitfall traps, left in the ground for 48 hours (see Chapter 9 for descriptions of these methods). The number of available mini-Winker sacks will usually be the limiting factor to the efficiency of this sampling method. The ALL Protocol recommends taking 20 samples. This implies that 20 mini-Winkler sacks will be needed to process all the samples at the same time. If 20 mini-Winkler sacks are available and can be run at the same time, then all samples can be processed in just over 48 hours. If fewer than 20 sacks are available, samples may be extracted one after the other. This will prolong the sampling process, since for every set of Winkler sacks used, 48 hours is needed for litter extraction.

Alternatively, a Berlese or Tullgren funnel (Chapter 9) may be used for extracting ants from the leaf litter, or the litter samples may be sorted by hand. Extraction using Berlese or Tullgren funnels should take the same length of time as that using Winklers, and hand sorting should also be completed in 48 hours.

During this 48-hour period, it is a good idea to do some general hand collecting in the area near the sampling transect, in order to collect a greater number of ant species. General collect-

ing is not standardized and therefore should not be part of a monitoring program, but it can be a valuable addition to an inventory. General hand collecting includes inspecting rotting logs, branches, and twigs on the ground; scraping soil; and visually searching for ants (see Chapter 9 for more details). When doing general collecting, one should be sure to record as much data as possible about where the specimens were collected, particularly distinguishing between ground and vegetation collections. The standardized protocol restricts sampling to ant species that live or forage in the leaf litter or on the ground. General collecting can add additional ant species from the vegetation.

At the end of the 48-hour period, the ants must be collected from the pitfall traps and Winkler sacks. This process may take from 2 to 4 hours for two people, depending on the ability of the researchers to distinguish tiny ants in the bottom of muddy cups. This is an important step, and it should be performed with great care so as not to miss any ants, some of which are nearly microscopic. Ant specimens should be placed in vials of alcohol and completely labeled with such information as the type of collection method, trap or sample number, date, and collector's name (Chapter 11). The steps involved in processing the specimens to prepare them for identification are covered in Chapter 11.

Beginning the Identification Process

Perhaps the most difficult part of incorporating ants into biodiversity programs is the identification process. Few people in the world are able to identify ants to species level, largely because of the lack of training and the poor state of tropical ant taxonomy. However, it is not impossible, and identification to genus and morphospecies can be done by most people after a little instruction and a lot of practice.

Is It an Ant?

The first step is to sort the ants from the other organisms collected in the pitfall traps and Winkler sacks (Chapter 11). Make sure that all organisms identified as ants really are ants (for help, see Chapters 5 and 11). All ants are classified into one family, the Formicidae, which is in the order Hymenoptera along with bees and wasps.

Ant Subfamilies and Genera

The next steps are to identify the ant specimens to subfamily and then to genus. There are 16 subfamilies of ants, with 296 genera and more than 9000 described species (Bolton 1994). Only a subset of these subfamilies, genera, and species are found in each biogeographic region of the world. Dichotomous taxonomic keys to ant subfamilies and genera are available in Hölldobler and Wilson (1990), Bolton (1994), and several other publications (Chapter 12). If these references are not available, consult the social insects Web site (http://research.amnh. org/entomology/social_insects/) for general pictorial keys and information on how to obtain these publications. These keys are fairly technical and require some knowledge of insect morphology. However, with practice and a little background reading on insect morphology, most researchers should be able to identify ants to genus. A more user-friendly, pictorial guide to ant genera of the world is currently being prepared.

Before attempting to identify an ant specimen to genus or species, it is best to become familiar with the taxonomic keys, the body parts of an ant (Fig. 5.1), and the morphological characters that are most frequently used to identify ants (Chapters 5 and 11).

Species Identifications

Identifying ants to species is much more difficult because taxonomic keys to species are scattered throughout the literature, many keys are out of date, and there are no keys for many regions of the world, particularly tropical areas (Chapter 12). The first step in species identification is to separate the ant specimens into morphospecies, or units that look different from one another (Chapter 11). Each morphospecies should be assigned a number so that specimens sorted later can be associated with similar previously encountered specimens. Morphospecies designations should be based on the traditional morphological characters used in ant taxonomy (Chapters 5 and 11).

It is unlikely that identification of all ant species at a site can be completed without some assistance. However, attempts should be made to identify as many of the specimens as possible, using publications that contain information on the ants of the area, especially those publications that contain taxonomic keys (see Chapter 12 for sources).

For a biodiversity inventory, the number of species (based on morphospecies) may be information enough. However, without knowing the scientific names of the ant species in an area, little can be inferred about the presence of particular species or the patterns of their distribution or diversity. Environmental management and monitoring studies require species identifications so that the presence of exotic species, rare or threatened species, or species specialized on particular climatic conditions can be recognized (Chapters 6 and 7). In addition, comparisons between sites, both locally and globally, require species identifications if the studies are not conducted by the same researchers.

Essential Collaboration with Ant Taxonomists

After identifying the ant specimens as far as possible, the next step is to contact local entomologists, some of whom may be familiar with ants, to provide taxonomic training or assist with identifications.

If no entomologists are available in the area, scientists trained in ant taxonomy may be able to help. The major ant collections of the world are listed in Chapter 12; most of these collections have ant taxonomists on staff. The social insects Web site also has lists of and links to ant taxonomists. Some ant taxonomists may be available to visit a site and provide training in collecting methods and ant identification. This approach is highly recommended—especially if funding can be provided for the taxonomist to travel to the site—since it can provide valuable training for the project participants and enable local researchers to identify ants on their own in the future. If training is neither desired or financially feasible, ant taxonomists can likely provide species identifications if specimens are sent to them. When seeking assistance from any of the taxonomists, be sure to contact them well in advance (at least two months) of sending the specimens for identification. It is important to keep in mind that these taxonomists are busy and that species identifications take considerable time. Allow at least three to twelve months for identifications to be made, depending on how many specimens must be identified.

The Value of Information Exchange

The exchange of ant collections and taxonomic expertise between local researchers conducting biodiversity inventories and ant specialists can be beneficial for both parties. Field researchers receive assistance with species identifications, enabling them to learn more about their local fauna. Biological information about particular species may also be provided by the ant specialists, and this helps the local researchers better understand and utilize the patterns of ant diversity in the management of their areas. Ant taxonomists also benefit by receiving ant specimens from taxonomic groups and geographic

areas of interest to them. The specimens and associated biological information that specialists receive from field projects contribute to systematic studies and taxonomic revisions, thus adding to our understanding of the biology and evolutionary history of ants and furthering our ability to identify particular species.

What to Do with Specimens

The ants collected in biodiversity studies are potentially valuable to taxonomists and local researchers, so they should be handled with care. A reference collection of the species from the site should be established at the local field station, university, or research institution. If possible, a few representatives from each species should be pinned and housed in a cool, dry collection case (see Chapter 11 for details). The pinned specimens will serve as a reference for future ant identifications. The remaining specimens may be stored in vials of alcohol.

Ant specimens should also be sent to those ant taxonomists who are working on particular groups of ants, regardless of whether their taxonomic assistance is needed (see Chapter 12 or the social insects Web site). These specimens may prove valuable in a taxonomic revision by providing needed material on poorly known species or additional data on geographic distributions.

Additional specimens should be deposited in major ant collections (see Chapter 12 for a list). Depositing ant specimens in national collections allows other researchers to examine them for taxonomic comparisons.

Data Output: How Best to Utilize the Information

Collecting and identifying ants provides data that can be used in furthering the goals of the biodiversity project. How the data are processed after collection is perhaps the most important part of the entire study. Careful consideration should be given to which methods of data analysis will best address the questions of each particular study. Several possible analytical methods are described in Chapter 13.

Analytical tools can be used throughout the study to determine the ultimate sample size needed to collect representatives of all ant species in an area (Chapter 10), to determine how long a monitoring program need be run, or to make adjustments to management practices.

The Importance of Training

Depending on their background knowledge of insects and ant biology, researchers involved in biodiversity studies may or may not feel qualified to carry out the standard protocol for ants on their own. Although this book should provide enough detail for almost anyone to utilize the methodology, some may feel that they need more direct training or assistance. As mentioned earlier, on-site training is recommended if the project has sufficient funding to bring in ant specialists. Such training provides a lasting knowledge base that will enable the project to carry out future inventories and follow through with ant identifications. It also enhances the researchers' ability to train others in the ant collection process.

The standard protocol for ants is also being taught as part of the multitaxa inventory and monitoring approach of the Smithsonian Institution/Monitoring and Assessment of Biodiversity (SI/MAB) training courses throughout the world. For more information on becoming a part of the Smithsonian network, readers should contact the SI/MAB Biodiversity Program of the Smithsonian Institution, 1100 Jefferson Drive SW, Suite 3123, MRC 705, Washington, DC 20560; www.si.edu/simab.

In addition to receiving training themselves, many researchers will eventually be required to train others to implement the protocol. Standardized protocols, such as that described in this book, are an invaluable part of biodiversity programs, and promotion of their use will enable data from biodiversity studies worldwide to be put into broader geographic and global contexts.

Promoting Ants in Biodiversity Programs

We hope that this book will help facilitate the incorporation of ground-dwelling ants into biodiversity programs throughout the world. The advantages of their inclusion should be clear from this chapter as well as several others in this book (e.g., Chapters 2, 4, and 6–8). Ants are a key taxon for biodiversity studies and a valuable addition to multitaxa programs. Their incorporation into biodiversity programs in diverse habitats and geographical regions will provide a global database of ant diversity that can be used by taxonomists, ecologists, and natural resource managers in any country. Knowledge of global patterns of ant diversity and responses of ant communities to local and global environmental change will assist with conservation planning worldwide.

Further Information and Assistance

The contributors to this book have joined together to form the ALL Group, a team of ant taxonomists and ecologists from throughout the world dedicated to the promotion of ants in biodiversity programs. Readers should feel free to contact the ant experts of the ALL Group via the social insects Web site (http://research.amnh.org/entomology/social_insects/) for help with incorporating ants into biodiversity programs. We will try to arrange assistance in training, obtaining equipment, identifying specimens, or other matters. Be sure to contact the Group well in advance of target dates. Further information about the ALL Group and ants in general is available on the social insects Web site.

Chapter 2

A Primer on Ant Ecology

Michael Kaspari

As the first chapter of this book suggests, ants are a taxon that offers much to those interested in long-term monitoring, inventory, and basic ecology. This chapter has two goals. The first is to introduce the reader to key ecological features of ants. Its target audience is those unfamiliar with ants but interested in adding them to monitoring and biodiversity studies. Its coverage will not be exhaustive, and as in any review, this chapter will have a distinct point of view. Other sources of information with different viewpoints are available in this volume and in the form of other reviews (Sudd and Franks 1987; Hölldobler and Wilson 1990; Andersen 1991a). Many topics basic to ant ecology (e.g., functional groups, patterns of species composition, dynamics, and interactions) are scarcely covered here because they are treated elsewhere

in this book (see Chapters 3, 4, and 8). My second goal is to challenge potential myrmecologists by emphasizing that a number of basic puzzles remain to be solved.

This chapter is organized around five topics: Colony Life, The Ant Niche, What Regulates Ant Populations?, What Regulates Ant Communities?, and Open Questions in Ant Ecology. Where appropriate I will make suggestions on how to apply this natural history to the design of a monitoring program. These tips will be greatly expanded upon in Chapter 9.

Colony Life

In an introductory chapter such as this, one inevitably glosses over much of the variation that scientists enjoy studying. Here I describe the behavior and composition of a generic ant colony, then move on to a few of the interesting variations on this theme.

Ants are eusocial organisms, characterized by cooperative brood care, overlapping generations of workers within the colony, and a highly developed caste system (Wilson 1971). Castes are groups of specialized colony members that perform different functions with corresponding differences in form. For example, if you were to dig into the soil of a large ant colony you would likely see hundreds of *workers* boiling out. The worker caste performs most of the colony's day-to-day tasks. These include collecting food, tending the young, and maintaining and defending the nest. Continued digging would reveal the colony's young—off-white eggs, larvae, and pupae—in small discrete chambers. All ants start as eggs, grow as larvae, and develop into adults as pupae; these immature stages are fed, groomed, and protected by workers. Depending on the time of year, you may encounter larger, winged ants—the male and female *alates* or *sexuals.* Alates are the reproductive phase of the colony and have little to do in the colony while they wait to fly off and mate. Nonetheless, as they represent the colony's reproductive future, they too are often vigorously defended. Eventually you will uncover the *queen* chamber and its occupant, surrounded by still more workers. The queen, once a winged female in another ant colony, is the center of colony life—often the largest ant in the colony, swollen with eggs and fat. The queen's central role is that of the egg-layer, that is, the mother of all the other colony members. No other members of the colony, although there may be thousands, produce eggs.

It has long been apparent that workers within a colony work together gathering food and defending the colony. This apparent cooperation and self-sacrifice was long held out as an example of virtue. How the castes work together and why workers forego reproduction has been a subject of constant fascination, as it seemed to be a potential exception to Darwin's theory of evolution by natural selection. Individuals were not selfish, but instead appeared to sacrifice themselves for the good of the colony.

One resolution to this paradox came through the study of the peculiar genetic system common to many social insects—*haplodiploidy* (Hamilton 1964, 1972; Alexander 1974; Trivers and Hare 1976). To see how this works, consider that humans differ between the sexes in the genetic makeup of our sex chromosomes. Females inherit two X chromosomes from their parents; males inherit an X chromosome from one parent and a Y chromosome from the other.

Sex in ants is determined in a fundamentally different way—there are no sex chromosomes. Queens, female alates, and workers have two pairs of each chromosome (i.e., are diploid); males have only one set and are haploid. Females and workers receive two sets of chromosomes through the conjoining of egg and sperm. Males are produced from unfertilized eggs.

This simple system has profound consequences for cooperation within the colony. Workers, it turns out, are closely interrelated, sharing 75% of their genetic makeup. Queens, like human mothers, share on average only 50% of their genetic makeup with each of their daughters. They also tend to live longer than their offspring in the insulated environment of the nest. In this situation, workers can advance the cause of their genes most by helping the queen produce more workers and reproductives. The best way to do that is to help keep the queen, and hence the whole colony, alive and functioning. If a worker engages in such "selfless" behavior as a vigorous and fatal defense of the nest or enabling the consumption of its own body by its nestmates, it is actually being selfish, insofar as it improves the chance of the colony surviving and its genes living on. This division of labor and social organization allows a high degree of behavioral sophistication. It may help account for the amazing proportion of biomass accounted for by social insects in the world's ecosystems (see Chapter 1).

This simple picture of colony life, although somewhat typical, ignores fascinating variation that occurs when workers are fertile (Peeters 1991) or when many queens contribute eggs to the same colony (Hölldobler and Wilson 1977). Both scenarios may enable "selfish" workers to prosper, and both are current, productive avenues of research in ant ecology.

The Ant's World—Life as a Colony of Tiny Organisms

It is often said that ecology and conservation biology suffer from a vertebrate bias (Wilson 1993). Humans choose to study organisms, such as birds and mammals, that experience the world as humans do. But ants live and interact in parts of the environment that are in many ways foreign to humans. An effective monitoring program requires a basic understanding of these differences in order to exploit them in the design of the project.

Most obviously, individual ants are small, with a dry weight typically much less than a gram. Their size allows ants entry into crevices and microenvironments (e.g., between soil particles or in the bark of the trees [Kaspari and Weiser 1999]). It allows ant colonies to exist on limited resources and to exploit the majority of the earth's other organisms, which are also small.

But small size has a cost. Small animals heat up and dry out more quickly (Hood and Tschinkel 1990; Kaspari 1993a). Ants, as ectotherms, are constrained to forage when it is warm enough, but not too warm. This results in a temperature "envelope" in which most ants forage at temperatures greater than 10°C and cease foraging much above 40°C, with an average peak foraging temperature of 30°C (Hölldobler and

Wilson 1990). Low humidity may also constrain foraging; the best time to collect ants in the deserts of North America is after summer rains (Schumacher and Whitford 1976). But even moisture is a two-edged sword, as standing drops of water are sticky and unmanageable to ants, and rain washes away chemical trails.

This interaction with their chemical environment introduces another profound way in which ant societies differ from human societies. Ants are filled with glands that open to the outside world, and these glands have three main functions. First, living in the soil requires a defense against fungal and bacterial pathogens; some glands produce secretions that help to keep the ant clean and disease-free. Second, ants are such a conspicuous part of their environment that they have many enemies. Ants—similar to other members of the insect order Hymenoptera (e.g., bees and wasps)—often defend themselves with their stings. Evolution has modified the ant sting in a variety of ways to produce defensive chemicals that are injected, dabbed, or sprayed on potential enemies and competitors.

Finally, some glands produce pheromones, allowing ants to communicate in sophisticated ways with others in the colony and with other colonies and species (Vander Meer and Alonso 1998). For example, each ant colony has an individual odor; ant queens use pheromones to control workers; ant workers use pheromones to leave scent trails to profitable food resources as well as to mark territories. Still other pheromones are released to alert the colony to dangers. In sum, chemicals are the main currency of communication among ants.

The Colony Life Cycle

The colony life cycle breaks down into three phases: *founding, growth,* and *reproduction.* Most ant colonies are founded when a newly mated queen flies off in search of a nest site. Most alates die during this journey, since they are an attractive source of food for a variety of predators (Whitcomb et al. 1973). Upon finding a nest site, a queen excavates or occupies a cavity in a plant or in the soil, where she lays eggs. The queen then depletes her own reserves and converts them into food for the first clutch of workers, either in the form of trophic eggs or salivary secretions.

Colonies enter the growth phase when the first clutch of workers matures. The queen's duties are reduced to egg production and pheromonal control of the colony. The workers take over the task of caring for young, foraging, nest maintenance, and defense. In this phase, colony growth is often exponential, as all resources are devoted to gathering food and raising more workers (Wilson 1971; Tschinkel 1993).

The length of the growth phase varies among species and is dependent on climate—cold slows brood development. The growth phase ends when colonies grow large enough to produce alates. This colony size threshold varies widely among species—some are sexually mature at 10 workers, others at 10,000 or 100,000. What drives this variation is still unclear, although one pattern that emerges is that the average colony size decreases as one approaches the equator (Kaspari and Vargo 1995) and passes from less productive to more productive environments (Kaspari et al., unpublished data). The determinants of size, growth rate, time to maturity, and other life history traits in ants are still only poorly known (Tschinkel 1991).

The reproductive phase of the colony's life cycle begins when attention is lavished on unfertilized eggs (destined to be male alates) and some fertilized eggs are raised, through extra nutrition, to be female alates (and ultimately queens of their own colonies). As alates are typically larger than workers, resources are diverted from worker production, and colonies may stop growing or even decrease in size.

These alates then fly off to mate with alates from other colonies. The males, after copulating, die. The females fly off in search of another nest site, completing the colony life cycle. Colony cycles are somewhat synchronized in habitats that show seasonal rainfall and temperature. Alate flights often occur at the beginning or end of the "benign season" (the warm season in the temperate zone, the rainy season in the tropics). Colonies may produce alates for as long as the queen lives, often for many decades.

This concludes a sketch of a "typical" ant colony. Most ant species have only one queen per colony, and the colony occupies a single nest. Colonies of some ant species, however, have multiple-queen nests for at least one part of their life cycle. This is a topic of enormous interest since it complicates the genetic-relatedness rules of ant behavior outlined previously. Further, many of these species nest not in one location, but in several nest chambers linked by long tunnels or runways that may stretch tens of meters.

Species with multiple queens and multiple nest sites often dominate habitats, owing to their high potential growth rate and large spatial extent. Such is the case for a handful of introduced ant species (*Linepithema humile,* the Argentine ant; *Pheidole megacephala,* the big-headed ant; *Solenopsis wagneri,* the fire ant). Species introductions are a problem worldwide, and these three ant species have been shown to have an enormous impact (Chapter 4; Williams 1994). They are a nuisance to humans, tending aphids on cultivated plants and invading households; the fire ants have a nasty sting. Introduced ant species are also scourges of their host ecosystems. Although native species perform a variety of ecosystem services (e.g., dispersing seeds) introduced species may not (Bond and Slingsby 1984). Freed of factors that limited their native populations, these species may outcompete and drive to extinction local assemblages of ants and arthropods (Porter and

Savignano 1990) and reduce wildlife populations (e.g., Allen et al. 1995).

The Ant Niche

The ecological niche of a species describes the roles it plays in an ecosystem. Describing a typical ant niche is as vexing as describing a typical ant colony. The variety of diets, nest sites, life spans, and associations of ants in any given habitat makes ants an attractive group for monitoring.

However, some general observations may be offered. Most ant colonies are relatively sessile, at most moving their colonies every two weeks, some not moving at all (Smallwood 1982). Ants derive their energy from other organisms— either plants (nectar, leaves, seeds) or other animals, alive or dead. From a central point, colonies send foragers through the environment, quickly recruit to new food resources, and just as quickly abandon them as the need arises. In this way ants collect and concentrate resources in the environment and are themselves predictable resources for those that exploit them. In many ways, ant colonies are decidedly "plant-like" (Andersen 1991a).

Species in an assemblage may vary along three niche axes of particular concern to those designing a monitoring program. These are the *nest niche,* the *food niche,* and the *temporal niche.*

The Nest Niche

Ant nests take a variety of forms. If you traveled along a gradient, from warm desert to moist tropical rainforest, the variety of nest sites that ants use would increase dramatically. In deserts ants tend to be soil nesters. Some soil-nesting species rarely breach the surface except to release alates (e.g., some *Acropyga* and *Neivamyrmex*). Many of these species are known only from male alates caught in light traps. The diversity and natural history of these species remain virtually unexplored (Lévieux 1976, 1983).

Other species nest in the ground but emerge from entrances to do at least some of their foraging above ground. This is probably what most people think of when they picture an ant colony, but even this simple idea is expressed in a number of ways. The nest entrance can be a discrete hole in the soil. In cooler deserts, particularly the tundra, soil nests are often found under stones: stones retain warmth longer than soil, and colonies in cool climates take advantage of these environmental hot spots to warm their brood (Brian and Brian 1951). In some upland meadows, almost every large, flat stone appears to have an ant colony underneath. Not surprisingly, then, stone nesting is less common where soil temperatures are warmer (e.g., hot deserts and lowland tropical rainforests; Brown 1973). In other cases ants bring the stones to the nest entrance, often paving a large gravel disk around the entrance.

Ants nest in a variety of soil types, from hard clay to loam to pure sand. Yet whereas a gardener or botanist can speak volumes on how pH, drainage, and other soil properties influence the plant community, ant ecologists can say little about how soil properties influence ant communities. For example, little is known, given the heterogeneity in soil and litter occurring at a given site, about the role that these differences may play in segregating species (but see Johnson 1992).

Moving from deserts into grasslands and savannas, most ants still nest in the soil. However, in dense grasslands, colonies may live in perpetual shade. A solution practiced by some species is to create a disk of bare soil around the nest entrance. Another solution, particularly in the Northern Hemisphere, is to form large thatch mounds that rise a foot or more above the surrounding grass. In woodlands, tree stumps and snags may also be hot spots for ant colonies. In each case, by avoiding or emerging from the shade, these nests prolong their residents' exposure to the sun.

As we enter forests, the ground becomes covered with a layer of woody debris, leaves, twigs, and fruits (e.g., acorns). In very dry woodlands, soil-nesting species still predominate. As the woods become moister, ants begin to nest in this litter. Litter-nesting ants may nest in cavities in twigs or fruit, between leaves, or in large, decaying logs (Herbers 1989; Kaspari 1993b; Byrne 1994). The fraction of species that nest in the litter is largest in the tropics (Wilson 1959). In the litter a single bit of hollow twig may house a colony of ten to a hundred workers; a scattering of leaves may provide meager shelter for a colony and its pupae, spread over several square meters; a large, rotten log may contain multiple colonies of a variety of species. To find these colonies, one need only crack some twigs, disturb some leaves, or cut into a rotten log. If the environment is seasonally cold or dry, these colonies may periodically move out of the litter and into the soil (Herbers 1985).

In tropical forests, a substantial portion of the local ant fauna will be found living in the plants themselves, from low herbs to the canopies of trees (Jeanne 1979). Some ants build nests out of chewed wood pulp. These "carton nests" are common in the tropics and may be found affixed to trees high in the canopy or in the understory, on the underside of leaves (Black 1987). Species such as the wood ants (*Camponotus*) may excavate a nest chamber in a partially rotting tree (look for regular columns of ants on the trunks of trees). Finally, a host of plants have evolved cavities and food bodies, providing ants food and shelter in exchange for protection from herbivores (Huxley and Cutler 1991, Chapter 4). Only a small subset of plants may have these cavities, but they are often used by species found nowhere else.

The Food Niche

The majority of ants appear to be opportunistic foragers, taking some combination of plant exudates, seeds, and animal matter, alive or dead. Some fraction of an ant assemblage, however, is more specialized in their diets. For example, in the warmer parts of the Americas, the Attini cut vegetation or collect dead insects or insect dung. This material is in turn used as a substrate on which to grow fungus, and this fungus is cultured and harvested for food (Weber 1972a, 1972b).

Other ants specialize to various extents on plant exudates (Tennant and Porter 1991; Tobin 1994). These exudates are obtained either directly from plant organs called nectaries or indirectly through such sucking insects as the Homoptera (Huxley and Cutler 1991). Given the enormous volume of ants in tropical canopies, an increasing body of evidence suggests that canopy ants live in a carbohydrate-rich and protein-poor environment (Tobin 1994; Davidson 1997; Kaspari and Yanoviak, in press).

Many ant genera include specialized predators (e.g., *Cerapachys, Neivamyrmex, Proceratium, Strumigenys, Thaumatomyrmex*) that feed on a restricted set of arthropods. Some "specialists" may have taxonomically narrow diets but feed on insects that are otherwise quite common (e.g., the ant specialists among the army ants *Neivamyrmex*) if not always apparent to the casual observer (e.g., the Collembola specialists in the genus *Strumigenys*).

Finally, ant species that make up a community may specialize to varying degrees on the size and density of a resource. Species with large workers often have access to a broader array of prey sizes (Kaspari 1996c). Likewise, as food comes in packages of different sizes, the larger, richer bits of food are often taken and defended successfully by species with large, aggressive colonies (Kaspari 1993b).

The Temporal Niche

Within an ant community, subsets of species may restrict their activity to some parts of the year or day. For example, *Prenolepis imparis,* which can forage at temperatures approaching

0°C, is a North American forest species commonly active in spring and fall, but rather inactive in summer (Talbot 1943; Fellers 1989).

Over a 24-hour period, the same patch of habitat may reveal very different parts of its ant fauna. Deserts (Whitford and Ettershank 1975; Bernstein 1979; Morton and Davidson 1988) and rainforests (Greenslade 1972; Kaspari 1993a), for example, often have distinctive nocturnal, crepuscular, and diurnal fauna. This segregation likely arises from a combination of physiological tolerances, competitive interactions, and predation risk (Whitford 1978; Orr 1992). However, the paucity of such studies leaves us with no reliable generalizations for an important question in ant monitoring: at what times of day do you sample? As a first approximation, desert (and perhaps rainforest) communities are likely have the most highly developed between-species segregation of daily activity. Desert and temperate communities likely have the most pronounced seasonal segregation. Bait studies in such habitats that fail to sample over the appropriate time intervals may underestimate species richness.

What Regulates Ant Populations?

A population is a collection of individuals of the same species found in a given area (e.g., the *Pogonomyrmex rugosus* population of the Jornada Experimental Range, southern New Mexico). In this section I briefly review what we know about factors that regulate ant populations, that is, factors that cause ant populations to increase, decrease, or stabilize (see also Chapter 3 for more examples).

Factors that regulate populations can be broken down into two groups (see also Chapter 3). *Resource-based* factors regulate populations by controlling the supply of resources as well as a colony's access to those resources. *Mortality-based* factors, in contrast, are those that kill and/or harvest parts of colonies. In short,

resource-based factors determine how fast populations can grow in a habitat; mortality-based factors determine the actual standing crop of a species. How these forces act together is a topic of ongoing research.

Resource-Based Regulation

Resource-based factors set the ability of colonies to grow and reproduce. They are of three types: *resources, conditions,* and *population interactions.* Resources are items actually used and depleted by ant colonies (e.g., food, nest sites). *Conditions* are abiotic factors that regulate access to resources (e.g., temperature, humidity). *Population interactions* describe how other populations in the habitat regulate access to available resources.

As habitats become more productive, they often have higher overall numbers of ant colonies. For example, net primary production measures the amount of photosynthesis in a habitat, in units of grams of carbon fixed per square meter per year. In the Americas, the density of ants increases from around 0.03 colonies/m^2 in the Colorado desert (<10 gC/m^2/y) to about 10 colonies/m^2 in an Ecuadorian rainforest (>1000 gC/m^2/y; Kaspari et al., 2000). But the increase is far from uniform. Access to resources is decreased by poor conditions (Andersen's "stressors"; see Chapter 3) and by competing populations of ants and other organisms.

Temperature is a preeminent condition for ant populations (Brown 1973). As a taxon, ants are thermophilic, shutting down in winter and avoiding cold shade (Brian and Brian 1951). Yet even in environments that are mostly cold year round, such as alpine tundra, ants are often common in direct sunshine (Heinze and Hölldobler 1994). Where it is warm year round (e.g., the tropical deserts, savannas, and rainforests) ants are a conspicuous part of the landscape.

Organisms or populations *compete* for resources when an individual or population grows at the expense of another individual or population. For example, as one colony grows, it may deplete the food supply sufficiently to make food unavailable to other nearby colonies. Likewise, if there are a limited number of hollow twigs or ant plants in a forest, whichever colony or population gets there first may exclude a second colony or population (Davidson et al. 1989; Longino 1991). Competition can be within a species (intraspecific) or between species (interspecific). In the former case, interactions within a species regulate the population; in the latter, interactions between species help regulate the population. Competition may occur indirectly, through the consumption of food or other resources, or quite directly, through the killing of other colonies. The distinction between competition and predation becomes fuzzy in ant-ant interactions (i.e., is a colony that kills a neighboring colony and carries away its pupae and food stores preying upon that colony, or simply getting rid of a competitor?).

Where there is abundant sunshine (or *insolation*), there is growing evidence that ant colonies compete with each other for resources such as nests and food. I can rank this evidence, in order of increasing confidence, as follows.

Many ant species are highly territorial. If colonies deplete resources and kill foundresses near established nests, this should result in a regular distribution of ant colonies in a homogenous landscape. Many studies have looked for these patterns, mapping out colonies in an area and testing the hypothesis that colonies are more dispersed than would be expected by chance. A summary of the evidence (Levings and Traniello 1981) suggests that territoriality is often, but by no means always, the case. At least one cautionary note for this type of evidence is sounded by Ryti and Case (1992).

Colony density and size can also affect competition between ants. Resources can be subdivided by a population in a number of ways. All the resources can be dominated by a single large colony, or they can be divided among many, smaller colonies. Put another way, as average colony size in a habitat increases, the density of colonies in that habitat should decrease. This inverse relationship between colony size and density is often observed over the course of a growing season. Early on, habitats are colonized by many foundresses. These foundresses raise broods that find, fight, usurp, and kill other colonies, until that same habitat is left with only a few victors (Ryti and Case 1988b; Tschinkel 1992; Adams and Tschinkel 1995). This is not always the case, however. In one study of tropical litter ants, there was little relationship between colony size and density (Kaspari 1996b); disturbance by rainfall and army ants may be sufficient to prevent colonies from saturating the environment.

If ants compete for resources or good conditions, then removal of one colony should benefit another. Experimental removal of ant colonies often results in the rapid use of the vacated site by foragers and nests of adjacent ant colonies (Davidson 1980; Andersen and Patel 1994).

If resources or conditions limit a colony's growth, then increasing resource availability should enhance that colony's survival or reproduction. This individual success should ultimately result in higher local population densities. This is a simple experiment, but it has rarely been performed by ant ecologists. Food addition studies in warm desert environments have yielded mixed results (Ryti and Case 1988a; Munger 1992). Ants in temperate environments closer to the poles may be more likely to respond to increases in food supply (Deslippe and Savolainen 1994). One reason may be that warm deserts often run on a seed economy, and seeds can be stored for long periods. "Harvester ants" may thus be better buffered against food shortages and more likely to respond only to prolonged periods of shortfall. Shortages of nest sites may also limit ant populations. In two Neotropical litter ant assemblages, nest densities doubled with the addition of bamboo twigs, but 75% of the nest sites remained unoccupied (Kaspari 1996b).

Clearly resources, climate, and competitors work together to regulate ant populations (Brown 1973). Even in resource-rich environments, cool-damp climates have few ants. A good example is provided by cool, temperate rainforests. In contrast, warm-dry environments appear to have conditions quite well suited to ants. In this case, the number of ants often appears to be set by rainfall—a good predictor of the seed crop on which desert ants feed (Morton and Davidson 1988).

Mortality-Based Regulation

A variety of factors kill and cull ant colonies. The death of a colony's queen generally spells the end for the colony, although it may still produce a final batch of alates from existing eggs (or eggs laid by workers). Colonies of some multiple-queen species may also adopt a new queen (e.g., Tschinkel and Howard 1978).

But as suggested previously, most queen deaths occur early in the colony's life cycle, when roving foundresses are a vulnerable (and nutritious) food to predators ranging from dragonflies to birds. After founding, more queens, newly ensconced in the nest, are killed by roving workers from mature colonies. Mature colonies die less frequently, but the causes of queen death in older colonies are obscure, in part because so many queens are high in the treetops or deep underground, and thus difficult to observe. Weather must play a role. Unable to get up and move quickly (a large colony of *Atta colombica* may take 8 days to execute a nest move; Porter and Bowers 1981), many ant species are probably susceptible to flooding.

Given their densities, biomass, and interactions, ant colonies are a conspicuous part of the environment (see Chapter 1). It should not be surprising that they have attracted their share of predators and parasites—many of which are other ant species (Kistner 1982). For example, some ant species are social parasites. They have queens that invade the nests of a host species, find the queen, and kill it, "adopting" the colony's workers to raise the intruder queen's eggs (Wilson 1984).

Ant populations are also regulated by harvesting. Just as regular pruning of a garden can keep individual plants in check, predators that drain a colony's resources by killing its workers can help regulate ant populations. In the boreal and temperate zones some ant species conduct "slave raids," stealing the pupae of other colonies (Topoff 1990). As the term implies, these pupae are carried off and raised by the raiding species to take over many of the colony tasks. Slave-raiding species are replaced (in an intriguing but as yet unexplained pattern) toward the tropics by army ants—nomadic raiding colonies of ants that kill and carry off pupae for immediate consumption. Over 20 army ant species may inhabit a single Neotropical forest. It is possible that their combined effects on the ant community are profound (Rettenmeyer et al. 1983; Kaspari 1996a). But do slave raiders and army ants keep colony densities lower than they would otherwise be? No one has yet performed the simple experiment of removing army ants and slave raiders—or building fences around their prey—to observe the response of the host species.

A host of other animals kill or harvest ant colonies. For example, almost every continent has a series of vertebrates (e.g., anteaters and lizards) that consume ants. It has been shown in rare instances that these predators regulate the distribution of their prey (e.g., where ant lions are common, ants are not; Gotelli 1993). Ant colonies also have live-in associates, including mites, nematodes, spiders, and beetles (Chapter 4; Kistner 1982; Hölldobler and Wilson 1990).

The impact of these associates on the colony's economy is also largely a mystery.

Finally, just the *risk* of parasitism may keep some ant colonies from growing faster. Recent work on the interactions between phorid flies and ants exemplifies this phenomenon. Phorids are tiny parasitic flies, many of which specialize on a single ant species or genus (Brown and Feener 1991a, 1991b; Brown 1993). Phorid flies search for their host ants (often following the odor plume of the ants themselves), hover, then zoom in to lay an egg somewhere on the worker ant's body (Porter et al. 1995a, 1995b; Feener et al. 1996). The worker falls over, stunned, then eventually returns to the colony and dies when the egg hatches and the ant serves as food for the developing maggot.

However, phorid flies first must *catch* worker ants. Therein lies the tale. Host ants often run and hide in the presence of their phorid parasite (Porter et al. 1995c). This reaction is so profound as to interfere with foraging, and perhaps swing the competitive balance away from the host ant to its phorid-free competitor (Feener 1981).

Ant ecologists have compiled a catalogue of parasites and predators, with effects ranging from killing the queens to frightening the workforce. But the effect of these parasites and predators on the number of ant colonies in an area is still largely unknown. Just because horned lizards in a desert bajada consume harvester ants for a living does not mean they play a meaningful role in limiting the number or size of those colonies. Put another way, there is not a single study of an ant population (let alone a community; see later in this chapter) in which all the predators and parasites have been enumerated and their impact on ant colonies has been quantified.

A Word on Patchiness

Density, the number of ant colonies per unit area, is an abstraction—ant colonies are never evenly distributed over the landscape. In fact, ant colonies can be quite patchy, a phenomenon long recognized (Wilson 1958) and one that continues to fascinate ant ecologists (Levings and Traniello 1981; Levings 1983; Kaspari 1996a, 1996b). For example, a single 1-m^2 patch of litter in a tropical forest may have 1–17 species nesting in it. The role that top-down and bottom-up forces play in creating this patchiness, and in creating the broader geographical trends of diversity, is a subject of ongoing research.

What Regulates Ant Communities?

An ecological community is a collection of species living in a given environment. Most monitoring programs have as one of their goals the description of an ecological community. Community descriptors can be grouped into those describing *form, function,* and *diversity. Form* describes the size, shape, and mass of an ant community. *Function* describes what the ants actually do to the ecosystem—what foods they eat, how much soil they turn over, what other populations they regulate. *Diversity* describes the composition, number, and taxonomic relationships of species between and across communities. Community form, function, and diversity vary in time and space in predictable ways.

Form

A community's biomass is the summed weights of all its species. Ants and termites may represent up to one-third of the total animal biomass in some tropical forests (Fittkau and Klinge 1973). This preponderance of ant biomass is especially high in the tropical canopy, where up to half the individuals may be ants (Stork and Blackburn 1993; note that this is not the same as saying that ant population densities are higher since the majority of these ants are workers from a few colonies).

All the species in a community sum to form a size distribution. What constitutes "size" in ants and other social insects is a bit complicated— ant species have a characteristic distribution of sizes of individual ants and a characteristic number of ants in the colony. Taken together, ant colonies represent some of the largest insects recorded (Kaspari, forthcoming).

Both the average size of ants and the number of ants per colony appear to decrease as one travels from the poles toward the equator (Cushman et al. 1993; Kaspari and Vargo 1995). Tropical ants and ant colonies tend to be smaller. The cause(s) of this pattern, shared with many other organisms, is not yet clear. It may be linked to the increasing use of litter nests in the tropics or to adaptations to living in richer, less seasonal environments (Kaspari and Byrne 1995; Kaspari et al. 2000a). Small size, as discussed earlier, has both costs and benefits. For example, smaller ants may be restricted, on average, to moister environments and cooler, moister times of the day (Hood and Tschinkel 1990; Kaspari 1993a) since they desiccate more quickly.

Even within colonies ants vary in size and shape beyond the obvious differences between workers and reproductives. Sometimes there are discrete worker castes in the colonies (e.g., "minors" and "majors"). In such cases, a persistent question in ant ecology is how and why this occurs, and how caste allocation varies with environment (Wilson 1985; Schmid-Hempel 1992; Kaspari and Byrne 1995). Relatively few species have such distinct forms: species typically show some continuous size variation in the size of the worker caste. Again, the question arises, is this variation a natural outcome of changing food supplies and energy demands within the colony, or is it "fine-tuning" by the colony, allowing larger ants to specialize on larger prey (Rissing 1987; Wetterer 1991)? Ant

ecologists still have not settled this issue, in part due to lack of data on how and when resources limit colony growth (Beshers and Traniello 1994; Kaspari and Byrne 1995).

However, if food is a limiting resource, we might expect the advantages of worker size variation to be greatest when a species has the environment all to itself. In other words, size variation should evolve to exploit the "empty niche space" left in the absence of other species. This appears to be the case with *Messor pergandei,* an ant of the desert southwest in North America. In assemblages with few species, *M. pergandei* workers vary greatly in size (Davidson 1978). In richer environments, with a larger number of ant species, much of that size variation disappears.

Function

Given their diversity and biomass, it is not surprising that ants play such a large role in the functioning of ecosystems. Many of these functions (e.g., seed dispersal) are discussed in more detail in Chapter 3. Here I discuss a few ways in which ants shape ecosystems as soil movers, as "keystone species," and, pathologically, as introduced species.

First, ants greatly affect the structure of their environments as "ecological engineers"— organisms that rearrange the environment in ways that affect other organisms (Lawton 1994). One way they do this is by moving and enriching soil—large ant colonies may excavate liters of soil in their lifetime, aerating the soil and incorporating litter from the surface in much the same way as do earthworms (Elmes 1991). Lesica and Kannowski (1998) suggest that hummocks in northern peatlands may be abandoned nests of *Formica podzolica,* based on similar elevated levels of soil nutrients.

Since ants bring food in from their entire foraging territory, they may serve to concentrate nutrients in the nest. However, this effect can vary from species to species. For example,

Haines (1978, 1983) has studied two species of leaf cutter ants, *Atta colombica* and *A. cephalotes*. Leaf cutters harvest vegetation and use the cut leaves as a substrate on which to grow fungus. The fungus is then harvested for food. Since leaf cutter colonies can consist of millions of workers, refuse disposal is a big job, handled differently by the two species. These habits, Haines suggests, predispose the two species to have very different effects on the nutrient recycling in the soil. *Atta cephalotes* stores its refuse underground; *A. colombica* dumps its refuse in large conical piles above ground. With *A. cephalotes* the refuse pile's nutrients are leached away deep underground, while with *A. colombica* the nutrients are retained near the surface of the soil and are more easily recycled by the plant community.

Some ants are likely keystone species—organisms that disproportionately impact their community (Paine 1968; Lawton 1994). One potential example is the army ant *Eciton burchelli*. Army ants are nomadic species, with hundreds of thousands of workers. Army ant colonies roam in search of prey, mainly arthropods, especially social insects. *E. burchelli* may be a keystone species for at least two reasons. First, a raid by *E. burchelli* creates a seething crowd of escaping arthropods just ahead of the raid front. These are easy prey for numerous species of birds that form mixed flocks and spend their lives following army ant swarms (Willis and Oniki 1978; Willis 1983), as well as other associates (Rettenmeyer 1962; Kistner 1982). Seond, there is some evidence that, by preying on large ant species, *E. burchelli* may open up opportunities for smaller ant species that escape predation (Franks and Bossert 1983).

The role of ants in ecosystems is clearly seen when introduced ants disrupt communities. Ants transported outside their native ecosystems can disrupt the ecosystems of their new homes. Accounts of two introduced species will make this clear. The first is *Linepithema humile*,

the Argentine ant, now common in warm-temperate habitats the world over (Bond and Slingsby 1984). In South Africa, the fynbos plant community is extraordinarily diverse. Many plants of the fynbos depend on native ants to disperse seeds to new habitats away from the parent. As *L. humile* gradually invades, it displaces the native ants but fails to disperse the seeds. As a result, many of the plants appear on their way to local extinction as *L. humile* continues to spread.

In North America, the fire ant, *Solenopsis wagneri* (formerly *S. invicta*), has occupied much of the southeastern United States. A host of studies has begun to assemble a picture of widespread ecosystem disruption. In Texas, *S. wagneri* makes up more than 99% of ants captured at infested sites. In infested sites, the number of common ant species has declined from an average of 13 to 4 species, and the number of other arthropod species has also declined (Camilo and Phillips 1990; Porter and Savignano 1990). In contrast, in *S. wagneri*'s native Brazil, it is found at 0.1–19% of ant baits, and it co-occurs with up to 48 species of ants (Fowler et al. 1990). Population densities of *S. wagneri* are at least four times higher in the United States than in Brazil (Porter et al. 1992).

Such population explosions of pest ants and destruction of native arthropod communities are repeated with *Wasmannia auropunctata* in the Galápagos (Clark et al. 1982; Lubin 1984), *L. humile* in California (Erickson 1971; Ward 1987), and a host of exotic species in Hawaii (Fluker and Beardsley 1970). One goal of this volume is to provide a means for better surveying this damage, and we hope that more scientists will be motivated to study the ecology of introduced species in order to slow or reverse their impact.

Diversity

Studies of diversity document the number and identity of species in a given area. As the world's habitats disappear, careful quantification of diversity has taken on new import. But diversity is also one of the most difficult things to measure unambiguously (see Chapter 13). Comparing diversities between areas demands a standard protocol, as species richness increases with the size of the area sampled and increased time spent sampling. A chief aim of this book is to summarize those protocols. Here I review a few of the major patterns in species richness and species composition that have been discovered in ants thus far.

The most striking pattern of species richness (the number of species in a given area and time) is its increase from the poles to the tropics. Jeanne (1979) was the first to study this trend in a standardized way. Jeanne investigated the intensity of ant predation along a transect from the New World temperate zone and tropics. The transect consisted of five forest sites, located in the northern and southeastern United States, tropical Mexico, Costa Rica, and Brazil. The same baits (a wasp larva in an open vial) were set out for a specified time in a variety of habitats: old field and intact forest, high and low in trees. A number of trends were apparent. First, species richness in each habitat increased from the temperate zone to the tropics. Richness, however, increased at different rates at different areas within a site. Arboreal ants were all but absent in the north but made up an increasing proportion of the ant fauna toward the tropics. In contrast, old fields made up a decreasing proportion of species richness compared with forests.

The latitudinal gradient has many causes. As we discussed earlier, tropical environments are more productive. Since plants form the bottom of trophic pyramids, more productive environments should be able to support more ants and hence more species. But, as we also discussed earlier, ants are thermophilic. Since ants do better at warm temperatures, temperature may regulate access to productivity (Brown 1973).

Put this way, the other two trends make sense. Arboreal nesting allows ants to be closer to the majority of a forest's productivity—in the canopy. However, the canopy lacks the soil's ability to buffer the colony from hard freezes. As average temperature becomes more amenable, arboreal nesting increases. Likewise, average temperature decreases toward the poles and temperature in the shade is always cooler than temperature in full sun. In cold environments, ant abundance and species richness may be proportionately higher in open areas than in the cool shade of a developed forest.

Species richness shows other patterns. For example, larger islands tend to have more species than smaller islands (Wilson 1961; Goldstein 1975). Species richness also tends to increase, but in an often sporadic way, after an area has been disturbed (see Chapter 7). In contrast, as pointed out earlier, introduced ant species can quickly "simplify" an ant assemblage, driving many native species to extinction.

Major puzzles in the patterns of species richness remain, however. For example, two studies examined the correlation between rainfall and harvester ant diversity in arid environments. Davidson, studying the deserts of North America, found a positive correlation (Davidson 1977a, 1977b). Since productivity in dry environments is largely limited by rainfall, this finding appears to support a productivity explanation for the trend. However, when the same techniques were applied to Australia's desert and shrubland ant faunas, no correlation existed (Morton and Davidson 1988). Species richness is obviously a complex variable shaped by a number of factors, including the unique history of an area (see Chapter 8).

The geographical distribution of species composition is covered by Ward (Chapter 8) and Brown (1973). Here I briefly review how species composition may vary in interesting ways within a habitat.

Much of the work on species composition has centered on the role that interspecific competition plays in arranging species across the landscape. For example, in many temperate communities a regular hierarchy is apparent (Wilson 1971; Savolainen and Vepsäläinen 1988; Savolainen 1990; Andersen 1992b). Hierarchies have been standardized in various ways but boil down to dominants and subordinates. The dominant species often form(s) large colonies. They are aggressively territorial and recruit quickly, in numbers, to food. Ant diversity and density are often low around dominant species. Subordinate species often form smaller colonies, are poor recruiters, and are found on the periphery of territories controlled by dominants. This pattern is exceptionally pronounced in boreal and north temperate habitats, in arid deserts and shrublands in Australia, and in simple communities (Hölldobler and Wilson 1990). A similar phenomenon may occur in tropical canopies, where a few species are dispersed in a mosaic throughout the treetops, making up 95% of the biomass and/or numbers (Majer 1976; Blackburn et al. 1990; Adams 1994; Tobin 1997).

This dominance hierarchy is by no means universal in ant communities (see Chapter 3). In the tropical litter, species show few strong positive or negative correlations in abundance predicted by strong interspecific competition by dominants (Kaspari 1996b). This may be due to a number of factors. Litter colonies may never reach a size that allows them to dominate their neighbors. Litter nests, rotting around their occupants, may prevent them from setting up a large, stable territory. Ants of the tropical canopy, on the other hand, form "islands," with narrow, defensible trunks leading to the lush canopy above (Hölldobler and Lumsden 1980). A single large colony can thus monopolize whole trees and keep out other species (Hölldobler 1983; Adams 1994). Similar phenomena, on smaller scales, occur in the ant plants of the tropical understory (Davidson et al. 1988, 1989).

Dominance hierarchies may also be muted when predators decrease a dominant's ability to monopolize resources. In a classic study, Feener (1981) studied two species of ants, *Solenopsis texana* and *Pheidole dentata,* with a parasitic phorid fly, *Apocephalus. Pheidole* have large-headed soldiers that are recruited to rich food supplies. These soldiers are also the preferred hosts of the flies. As a consequence, when the flies are present, the *Pheidole* soldiers run away in a panic, leaving the *Pheidole* colonies outmatched at food resources compared with *Solenopsis.* A phorid fly may thus tilt the balance in competition among these two common and conspicuous ants of the Texas litter.

In sum, although interspecific competition may play a large role in shaping patterns of species composition, it is by no means common everywhere. Its effects are modified by factors such as the architecture of the environment and the presence of predators.

Open Questions in Ant Ecology

This chapter has presented one view of the status of ant ecology. As has become clear by this point, large gaps remain in our understanding of these important insects. Following are just a few of the unanswered questions.

What kills ant adult colonies? We know that the majority of ant colonies die at the foundress stage. But how do abiotic factors (e.g., floods, drought, cold spells) and biotic factors (e.g., army ants, parasites, slave raiders, viruses, and predators) combine to kill colonies that have reached maturity?

What are the impacts, and possible means of control, of introduced ant species? Wherever they have been studied, introduced ants such as the imported fire ant and the Argentine ant have disrupted ecosystems. What regulates the spread and final distribution of imported ant species? Are their effects ameliorated over time? Can we safely introduce biocontrol?

How does mating system influence interactions? Under what circumstances do large, multiple-queen colonies evolve and occupy landscapes (Davidson 1998; Holway et al. 1998)? Why aren't they more common?

What do canopy ants, litter ants, and subterranean ants do? These forms are all integral parts of ant assemblages, but their inaccessibility has precluded extensive study. Is their biology fundamentally different from that of the soil-nesting ants that have been the subject of most ant research?

What factors regulate ant populations? How do resources and mortality combine to determine long-term population trends in ants? How does this answer differ as we go from boreal to tropical forests, from deserts to grasslands? Does this answer differ fundamentally between regions (e.g., Australia, Asia, North America) with differing histories and taxonomic contributions?

How interchangeable are species? We know surprisingly little natural history about most ants, especially those of the tropics. To what extent, in species-rich communities, do the 30 or so *Pheidole* or *Camponotus* species do much the same thing? What roles do ants play in ecosystems?

How are current environmental changes (climate change, habitat destruction) reflected in changes in ant abundance and diversity? Are ants sensitive indicators of environmental change? Or do their lifestyles buffer them against anthropogenic change?

Tips for Field Work

One never wants to wreak more havoc on populations than is necessary to satisfy the needs of a given scientific study. Thankfully, collecting workers from large colonies is akin to pruning a bush or scraping off some skin cells—you typically leave the colony (the queen and most workers) behind.

Choose a survey period that maximizes ant activity—typically one during which temperatures are high and rainfall is plentiful. Desert and tropical assemblages are often most active during the wet season. Early summer is often a good time to sample most other temperate zone assemblages. Mediterranean climates, where it rains most in the winter and is warm in the summer, often require spring sampling. Avoid collecting when there is standing water or when the vegetation is wet. This can slow ant activity, especially among small ants.

Be careful how you handle baits, pitfall cups, or anything with which an ant will come into contact. Avoid wearing perfumes, colognes, and insect repellant when sampling ants.

If one goal is to monitor colony densities, then species with multiple-nest entrances may be overcounted. There are different ways to compensate for this tendency. One is ignore conspecific nests within a given distance of each other (e.g., 1 m). Another is to perform "transplant experiments" (i.e., if a worker from one nest entrance is vigorously attacked when placed near another nest entrance, it is likely that the two are from different colonies).

If you use bait in your protocol, use small baits that are hard to monopolize by a large, aggressive colony. We have found that short-bread cookies (made with flour, sugar, eggs, and nutmeats) attract the widest variety of ants (compared with peanut butter, tuna, and sugar water). Even Attini (fungus-growing ants) and *Strumigenys* (Collembola specialists) stop to carry off shortbread cookie crumbs. On the dark rainforest floor, the investigator often sees the pale yellow cookie crumb move long before she or he sees the ant!

If possible, spread sampling effort out over the hours of the day. One way to do this is to use passive sampling techniques such as pitfall traps.

Patchiness is a problem when putting together a monitoring program. Ant densities can vary so much from 1-m^2 plot to 1-m^2 plot, or from valley to valley, that it is important to locate sample plots randomly and to have enough of them to account for this variability.

In sum, the ecology of ants—small, colonial, sessile, and chemosensory organisms—differs in basic ways from the ecology of vertebrates and hence from our own world. The impact of ants in today's ecosystems is profound and their presence is ubiquitous. For all these reasons, adding ants to the list of taxa surveyed in a monitoring program is a wise investment. A careful consideration of ant life history during the study's design phase will maximize the returns on that investment.

Chapter 3

Global Ecology of Rainforest Ants

Functional Groups in Relation to Environmental Stress and Disturbance

Alan N. Andersen

This chapter deals with global ecology—the analysis and synthesis of ecological patterns and processes on a global scale (Cowling and Midgely 1996), referred to by Brown (1995) as *macroecology*. Global ecology does not address details of community composition and dynamics in any particular place, but instead provides a broad framework for doing so.

Study of the global ecology of rainforest ants seeks to understand how the structure and function of ant communities vary between rainforest and other biomes, among different rainforest types, among different strata within rainforest, and in response to disturbance. This approach requires a predictive understanding of the responses of rainforest ants to environmental stress and disturbance, where stress is defined, following Grime (1979), as any factor limiting productivity, and disturbance as any factor removing biomass. A key to such an understanding is the identification of functional groups that transcend taxonomic and biogeographic boundaries and respond predictably to stress and disturbance (Lavorel et al. 1997; Smith et al. 1997).

Principles of Stress and Disturbance

The primary stressors for ants are the following:

1. *Low temperature*. I consider low temperature to be the primary stress controlling global patterns of ant productivity and community structure (Andersen 1995). From an ant's perspective, temperature is a product

of both climate (which controls ambient temperature) and habitat structure (which determines the degree of insolation of the foraging surface and therefore microclimate). Low-temperature stress is high in cool and shaded habitats, moderate in cool and open or warm and shaded habitats, and low in warm and open habitats (Andersen 1995).

2. *Nest site availability.* The availability of nest sites (the range of types and their abundance) exerts an important influence on ant productivity and community structure. The range of types of nest sites varies with the structural complexity of the habitat, and this range constrains the types of ants that can occur there. Structurally complex habitats, such as lowland tropical rainforests, offer nest sites (e.g., leaf litter, rotting logs, epiphytes, myrmecophytes) that are often not available in other habitats, and therefore they support functional types of ants (e.g., cryptic, myrmecophytic, and other arboreal species) that are often uncommon or absent elsewhere (Wilson 1987; Benson and Harada 1988; Byrne 1994). Herbers (1989) considered the abundance of nest sites such as preformed plant cavities in acorns and twigs to be a key limiting factor in temperate forests of the United States. In structurally simple habitats, where most ant species nest in soil, soil type has a major influence on ant productivity and community structure. Throughout Australia, for example, the highest degree of ant richness and abundance is often found on sandy soils, and the lowest on heavily textured soils (Greenslade 1979; Andersen and Spain 1996), reflecting differences between the substrates as nest sites.

3. *Food supply.* Food availability is obviously a critical determinant of the distributions of species with specialized diets, such as seed harvesters and specialist predators. However, most ant species are scavengers, generalist predators, collectors of honeydew, or a combination of these, and the extent to which overall ant productivity is limited by food supply is not clear (Kaspari 1996b). There is no clear global relationship between primary productivity on the one hand and the productivity (reflected in either abundance or species richness) of ants on the other. Food resources often do not appear to be limiting in local ant communities (Byrne 1994), and it appears that factors such as temperature (e.g., insolation of foraging surfaces) and nest site availability (e.g., soil type) are more important (Kaspari 1996b), except in the most unproductive habitats, such as true deserts (Marsh 1986).

4. *Microhabitat structure and resource capture.* The structural complexity of the foraging surface exerts a major influence on the ability of ant species to capture food resources. For example, leaf litter on the ground reduces the efficiency with which resources can be located, retrieved, and defended by epigaeic ants. This factor has a major effect on ant community structure and possibly also influences overall ant productivity.

Given that disturbance is defined as the removal of biomass, for most animals it is synonymous with mortality. Ants, however, are modular organisms, and many "modules" (individual ants) can be lost without necessarily threatening the reproductive unit (the colony), in a manner analogous to the effects of herbivory on plants (Andersen 1991a). Therefore, combined with the protection provided by nests, especially those in the soil, habitat disturbance is often not much of a disturbance to ants at all, unless it is so severe that it causes widespread destruction of colonies. The major effects of habitat disturbance are often indirect and stress-related, influencing habitat structure, microclimate, and food supplies (Andersen 1995). The importance of predation as a disturbance of ant communities has been little studied. Despite a wide range of animal species that feed on ants, some exclusively so (Redford 1987; Abensperg-Traun and Steven 1997), predation is not generally regarded as a major force structuring ant communities. However, there is increasing evidence that predation by other insects (Gotelli 1996) and more particularly parasitism by phorid flies (Feener 1981; Orr 1992; Porter et al. 1995c) can be an important factor in regulating foraging in some ant species, with significant effects on community dynamics.

From a global perspective, environments can be classified according to the relative importance of stress and disturbance as factors driving community structure, following the nomenclature of Grime (1979). Environments subject to severe stress or disturbance are characterized by highly specialized stress-tolerant species and unspecialized ruderal species, respectively. At very low levels of stress and disturbance, competition becomes the primary factor regulating community structure, and highly competitive species predominate. Three primary types of communities—stress-tolerant, ruderal, and competitive—can therefore be recognized in relation to stress and disturbance (Andersen 1991a, 1995). At intermediate levels of stress and disturbance, a variety of secondary community types can also be identified (Fig. 3.1).

Just as the degree of disturbance from a vegetation perspective does not necessarily reflect the level of disturbance to ant communities, the same environmental conditions can represent very different levels of stress from plant and ant perspectives, and can therefore support very different structural types of communities (Andersen 1995). For example, hot and open environments represent low levels of stress for ants and

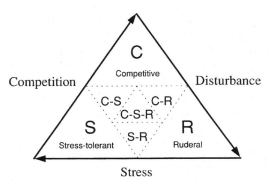

Figure 3.1. Classification of communities in relation to stress and disturbance, following the nomenclature of Grime (1979). See text for details. Modified from Andersen (1995).

support competitive ant communities. However, these same conditions are stressful for plants (i.e., primary productivity is low), and such environments support stress-tolerant plant communities, dominated by taxa such as cactuses in North America and hummock grasses in Australia. Environments support structurally analogous plant and ant communities only when ants and plants respond similarly to limiting factors. An example of this is the ground layer of rainforests, which supports stress-tolerant ant and plant communities because both ants and plants are limited by low levels of sunlight.

Ant Functional Groups

Global community ecology requires the identification of functional groups that transcend taxonomic and biogeographic boundaries and vary predictably in response to stress and disturbance. Such groups have been identified for ants based on Australian studies (Greenslade 1978; Andersen 1995, 1997a). There are seven such ant functional groups, and their major representatives in Australia and the New World are listed in Table 3.1. A generalized model of the relationships of these groups to each other, and to environmental stress and disturbance, is

Table 3.1 Ant Functional Groups in Relation to Stress and Disturbance, with Major Representatives in Australia and the New World[a]

Functional Group	Australia	New World
1. Dominant Dolichoderinae	*Anonychomyrma, Froggattella, Iridomyrmex, Papyrius, Philidris*	*Azteca, Forelius, Linepithema, Liometopum*
2. Subordinate Camponotini	*Calomyrmex, Camponotus, Opisthopsis, Polyrhachis*	*Camponotus*
3. Climate specialists		
a. Hot	*Melophorus, Meranoplus, Monomorium* (part)	*Pogonomyrmex, Solenopsis s.s, Myrmecocystus*
b. Cold	*Monomorium* (part), *Notoncus, Prolasius, Stigmacros*	*Formica* (part), *Lasius, Leptothorax, Stenamma, Lasiophanes*
c. Tropical	Many taxa	Many taxa
4. Cryptic species	Very many small myrmicines and ponerines, including *Hypoponera*, most Dacetonini, and *Solenopsis* (*Diplorhoptrum*)	Very many small myrmicines and ponerines, including *Hypoponera*, most Dacetonini, and *Solenopsis* (*Diplorhoptrum*)
5. Opportunists	*Paratrechina, Rhytidoponera, Tetramorium*	*Dorymyrmex, Formica* (*fusca* gp.), *Myrmica, Paratrechina*
6. Generalized Myrmicinae	*Crematogaster, Monomorium, Pheidole*	*Crematogaster, Monomorium, Pheidole*
7. Specialist Predators	*Bothroponera, Cerapachys, Leptogenys, Myrmecia*	*Dinoponera, Leptogenys, Pachycondyla, Polyergus*

[a]See text and Table 5.1 for details.

shown in Fig. 3.2. The seven functional groups are as the following:

1. *Dominant Dolichoderinae.* From a global perspective, competitively dominant taxa are by definition those that predominate in environments experiencing low levels of stress and disturbance. For ants, such environments are hot and open ones, and these are often dominated both numerically and functionally by highly aggressive dolichoderines. This is particularly true in Australia, where *Iridomyrmex* and other dolichoderines dominate the continental ant fauna to an extent unparalleled elsewhere. However, it is also true for warmer regions of the New World, where *Forelius, Linepithema,* and *Liometopum* are behaviorally dominant ants in open habitats, and *Azteca* and *Dolichoderus* are highly dominant in the canopies of rainforest. It is important to appreciate that global dominance (where *global* defines the spatial scale on which dominance is considered) does not at all imply universal dominance (Andersen 1997b). Dolichoderines are not at all universally distributed, and they are often absent entirely from even moderately stressful habitats.

2. *Subordinate Camponotini.* Camponotine formicines, especially species of *Camponotus,* are also very often diverse and abundant in rich ant communities. Most are behaviorally submissive to dominant dolichoderines, and many are ecologically segregated from them owing to their large body size and often nocturnal foraging.

3. *Climate specialists.* These taxa have distributions heavily centered on either arid zones (hot climate specialists), the humid tropics

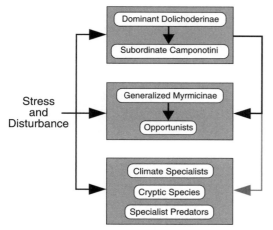

Figure 3.2. Functional group model of ant community organization in relation to environmental stress (factors limiting productivity) and disturbance (factors removing biomass). Arrows indicate direction and strength of influence. See text for details.

(tropical climate specialists), or cool-temperate regions (cold climate specialists). Both cold and tropical climate specialists are characteristic of habitats where the abundance of dominant dolichoderines is low, and, aside from their habitat tolerances, they are often unspecialized ants (army and fungus-growing ants are obvious exceptions). Hot climate specialists, on the other hand, are characteristic of sites where dominant dolichoderines are most abundant, and they possess a range of physiological, morphological, and behavioral specializations relating to their foraging ecology, which reduce their interaction with other ants. They include thermophilic taxa (such as species of *Cataglyphis, Melophorus, Myrmecocystus,* and *Ocymyrmex;* Snelling 1976; Marsh 1985; Christian and Morton 1992; Wehner et al. 1992) and specialist seed harvesters (including species of *Messor, Monomorium,* and *Pogonomyrmex;* Morton and Davidson 1988;

Andersen 1991b; Medel and Vásquez 1994), which feature in virtually all of the world's desert ant communities. Although species of *Forelius* have been described as dominant dolichoderines (Andersen 1997a), they might also be regarded as hot climate specialists (Bestelmeyer 1997).

4. *Cryptic species.* These are small to minute species, predominantly myrmicines and ponerines, that nest and forage primarily within soil, litter, and rotting logs. They are most diverse and abundant in forested habitats and are a major component of leaf litter ants in rainforest.

5. *Opportunists.* These are unspecialized, poorly competitive, ruderal species (Grime 1979), whose distributions appear to be strongly influenced by competition from other ants. They often have very wide habitat distributions, but predominate only at sites where stress or disturbance severely limit ant productivity and diversity, and therefore where behavioral dominance is low.

6. *Generalized Myrmicinae.* Species of *Crematogaster, Monomorium,* and *Pheidole* are ubiquitous members of ant communities throughout the warmer regions of the world, and they are often among the most abundant ants. As will be discussed later in this chapter, there is often competitive tension between them and dominant dolichoderines, including in tropical rainforest.

7. *Specialist predators.* This group comprises medium-sized to large species that are specialist predators of other arthropods. They include solitary foragers, such as species of *Pachycondyla,* as well as group raiders, such as species of *Leptogenys.* Except for direct predation, they tend to have little interaction with other ants owing to their specialized diets and typically low population densities.

Distribution of Behavioral Dominance

Globally, behavioral dominance becomes increasingly important to community structure with decreasing stress and disturbance. This trend is illustrated by ant behavior at tuna baits along an environmental gradient in southeastern Arizona, where monopolization by large numbers of behaviorally dominant species was greatest in desert (warm and open) habitats and least in forest (cool and shady) habitats (Fig. 3.3; Andersen 1997a).

The behaviorally dominant ants in warm regions are primarily Dominant Dolichoderinae and Generalized Myrmicinae, and, as previously mentioned, in open habitats there is often competitive tension between them. Dominant dolichoderines are strongly associated with hot, open habitats, such as deserts, Mediterranean ecosystems, and the canopies of tropical rainforests. Generalized myrmicines, by comparison, are far more shade tolerant, with *Pheidole* being a numerically dominant genus on the rainforest floor throughout the tropical world (Chapter 8). Globally, I consider Generalized myrmicines to be competitively subdominant ants (Andersen 1995) for the following reasons (Table 3.2):

1. They are considerably more stress tolerant than Dominant dolichoderines.
2. Whereas Dominant dolichoderines typically have large territories and individuals exhibit extremely high rates of activity, territory size tends to be more restricted in

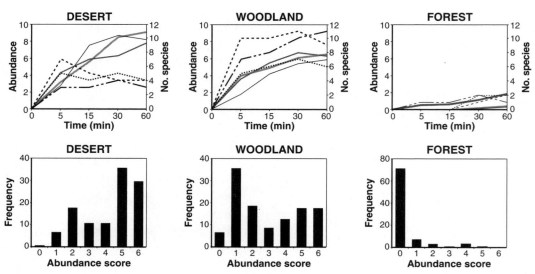

Figure 3.3. Behavior of ants at tuna baits at desert, woodland, and forest sites in southeastern Arizona, illustrating high, moderate, and low levels of behavioral dominance, respectively. Ant abundance (top; solid lines) at desert sites increases rapidly, reaching saturation levels after 30 minutes. Species richness (top; dotted lines), however, levels off after 5 minutes owing to competitive exclusion. Ant abundance is lower at woodland sites, but species richness continues to increase with time (local species richness is similar at desert and woodland sites). Both abundance and richness are very low at forest sites. Ant abundance scores (bottom) were usually either 5 (>20 ants) or 6 (>50 ants) at desert sites, fairly evenly distributed at woodland sites, and usually 0 (no ants) at forest sites. Data from Andersen (1997a).

Table 3.2 Generalized Myrmicines as Subdominant Ants to Dominant Dolichoderines[a]

Characteristic	Dominant Dolichoderinae	Generalized Myrmicinae
Primary distribution	Low stress	Moderate stress
Territory size	Large	Restricted
Rates of foraging activity	Very high	Moderate
Resource monopoly	Aggressive displacement	Occupation and defense

[a]See text for details.

Generalized myrmicines, and rates of activity are more moderate.

3. Dominant dolichoderines actively displace other ants from food sources, whereas Generalized myrmicines often rely more on stout defense of food sources they have initially occupied (Andersen et al. 1991).

In cooler parts of the world, Dominant dolichoderines are mostly absent, and the abundance of Generalized myrmicines is greatly reduced. Throughout the Palearctic and Nearctic, the behaviorally dominant ants of cool-temperate regions are mound-building formicines (Cold climate specialists)—species of *Formica* and to a lesser extent *Lasius* (Creighton 1950; Rosengren and Pamilo 1983; Savolainen and Vepsäläinen 1988). It seems likely that their behavioral dominance in such cool climates is related to the thermoregulatory properties of their nests (Hölldobler and Wilson 1990). For example, with air temperature less than 14°C, *Formica polyctena* can achieve nest temperatures of up to 25°C (Coenen-Stass et al. 1980).

The relative importance of behavioral dominance varies markedly within the rainforest in response to increasing stress. Most behaviorally dominant taxa that occur in the tropics are arboreal, a habitat in which they can exploit direct sunlight. Such taxa include Dominant dolichoderines (e.g., *Azteca, Dolichoderus, Philidris;* Greenslade 1971; Adis et al. 1984; Tobin 1991;

Shattuck 1992b), Generalized myrmicines (e.g., *Crematogaster;* Greenslade 1971; Majer 1976; Adis et al. 1984), Tropical climate specialists (e.g., *Myrmicaria, Oecophylla;* Greenslade 1971; Majer 1976; Stork 1991), and Subordinate camponotines (e.g., *Camponotus;* Wilson 1987). The canopy is the most productive microhabitat for both ants and plants in tropical rainforest, and there is increasing evidence that behaviorally dominant ants are predominantly primary consumers, being sustained by plant and homopteran exudates (Tobin 1994; Davidson and Patrell-Kim 1996).

The abundance of behaviorally dominant ants in rainforest decreases with increasing latitude and altitude, with Dominant dolichoderines virtually being restricted to the lowland tropics. On the rainforest floor, the heavy shade and litter represent considerable stresses for ants, and, as discussed by Kaspari (Chapter 2), behavioral dominance is relatively poorly developed, even in the lowland tropics.

Functional Group Composition

Consistent patterns of functional group composition can be recognized in relation to climate and vegetation (i.e., environmental stress). Functional group composition varies between climatic zones and, within any particular zone, varies systematically with vegetation type (Andersen 1995, 1997a). For example, in monsoonal northwestern Australia (Fig. 3.4a–c) the

Figure 3.4. Effects of vegetation on functional group composition in the monsoonal tropics of northern Australia (a–c; data from Andersen 1991c; Andersen and Reichel 1994) and in cool-temperate southern Australia (d–f; data from Andersen 1986a, 1986b). Functional groups: CCS, Cold climate specialists; CS, Cryptic species; DD, Dominant Dolichoderinae; GM, Generalized myrmicines; HCS, Hot climate specialists; OPP, Opportunists.

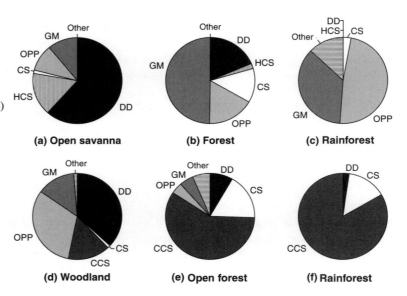

(a) Open savanna   (b) Forest   (c) Rainforest

(d) Woodland   (e) Open forest   (f) Rainforest

predominant vegetation is savanna, and functional group composition is similar to that in the arid zone (predominantly Dominant dolichoderines, Hot climate specialists, and Generalized myrmicines; Fig. 3.4a). The long-term absence of fire increases the structural complexity of the vegetation (Andersen 1996), thereby markedly reducing insolation at the soil surface. This dramatically reduces the abundance of Dominant dolichoderines and Hot climate specialists, and increases the abundance of Generalized myrmicines (Fig. 3.4b; Andersen 1991c).

In local patches of monsoonal rainforest, where insolation at the soil surface is even lower, Dominant dolichoderines and Hot climate specialists are absent altogether, and most ants are either Generalized myrmicines or Opportunists (Fig. 3.4c; see also Andersen and Majer 1991; Reichel and Andersen 1996). In cool-temperate southern Australia (Fig. 3.4d–f), the abundance of Dominant dolichoderines and Generalized myrmicines is generally low, and Opportunists and Cold climate specialists are usually among

the most common ants. Dominant dolichoderines and Generalized myrmicines are usually only abundant in open habitats (Fig. 3.4d), and the relative abundances of cold climate specialists and cryptic species increase with decreasing insolation (Fig. 3.4e,f).

The ground-foraging ant faunas of different rainforest types have distinctive functional group signatures. The lowland tropics feature Generalized myrmicines (particularly *Pheidole*), Cryptic species, Tropical climate specialists (including army and leaf cutter ants), and Specialist predators (primarily large ponerines; Chapter 8). With increasing elevation or latitude, the diversity and abundance of cryptic species and particularly Generalized myrmicines and Specialist predators declines, and Tropical climate specialists are replaced by Cold climate specialists (including *Stenamma* in the New World). The faunas of cool-temperate rainforests are composed almost entirely of Cold climate specialists (including *Lasius, Leptothorax, Prenolepis,* and *Stenamma* in the

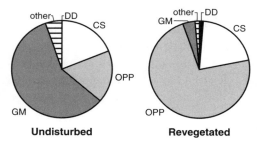

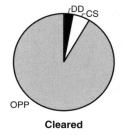

Figure 3.5. Effects of disturbance on functional group composition of rainforest ants in the humid tropics of Queensland. Functional groups as in Fig. 3.4. Data from King et al. (1998).

north, and *Lasiophanes, Notoncus, Prolasius,* and certain *Monomorium* in the south), Cryptic species (e.g., *Hypoponera*), and Opportunists (e.g., *Paratrechina, Rhytidoponera,* and the *fusca* group of *Formica*).

Functional group composition responds predictably to habitat disturbance in temperate and semiarid regions (Andersen 1990; Bestelmeyer and Wiens 1996), but the effects of disturbance on functional group composition of tropical rainforest ant communities have been poorly documented. In temperate southeastern Australia, for example, disturbance typically results in the proliferation of Opportunists, especially small species of *Rhytidoponera* (Andersen 1988, 1990; Andersen and McKaige 1987). Such a proliferation of Opportunists, especially species of *Formica* (*fusca* group) and *Myrmica,* following disturbance is also characteristic of cool-temperate regions in the Northern Hemisphere (Brian 1964; Gallé 1991; Andersen 1997a). Results from Queensland (Fig. 3.5) indicate that a proliferation of Opportunists (species of *Paratrechina* and *Rhytidoponera*) is also a characteristic response to severe disturbance in humid tropical Australia. This also appears to be true in the Solomon Islands, where tree clearing favors opportunist species of *Cardiocondyla, Paratrechina, Tapinoma,* and *Tetramorium* (Greenslade and Greenslade 1977). Aside from arbo-

real taxa, Specialist predators and Cryptic species were especially sensitive to tree clearing in the latter study. Cryptic species also appear to be especially sensitive to tree clearing in the neotropics (Majer et al. 1997), where edge effects can be manifest for up to 200 m into the forest.

Conclusion

In any functional group analysis there is an inevitable trade-off between generality and precision, and the broad-scale predictive power of a global scheme will inevitably be inadequate for a detailed understanding of the dynamics of particular communities (Andersen 1997b). However, a global ecology based on functional groups in relation to stress and disturbance provides a predictive framework for analyzing broad patterns of (1) community composition and behavioral dominance within and between rainforest types, and (2) the responses of rainforest ant communities to disturbance. Unfortunately, even such coarse-scale analyses are highly constrained by a patchy geographic coverage of relevant studies (e.g., very little has been published from Africa) and a paucity of information on the effects of habitat disturbance (other than tree clearing). Nevertheless, there appears to be substantial convergence between biogeographic regions in the distribution of behavioral dominance within and between rainforest types and in changes in functional group composition in relation to stress (primarily temperature) and disturbance. To the extent that these patterns are confirmed by further studies, global functional groups are a valuable tool for understanding the dynamics of rainforest ant communities.

ACKNOWLEDGMENTS

I thank all participants in the Ants of the Leaf Litter conference for stimulating discussions on rainforest ant community ecology. I am particularly grateful to Brandon Bestelmeyer and John Wiens for their comments on the manuscript. This chapter is CSIRO Tropical Ecosystems Research Centre contribution no. 937.

Chapter 4

The Interactions of Ants
with Other Organisms

Ted R. Schultz and Terrence P. McGlynn

There is little doubt that the common ancestor of all ants was eusocial and that it maintained some sort of stable nest environment. From this auspicious groundplan a broad variety of complex behaviors has evolved. Coupled with the sheer abundance of the Formicidae, this behavioral diversity has produced a spectacular array of interactions between ants and other organisms. Properly understood, these interactions could be used to predict ecological conditions within a given habitat by the presence of a particular ant species, with the goal of using ground-dwelling ants as indicators of biodiversity.

Two broad factors work against the realization of this goal. First, given our current understanding of ecological processes, we are unable to draw reliable inferences about general ecological conditions from knowledge of particular interactions between species. A more direct approach would be to establish empirically correlations between the presence of particular ant species and particular ecological conditions, and to use these correlations as predictors. This practical goal will likely be expedited if we pay special attention to ant species that participate in precise, obligate interactions with other species, or in which interactions are very complex, since we might expect such species to be most sensitive to general ecological conditions.

A second factor that mitigates against using the interactions of ants with other organisms for inferring biodiversity is our poor understanding of these interactions. Knowledge of ant biology relies heavily on the intensive study of only a few species. In many cases the diets of entire genera are unknown. Many plants are known to possess specialized structures for the housing of ants, yet the identities of their ant guests remain mysterious. What knowledge there is exists mostly in the form of brief notes in the literature, although repeated attempts have been made to collate subsets of this information (see references cited later in this chapter).

This chapter serves as a brief and limited introduction to the complex and fascinating subject of interactions between ants and other organisms. In discussing such interactions, we will employ two broad categories: trophic interactions, in which one organism is eaten by the other, and symbioses, in which the organisms coexist for an extended period of time. We will respect the traditional subdivisions of the latter category: (1) parasitism, in which one partner benefits at the expense of the other; (2) commensalism, in which one partner benefits and the other is neither harmed nor benefited; and (3) mutualism, in which both parties benefit. Symbioses may be further distinguished as either facultative or obligate on the part of one or both symbionts, depending on whether it is possible for that partner to survive outside the relationship. Our ability to place the variety of interactions described in the following sections into these categories is limited by our ignorance of the details of particular cases as well as by the artificiality of these groupings.

Interactions with Plants

Trophic Interactions

Myrmecological orthodoxy regards ants as carnivores, and certainly few ants are entirely herbivorous. However, as pointed out by Tobin (1994), nectar and other plant products play an important and generally underappreciated nutritional role in the diets of many ant species, especially those of the adults. Certainly many otherwise carnivorous ants are attracted to floral and extrafloral nectaries, and some (e.g., species of *Solenopsis* [Tennant and Porter 1991]

and *Atta* [Quinlan and Cherrett 1979]) are known to feed on plant sap and fruit juices. Some ants (e.g., the North American *Messor pergandei*) rely entirely on seeds for nourishment, and many more (including species in *Monomorium, Pheidole,* and *Pogonomyrmex*) rely heavily on seeds. Aside from such obligate "harvesting ants," many ant species are occasional seed consumers (e.g., Beattie 1985; Kaspari 1993b), and many more forage for seeds bearing elaiosomes, ant-attractive nutritive attachments manufactured by the plant to encourage seed dispersal (Handel et al. 1981; Handel and Beattie 1990a, 1990b). As discussed in more detail subsequently, ants also consume specialized food bodies produced by plants, such as Beltian bodies in the New World *Acacia* and Müllerian bodies in the New World *Cecropia.*

From the standpoint of ecological energetics, the Neotropical leaf-cutting ants could be regarded as herbivorous (Stradling 1978), since they harvest an estimated 15% of all fresh vegetation (Cherrett 1986) in the Neotropics. Likewise, ants that rely on homopteran-produced honeydew could be regarded as essentially herbivorous, since homopteran tending is bioenergetically comparable to collecting plant fluids directly (Tobin 1994). However, both leaf-cutting and homopteran-tending ants acquire their nutrition via symbiotic intermediates (fungi and homopterans)—an important consideration from the species interaction perspective taken here.

Symbiotic Interactions

Ants participate in symbioses with over 465 plant species in over 52 families (Jolivet 1996) and, not surprisingly, the literature of ant-plant symbioses is vast (for reviews, see Bailey 1922b; Bequaert 1922; Wheeler 1942; Buckley 1982a, 1982b; Beattie 1985; Hölldobler and Wilson 1990; Huxley and Cutler 1991; Jolivet 1996). Whether the majority of these symbioses are mutualisms, in which both partners benefit, or whether they are beneficial only to the ants was formerly a matter of argument among both myrmecologists and botanists (e.g., Belt 1874; Schimper 1888, 1898; von Ihering 1891; Rettig 1904; Wheeler 1913, 1942; Skwarra 1934; Brown 1960). Largely because of recent experimental work (e.g., Janzen 1966, 1967; Davidson et al. 1988), the majority of ant-plant symbioses are currently regarded as true mutualisms, in which ants obtain shelter, nourishment, or both and plants obtain protection against both arthropod and vertebrate herbivores. In some cases, plants may also obtain nutrients from ant waste materials and soil, gain protection from competing plants (which are removed by the ants), have their seeds dispersed, and, in rare cases, even get pollinated.

Cases of commensalism, in which ant colonies gain shelter but neither harm nor benefit the host, certainly exist. For instance, many ants occupy hollow stems (e.g., *Camponotus* and *Crematogaster* species), abandoned insect galls (e.g., *Leptothorax* species), and the tangled roots of epiphytes (e.g., *Anochetus* and *Strumigenys* species). Such associations presumably served as evolutionary precursors for mutualisms in which plants receive protection from herbivory in exchange for supplying cavities favoring ant occupation, including hollow branches, stems, and thorns; hollow pseudobulbs; or pouchlike domatia on leaves and petioles. In many cases, plants provide food as well as shelter, including carbohydrate-laden extrafloral nectaries and fatty or proteinaceous pearl bodies. According to O'Dowd (1982), the latter are produced by American, Asian, and African plants in over 50 genera in 19 families. Alternatively, plants may provide food but not shelter, encouraging frequent visitations by a variety of ant species and, presumably, many of the herbivore-repelling benefits that such visitations afford. Schupp and Feener (1991) showed

that more than a third of the plants in a Panamanian forest may practice such a strategy.

The classic case of ant-plant mutualism is that of the New World members of the genus *Acacia* (Leguminosae), which produce both extrafloral nectaries and proteinaceous Beltian bodies. Known for their painful stings, ants in the genus *Pseudomyrmex* occupy the hollow thorns, repelling arthropods (Coleoptera, Hemiptera, Homoptera, Lepidoptera) as well as large browsing mammals. The ants also kill other plants growing within a certain radius around the occupied *Acacia* (Janzen 1966, 1967). Another relatively well-studied case of mutualism is that of *Cecropia* (Moraceae), in which ants (primarily *Azteca* species, including six obligate *Cecropia*-dwellers, but also species of *Camponotus, Crematogaster, Pachycondyla,* and other genera) occupy the plants' hollow stems, excavating entrance holes in preformed, weakened areas present in the walls of internodes. *Cecropia*-dwelling *Azteca* species are particularly well known for their ferocity, reacting aggressively to any disturbance to their host plant. The plant provides nourishment to the ants in the form of glycogen-rich Müllerian bodies growing on a pad (trichilium) at the base of the leaf petiole (Bailey 1922a; Rickson 1971; Longino 1991). A third example is that of the hollow pseudobulbs of *Hydnophytum, Myrmecodia,* and other species of the Hydnophytinae (Rubiaceae) of the Far Eastern tropics. These plants possess swollen tubers with preformed, often complex chambers that in some species are nearly always occupied by ants (usually *Iridomyrmex* species). The plants obtain nutrition from ant excrement and debris absorbed by "warted" surfaces found in some of the chambers (Miehe 1911a, 1911b; Bequaert 1922; Huxley 1978; Jebb 1991).

Some ants cultivate and occupy "ant gardens," clusters of epiphytes planted in the branches of trees on soil and carton provided by the ants (Ule 1902; Kleinfeldt 1978, 1986;

Buckley 1982a; Davidson 1988). The ants, predominantly species in the genera *Camponotus, Crematogaster,* and *Solenopsis,* obtain nutrition from extrafloral nectaries, elaiosomes, and fruit sap; the plants gain protection from herbivores, including leaf-cutting attines (Weber 1943). Some ant species—including *Camponotus femoratus, Crematogaster parabiotica,* and *Monacis debilis*—may be obligate ant-garden nesters.

As already mentioned, ants are important dispersers of seeds (Beattie 1985) and are frequently encouraged in this role by attractive and nutritious elaiosomes provided by plants (Handel et al. 1981; Handel and Beattie 1990a, 1990b). Ants may also exercise an underappreciated influence on seed germination. Oliveira et al. (1995) showed that the South American fungus-growing ant *Mycocepurus goeldii* (Myrmicinae) significantly enhanced germination of the seeds of the tree *Hymenaea courbaril* (Caesalpiniaceae) by removing fruit pulp and thereby reducing fungal infestation. Although ant pollination may be important for some plants in some habitats (Peakall et al. 1991), it has been suggested that, because the antibiotic secretions of the metapleural gland demonstrably inhibit normal pollen function, ants are unlikely to be recruited into insect-plant pollination symbioses (Iwanami and Iwadare 1978; Beattie et al. 1984, 1985, 1986).

Interactions with Animals

Trophic Associations

ANTS AS PREDATORS. The ancestral ant was very likely a generalized predator. Arising from this lifestyle, highly specialized predation has evolved in many ant groups. For instance, diverse groups of ants—including *Acanthostichus, Cylindromyrmex,* and *Eurhopalothrix heliscata*—have independently specialized on termites (Brown 1975; Wilson and Brown 1984;

Overal and Bandeira 1985). Some myrmicine ants in the genera *Carebara, Carebarella, Erebomyrma, Liomyrmex, Paedalgus,* and *Solenopsis* are known to make their nests in close proximity to those of termites, and it is assumed that they steal termite eggs and brood for food (Forel 1901; Wheeler 1914, 1936; Wilson 1962b; Ettershank 1966). Species of *Discothyrea, Proceratium* (Ponerinae), and *Stegomyrmex* (Myrmicinae) prey on arthropod eggs (Brown 1974f, 1979; Diniz and Brandão 1993). A variety of species in the Dacetonini (Myrmicinae) prey on *Collembola* (Wilson 1953; Masuko 1984). Species of the Neotropical genus *Thaumatomyrmex* use their bizarre, pitchfork-like mandibles to remove the repellent hairs of what is apparently their sole prey item, millipedes in the family Polyxenidae (Brandão et al. 1991). A number of *Leptogenys* species specialize on isopods (pillbugs); at least one specializes on Dermaptera (earwigs) (Steghaus-Kovac and Maschwitz 1993). Adult workers of the Japanese myrmicine species *Myrmecina graminicola nipponica* and *M. flava* capture oribatid mites, skillfully tearing a hole in the highly sclerotized integument; the larvae then feed by inserting their peculiarly elongate heads into these holes (Masuko 1995). Finally, some ant species are specialized predators on other ants, including species of *Cerapachys* and *Neivamyrmex* (Wheeler 1918; Rettenmeyer 1963).

ANTS AS PREY. Many ant species represent predictable food sources for predators because of their large numbers, their tendency to forage in trails, and their long-lived, stable, usually stationary nests. In what is no doubt a continuing evolutionary arms race, ants have adopted numerous defenses, including repellent chemicals and soldier castes, to discourage predators, while predators have acquired methods of overcoming such defenses, becoming increasingly specialized in the process. Such predators include assassin bugs (Reduviidae), ground beetle larvae (Carabidae), rove beetles (Staphylinidae), ant lions (Myrmeleontidae), and worm lions (Diptera: Rhagionidae: *Vermileo*). Sphecid wasps in the genera *Aphilanthops* and *Clypeadon* provision their nests exclusively with ants (Evans 1962, 1977). In Costa Rica, windscorpions (Solifugae) run along and within nocturnal foraging columns of the leaf-cutting ant *Atta cephalotes,* probably preying on single workers (Bukowski 1991). Spiders that specialize on ants are often striking visual mimics of their prey, presumably as camouflage against small vertebrate predators that have learned to avoid the ants (Oliveira and Sazima 1984; Oliveira 1988). Although most vertebrates studiously avoid ants because of their stings and noxious chemical deterrents, vertebrate ant specialists include anteaters and some toads, lizards, snakes, and birds (Bequaert 1922; Weber 1972b).

Symbioses

Symbioses between ants and other animals (particularly arthropods) constitute a vast and fascinating subject, the far-flung literature of which has yet to be exhaustively catalogued (but see Kistner 1979, 1982, and Hölldobler and Wilson 1990 for excellent reviews).

ANT-HOMOPTERAN ASSOCIATIONS. The tending of homopterans by ants is well known, no doubt because of the striking parallel with the husbandry of cattle by humans. The majority of homopteran-tending ant species occur in the subfamilies Dolichoderinae, Formicinae, and Myrmicinae, although some ponerines (notably in the Ecatommini) also obtain significant nutrition through such interactions. In addition to Aphidae, ants also tend homopterans in the families Cercopidae, Cicadellidae, Coccidae, Fulgoridae, Membracidae, Pseudococcidae, and Psyllidae. Ants feed on "honeydew," a by-product of homopteran phloem-feeding consisting largely of carbohydrate but also containing amino acids that in some cases are added by the homopteran (Dixon 1985). In return, the ants protect homopterans from predators and parasitoids.

Most ant-homopteran associations are facultative mutualisms. However, ants in the North American genus *Acanthomyops* (Formicinae) appear to be obligately dependent on their root coccid symbionts (Wing 1968), as are species of the pantropical genus *Acropyga* (Formicinae). A virgin queen of an *Acropyga* species departs on her nuptial flight carrying in her mandibles a coccid symbiont to serve as the parthenogenetic progenitor of a future "herd" (Silvestri 1925; Wheeler 1935; Bünzli 1935; Brown 1945; Buschinger et al. 1987). This behavior has also been observed in a Sumatran *Cladomyrma* species (Roepke 1930). The Malaysian ant *Hypoclinea cuspidatus* (Dolichoderinae), an obligate symbiont of the mealybug *Malaicoccus formicarii* (Pseudococcidae), is a true "nomadic herdsman." The entire colony (which can consist of more than 10,000 workers and 5000 pseudococcids) is constantly on the move as old feeding sites are depleted and new ones required (Maschwitz and Hänel 1985).

Ants also tend caterpillars in the butterfly family Lycaenidae (Hinton 1951; Atsatt 1981; Pierce 1987). In this case, nourishment is provided to the ants via specialized glands, and, as in the homopteran case, protection from predation and parasitism is provided by the ants (Malicky 1969; Pierce and Mead 1981; Pierce and Easteal 1986).

GUESTS IN ANT NESTS. Many thousands of arthropod species make their homes and/or earn their livings in the stable environment afforded in or near ant nests, including members of the Acari (Chelicerata), Araneae, Collembola, Diplopoda, Isopoda (Crustacea), Pseudoscorpionida, and, within the insects, of the orders Blattaria, Coleoptera, Diptera, Homoptera, Hymenoptera,

Lepidoptera, Neuroptera, Orthoptera, Psocoptera, and Thysanura. Some parasitic symbionts simply steal food from ant foragers. For example, species of the Old World calliphorid fly genus *Bengalia* dart into columns and snatch away the food of various ant species (e.g., *Bothroponera, Camponotus, Dorylus, Leptogenys, Technomyrmex* species) (Bequaert 1922; Maschwitz and Schönegge 1980). Mosquitos of the genus *Malaya* (=*Harpagomyia*) are able to solicit regurgitated droplets from various Asian and African *Crematogaster* species (Jacobson 1909; Farquharson 1918; Wheeler 1928).

Many symbionts live inside ant nests, gaining their nourishment by feeding on refuse in the nest middens, by stealing the food of the ants, by preying on adult ants or brood, or by preying on other symbionts. In the very large nests of some ant species, remarkably large numbers of such "ant guests" can be found. For example, in one large refuse chamber within a four-year-old nest of *Atta sexdens rubropilosa,* Autuori (1942) found adult forms of 1491 Coleoptera, 56 Hemiptera, 40 Mollusca, 15 Diptera, 4 Reptilia, and 1 pseudoscorpion. In a study of 150 army ant colonies, Rettenmeyer (1962) collected 8000 mites, 2400 phorid flies, 1100 limulodid beetles, 300 staphylinid beetles, 300 Collembola, 170 Thysanura, 150 Diplopoda, 140 hysterid beetles, and 6 diapriid wasps.

Facultative ant-nest symbionts, which are also found living without ants, are typically species that are predisposed to soil and leaf litter environments, such as oribatid mites and Collembola. For example, pyrgodesmid millipedes are frequently found in the refuse piles of the nests of lesser attines, e.g., *Mycetarotes parallelus.* In contrast, obligate symbionts of ants, presumably derived from facultative ancestors, are found only in ant nests, and often only in the nests of particular ant species. Mites of the genus *Antennophorus* (Antennophoridae), for example, live on the body surfaces of ants in the closely related genera *Acanthomyops* (Formi-

cinae) and *Lasius* and obtain nourishment by stealing drops of food during trophallaxis or by actively soliciting such droplets by mimicking the tactile signals used by ants for this purpose (Janet 1897; Wasmann 1902; Karawajew 1906; Wheeler 1910). Many other ant-nest symbionts, including the thysanuran *Atelura formicaria* and the hysterid beetle *Hetaerius brunneipennis,* steal or successfully solicit regurgitated food (Wheeler 1908).

The pseudoscorpion *Sphenochernes schulzi* lives in nests of the Argentinean fungus-growing ant *Acromyrmex lundi,* where it apparently feeds on worker ants by first immobilizing them with injected poison, then imbibing their hemolymph (Turk 1953). The third instar of the lycaenid caterpillar *Maculinea teleius* (Lepidoptera) follows ant pheromone trails and enters nests of *Myrmica rubra,* where it feeds on the brood (Chapman 1920; Malicky 1969; Schroth and Maschwitz 1984). Other examples of nest symbionts that have acquired the ability to follow ant pheromone trails include the milichiid fly *Pholeomyia decorior,* a symbiont of the fungus-grower *Trachymyrmex septentrionalis* (Sabrosky 1959) and the cockroach *Attaphila fungicola,* resident in nests of *Atta texana* (Moser 1964). In an example of extreme integration, the staphylinid beetle *Lomechusa stumosa* (Staphylinidae: Aleocharinae) possesses specialized "appeasement glands" at the tip of its abdomen containing a proteinaceous substance that seems to exercise a calmative effect on its ant host, the European *Formica sanguinea.* Once incorporated into the nest, it preys on the ant brood and obtains regurgitated liquid food from workers (Hölldobler 1967, 1968; Hölldobler and Wilson 1990).

Numerous parasitoid species prey upon ants, including species of the hymenopteran families Diapriidae (Masner 1976; Huggert and Masner 1983) and Eucharitidae (Clausen 1940a, 1940b, 1940c, 1941; Heraty and Darling 1984; Heraty 1985, 1986). Flies (Diptera) of the family

Phoridae are particularly important ant parasitoids (Borgmeier 1963; Feener 1981; Feener and Moss 1990; Brown 1993); most phorid larvae are internal parasites of their ant hosts, but larvae of at least one species are free-living in nests of the European *Plagiolepis pygmaea,* receiving regurgitated liquid from worker ants (LeMasne 1941).

Predatory velvet worms (Phylum Onychophora) have been discovered in rainforest *Pheidole* nests, but it is unknown whether they feed upon the ants (McGlynn and Kelley 1999). In what may be a mutualistic association, the earthworm *Dendrodrilus rubidus* (Phylum Annelida) is found in nests of the European red wood ant *Formica aquilonia* (Laakso and Setälä 1997). Adult snakes in the families Colubridae, Elapidae, and Leptotyphlopidae, and lizards in the families Amphisbaenidae and Teiidae, live in nests of species of the leaf-cutting ant genera *Acromyrmex* and *Atta* and/or use the nests as oviposition sites. In some cases these associations are obligate. Some of these snakes are capable of following ant pheromone trails and may utilize the ants or brood as food (Goeldi 1897; Autuori 1942; Gallardo 1951; Vaz-Ferreira et al. 1970, 1973; Weber 1972b; Brandão and Vanzolini 1985).

Internal metazoan parasites of ants are known to include nematodes, trematodes, and cestodes. Protozoan ant parasites include (class) Microsporidea (phylum Cnidospora), known from *Leptothorax, Myrmecia, Pheidole,* and *Solenopsis* species, and (class) Neogregarinida (phylum Apicomplexa), known from *Leptothorax* and *Solenopsis* species (Hölldobler 1929, 1933; Gösswald 1932; Allen and Buren 1974; Allen and Silveira-Guido 1974; Jouvenaz and Anthony 1979; Espadaler 1982; Buschinger and Winter 1983; Jouvenaz 1986; Crosland 1988; Buschinger et al. 1995).

SYMBIOSES BETWEEN ANTS. Finally, ant species may enter into varying degrees of symbiosis with each other. For instance, "thief ants" of *Solenopsis* subgenus *Diplorhoptrum* live in the walls of the nests of larger ant species and steal their food and larvae. In a more derived case, *Megalomyrmex symmetochus* is found within the nests of the fungus-growing ant *Sericomyrmex amabilis* (Wheeler 1925) and has also been reported from the nest of an unidentified *Trachymyrmex* species. The queen and brood occupy the fungus gardens, ignored by their hosts and apparently feeding on the fungus. The closely related *M. silvestrii* and an undescribed *Megalomyrmex* species parasitize other fungus-growing ants (Brandão 1990; J. Wetterer, pers. comm.).

In cases of social parasitism, ants of one species utilize the work force of another colony in order to raise their own brood. Workers of "slavemaking" species raid colonies, steal brood, and raise them as slaves. In other cases, parasitic foundress queens enter established colonies and take over, by either killing or dominating the host colony queen. Thereafter, the host workers aid in raising the interloping queen's offspring, which ultimately supplant the former inhabitants. In the most extreme cases, the parasite queen produces only sexual brood, having lost the ability to produce a worker caste. For example, the attine ant *Pseudoatta argentina* parasitizes nests of *Acromyrmex lundi.* Its exclusively sexual brood is reared by the *Acromyrmex* workers, and upon maturity they depart and mate, and the queens find new *A. lundi* nests to parasitize (Gallardo 1929).

Associations with Fungi

In general, ants avoid associations with fungi. Indeed ants have evolved at least two important characteristics for discouraging the presence of fungi (as well as bacteria) in their nests: elaborate grooming behaviors (Wilson 1962a) and the antiseptic-secreting metapleural gland (Maschwitz et al. 1970; Maschwitz 1974; Beattie et al. 1986). Little is known of fungal diseases of ants, although ant-pathogenic fungi have been described in the families Clavicipitaceae (Thaxter 1888; Rogerson 1970), Hyphomycetales (Balazy et al. 1986), and Laboulbeniales (Thaxter 1908; Bequaert 1922); an unidentified unicellular fungal pathogen has been reported from the hemolymph of *Solenopsis wagneri* (Jouvenaz et al. 1977).

Outside the tribe Attini (Myrmicinae), no confirmed examples of fungivory are currently known in ants. Although the infrabuccal pockets of many ant species contain fungal filaments, the digestive tract does not; rather, the infrabuccal pocket serves as a temporary repository for fungi accumulated during the cleaning of the body, and the infrabuccal pellet is discarded in the refuse heap (Bailey 1920). In a rare exception, fungal filaments have been discovered within the digestive tracts of ants in the myrmicine tribe Cephalotini, *Cephalotes atratus* and *Zacryptocerus clypeatus* (Caetano and Cruz-Landim 1985; Caetano 1989; Kane 1995), but in this case the fungus is thought to serve as a digestive tract symbiont.

The northern European *Lasius fuliginosus* (Formicinae) constructs carton nests in hollow tree trunks and in the soil, cementing the carton with regurgitated sugary liquid. The fungus *Cladosporium myrmecophilum* is found growing only on this cemented carton (Lagerheim 1900); however, the ants do not consume it (Maschwitz and Hölldobler 1970). Fungi have been reported growing in epiphytic Rubiaceae that are inhabited by *Iridomyrmex* ants (Miehe 1911b; Bequaert 1922; Huxley 1978), within hollow stems of South American *Hirtella* (Chrysobalanaceae) occupied by *Allomerus* ants (Dumpert 1981), and on carton nests constructed by *Crematogaster* species in Nigeria (Farquharson 1914), but, again, there is little reason to believe that any of these ants is fungivorous.

Ants are thought to disperse the spores of mycorrhizal fungi in the order Glomales, fami-ly Endogonaceae, including those of the genus *Glomus* (McIlveen and Cole 1976; Allen et al. 1984; Friese and Allen 1988, 1993; Janos 1993). Since glomalean spores are large (50–800 μm in diameter) and rich in lipids, and since spores are often found associated with roots in the chambers of soil-nesting ants, it has been suggested that some ants may consume them (D. P. Janos, pers. comm.). Went et al. (1972) found mycelium of various fungi in the refuse chambers of desert harvester ants in the genera *Veromessor* and *Manica,* and they reported that *M. hunteri* larvae consumed an unidentified fungus offered to them in artificial culture. Perhaps the best support for ant fungivory outside the Attini comes from observations made in Malaysia and Indonesia. In one case, a Malaysian *Prenolepis* species was photographed carrying off pieces of an unidentified basidiomycete fruiting body (Rosciszewski 1995). In a second case, fungal tissue fragments made up 50–80% of all food items carried to the nest by the giant forest ant *Camponotus gigas* in Borneo (Orr and Charles 1994; Levy 1996; S. Yamane, pers. comm.)

One group of exclusively New World ants, the Attini (Myrmicinae), are obligately fungivorous, with the fungus constituting the sole source of nourishment for the larvae and the dominant source for the adults (Barrer and Cherrett 1972; Littledyke and Cherrett 1976; Quinlan and Cherrett 1979). Attine ants possess an elaborate array of behaviors for cultivating fungus gardens. The less-derived species cultivate their fungi on insect frass, seeds, and other organic detritus obtained from foraging in the leaf litter. The derived "higher" attines, including *Acromyrmex* and *Atta* species, cultivate their fungi on fresh vegetation, including leaves and flowers cut for that purpose. The identity of the attine fungal symbiont has been the source of speculation for over a century, although most researchers agreed that attines cultivated one or more species within the families Agaricaceae or

Lepiotaceae (subdivision Basidiomycotina, order Agaricales). It is now known that most Attini cultivate lepiotaceous fungi, although some species within the genus *Apterostigma* cultivate a distantly related fungus in the Agaricales, closely related to the genus *Gerronema* in the family Tricholomataceae (Chapela et al. 1994; Moncalvo et al. 2000).

In a striking parallel with the ant-coccid association of *Acropyga* and *Cladomyrma* species already described, virgin attine queens carry a pellet of the natal nest fungus garden within their infrabuccal pockets and use this to start their new gardens following colony founding. This clonal propagation leads to the expectation of fungal lineages that closely parallel the lineages of their ant hosts. However, at least in the lower attines, this expectation is not borne out. Instead, the fungal cultivars of many lower attine ants are more closely related to free-living species than they are to other attine fungi, indicating that some fungus-growing ants occasionally replace their resident fungal cultivars with free-living stocks. Furthermore, within a given geographic area distantly related ants—in some cases species in different genera—may cultivate the same fungal clones, indicating that some fungus-growing ants occasionally replace their resident cultivars with cultivars acquired from the gardens of other ant colonies (Mueller et al., 1998).

Introduced Ant Species

The most widespread ants have been called "tramp" species; their geographic spread is tied with human activity (Table 4.1; Passera 1994). The ants belong to a variety of functional groups (Chapter 3) and use a variety of strategies to fit into widely variable habitat types. They are most frequently encountered in urban environments, in disturbed areas, and on oceanic islands (Lieberburg et al. 1975; Clark et al. 1982; Brandão and Paiva 1994; Passera 1994).

Table 4.1 Major Exotic Tramp and Invasive Ant Species[a]

Species	Tramp Species	Invasive Species
Subfamily Dolichoderinae		
Linepithema humile (Mayr)	Yes	Yes
Tapinoma melanocephalum (Fabricius)	Yes	
Technomyrmex albipes (Smith)	Yes	
Subfamily Formicinae		
Anopolepis gracilipes (F. Smith)	Yes	Yes
Paratrechina. fulva (Mayr)		Yes
P. longicornis (Latreille)	Yes	Yes
P. vaga (Forel)	Yes	
Subfamily Myrmicinae		
Cardiocondyla emeryi Forel	Yes	
C. nuda (Mayr)	Yes	
C. venustula W. M. Wheeler	Yes	
C. wroughtoni Forel	Yes	
Monomorium destructor (Jerdon)	Yes	
M. floricola (Jerdon)	Yes	
M. pharaonis (Linnaeus)	Yes	
Pheidole megacephala (Fabricius)	Yes	Yes
Quadristruma emmae (Emery)	Yes	
Solenopsis geminata (Fabricius)		Yes
S. richteri Forel	Yes	
S. wagneri (*invicta*) Santschi		Yes
Tetramorium bicarinatum (Nylander)	Yes	
T. caespitum (Linnaeus)	Yes	
T. lanuginosum Mayr	Yes	
T. pacificum Mayr	Yes	
T. simillimum (Smith)	Yes	
Trichoscapa membranifera (Emery)	Yes	
Wasmannia auropunctata (Roger)	Yes	Yes
Subfamily Ponerinae		
Hypoponera eduardi (Forel)	Yes	
H. opaciceps (Mayr)	Yes	
H. punctatissima (Roger)	Yes	

[a]Tramp ants are closely associated with human activity and often nest in human structures. Invasive species move into natural habitats (either disturbed or undisturbed) and outcompete native ant species. From McGlynn (1999b).

Although they are not as frequently encountered in undisturbed continental habitats, in tropical and subtropical areas they can be encountered anywhere.

The five most widespread ant species are the pharaoh's ant (*Monomorium pharaonis*), Argentine ant (*Linepithema humile*), crazy ant (*Paratrechina longicornis*), ghost ant (*Tapinoma melanocephalum*), and big-headed ant (*Pheidole megacephala*). Although these are the only species known to appear in every nonpolar biogeographic region (McGlynn 1999a), they have no functional group or taxonomic commonalities but do excel as human commensalist species. Although some species are dominant and have been known to maintain absolute territories (Haskins and Haskins 1965; Crowell 1968; Lieberburg et al. 1975; Holway 1995), others are opportunistic or cryptic, and are capable of coexisting with nondominant ant species (Hölldobler and Wilson 1990; Delabie et al. 1995).

Where dominant species (for example, *P. megacephala, L. humile, Wasmannia auropunctata*) are introduced, their impact upon native ants is obvious and drastic. Invasive ants exclude competing species from food resources and are known to raid heterospecific nests (Clark et al. 1982; de Kock and Giliomee 1989; Brandão and Paiva 1994). Studies of invaded areas show that noncryptic aboveground foraging ants are the most severely affected (Holway 1995; Human and Gordon 1996). In at least one instance, invasive ants have disrupted ant-plant mutualisms (Bond and Slingsby 1984). Clearly the areas that contain these ants will have a reduced native ant diversity. At the ecosystem level, at least one invasive ant (*Solenopsis wagneri* Santschi, formerly *S. invicta* Buren) has decimated areas where it is introduced, affecting ecological interactions at the levels of soil cycling, fruit decomposition, and the biodiversity of the terrestrial arthropod community in general (Porter and Savignano 1990; Vinson 1991).

The long-term effects of introduced ants are not well understood. Accounts from the West Indies of invasions of a exotic ants dating back to the early nineteenth century demonstrate that there is a turnover in the species composition of the introduced ant fauna (Haskins and Haskins 1965). Wilson and Taylor (1967) suggest that the species composition of invasive ants on a given island changes over time. Invasive species are generally more successful in disturbed areas and do not create a monospecific stand of ant colonies in most localities.

Introduced ant species can serve as an excellent bioindicator for assessing the status of an ant community. The impact of human use may be indicated by the presence of introduced ants before any long-term community effects are observed. An excellent case study is in the Galápagos Islands (Clark et al. 1982), where the foci for the spread of the invader *W. auropunctata* were located in cities and campsites. As activity spreads throughout many of the Galápagos Islands, the introduced ant is marching in file with human activity.

ACKNOWLEDGMENTS

We thank Beth Norden for extensive bibliographic research. During the writing of this chapter, TRS was supported by a Smithsonian Institution Scholarly Studies grant and by National Science Foundation Award DEB-9707209; TPM was supported by the National Science Foundation Division of International Programs.

Chapter 5

Diversity of Ants

William L. Brown Jr.

This chapter provides a background of taxonomic information for those who may be sampling ant diversity or dealing with other myrmecological matters (such as ant ecology) that demand some understanding of the challenge of identification of ant species. It is assumed that ants are chosen for biodiversity study mainly because they are ubiquitous and easily sampled, but also because identifying them is practicable as compared with identification of such other teeming taxa as mites and collembolans.

Although the last few years have seen great leaps forward in ant taxonomy—especially publication of *The Ants* (Hölldobler and Wilson 1990), *Identification Guide to the Ant Genera of the World* (Bolton 1994), and *A New General*

Catalogue of the Ants of the World (Bolton 1995b)—huge gaps still remain in our ability to identify given ant species with dispatch and confidence. The main challenges involve such immense and unrevised genera as *Camponotus, Crematogaster,* and *Pheidole,* plus various smaller but nevertheless dominant ground-dwelling taxa (Chapter 8). Some genera (e.g., *Pachycondyla,* New World *Pheidole,* the genera of the dacetonines) are currently under revision, as indicated by "forthcoming" in the list of identification aids in Chapter 12.

Ant Genera: An Overview

Table 5.1 summarizes general information about ant genera; some species may not conform to these generalizations. Numbers of

---

The author died shortly after completing this chapter.

Table 5.1 Distribution, Biology, and Ecology of the Ant Genera of the World[a]

Genus	Subfamily	Tribe	No. Species	Distribution	Habitat	Microhabitat	Biology	Functional Group[b]	Keys
Acanthognathus	Myrmicinae		6	Neotropical		Nesting in litter	Predators of several arthropods	SP	Brown and Kempf (1969)
Acanthomyops	Formicinae		16	Nearctic		Subterranean	Tend homopterans	C	Wing (1968)
Acanthomyrmex	Myrmicinae	Myrmecinini	11	Sri Lanka to Melanesia	Mesic forest		Seed harvesters, especially of ficus	TCS	Moffett (1986)
Acanthoponera	Ponerinae	Ectatommini	4	Neotropical	Mesic forest, savanna	Arboreal	Predators	?TCS	Brown (1958)
Acanthostichus	Cerapachyinae	Acanthostichini	11	Neotropical, S Nearctic		Subterranean	Mass foraging predators of termites	C	MacKay (1996)
Acromyrmex	Myrmicinae	Attini	26	Neotropical, S Nearctic		Nesting in ground	Cultivators of fungi	TCS	Goncalves (1961); Fowler (1988) (*Moellerius*)
Acropyga	Formicinae	Plagiolepidini	56	Worldwide in tropics and warm temperate		Hypogaeic	Tend coccids	CS	Weber (1944) (Neotropical); Terayama (1985b) (Taiwan, Japan)
Adelomyrmex	Myrmicinae	?Myrmecinini	8	Neotropical, Indo-Australian			Predators of mites	TCS	Smith (1947a,b) (*Apsychomyrmex*, Neotropical)
Adetomyrma	Ponerinae	Ponerini	1	S Madagascar			Predators	?	Ward (1994)
Adlerzia	Myrmicinae	Pheidologetonini	1	Australia				?C	Brown (1952)
Aenictogiton	Dorylinae	Aenictogetini	7	C Africa				TCS	Santschi (1923b); Brown (1975)
Aenictus	Dorylinae	Aenictini	100	E Mediterranean to E Australia			Army ants	TCS	Wilson (1964) (Indo-Australian); Gotwald (1982); Terayama and Yamane (1989) (Sumatra)
Afroxyridris	Myrmicinae	Pheidologetonini	1	C, W Africa				?C	Belshaw and Bolton (1994b)
Agraulomyrmex	Formicinae	Plagiolepidini	2	Afrotropical				?	Prins (1983)
Alloformica	Formicinae	Formicini	3	Tadzhikistan	Steppe	Epigaeic	Generalized foragers, visit flowers	?	Dlussky (1969: 219); Agosti (1994b)

Genus	Subfamily	Tribe	No. of species	Distribution	Vegetation	Nesting	Habits	Code	References
Allomerus	Myrmicinae	Solenopsidini	3	Neotropical		Mostly nesting in plant cavities		?C	Wheeler and Mann (1942); Kempf (1975b); Bolton (1987)
Amblyopone	Ponerinae	Amblyoponini	64	World temperate and tropical			Predators, esp. of Chilopoda	C	Brown (1960); Baroni Urbani (1978a) (Mediterranean); Taylor (1978c) (Melanesia); Terayama (1987) (Taiwan); Lattke (1991) (Neotropical)
Ancyridris	Myrmicinae	Stenammini	2	Melanesia	Wet forest			?	
Anergates	Myrmicinae	Tetramoriini	1	Palearctic, adventive in Nearctic			Workerless parasites of *Tetramorium caespitum*	?	Ettershank (1966)
Aneuretus	Dolichoderinae	Aneuretini	1	Sri Lanka	Mesic forest		Predators	?	Wilson et al. (1956)
Anillidris	Dolichoderinae	Dolichoderini	1	S Neotropical		Hypogaeic		?	Santschi (1936, 1937)
Anillomyrma	Myrmicinae	Solenopsidini	2	Afrotropical, Indomalayan			Cryptic foragers	C	Bolton (1987)
Anisopheidole	Myrmicinae	Pheidologetonini	1	Australia				?C	Ettershank (1966)
Ankylomyrma	Myrmicinae	Leptothoracini	1	Afrotropical		Arboreal		TCS	Bolton (1973a)
Anochetus	Ponerinae	Odontomachini	27	World tropics and warm temperate, except Nearctic		Nesting in rotten logs	Predators	SP	Brown (1978) (world); Lattke (1986) (Neotropical); Wang (1993) (China)
Anomalomyrma	Leptanillinae	Anomalomyrmini	4	Indomalayan			Cryptic predators	C	Taylor (1990a)
Anonychomyrma	Dolichoderinae	Dolichoderini	24	Indomalayan, Australia		Epigaeic and arboreal		DD	Shattuck (1992a, 1992b)
Anoplolepis	Formicinae	Plagiolepidini	22	Afrotropical, S Palearctic, adventive in Indomalayan		Epigaeic	Foragers	CS	Prins (1982) (custodiens group, partial)
Antichthonidris	Myrmicinae	Solenopsidini	2	S Neotropical				?	Snelling (1975)
Aphaenogaster	Myrmicinae	Pheidolini	142	World except Afrotropical and South America			Generalized foragers	O	Smith (1961) (Niugini); Arnol'di (1976a) (USSR); Umphrey (1996) (Nearctic)
Aphomomyrmex	Formicinae	Brachymyrmecini	1	Afrotropical		Arboreal, nesting in plant cavities		TCS	Snelling (1979b)
Apomyrma	?Leptanillinae	Apomyrmini	1	Afrotropical		Hypogaeic	Predators of geophilomorph Chilopoda	C	Brown et al. (1970)
Apterostigma	Myrmicinae	Attini	34	Neotropical		Nesting in ground	Fungus cultivators	TCS	Lattke (1997)

Continued on next page

Table 5.1 continued

Genus	Subfamily	Tribe	No. Species	Distribution	Habitat	Microhabitat	Biology	Functional Group[b]	Keys
Asketogenys	Myrmicinae	Dacetini	1	Indomalayan			Predators		Brown (1972)
Asphinctanilloides	Leptanilloidinae	Leptanilloidini	3	Neotropical, Amazon basin, and Atlantic forest		Hypogaeic	Nomadic	C	Brandão et al. (1999)
Asphinctopone	Ponerinae	Ponerini	1	Afrotropical	Rainforest floor		Predators	C	Brown (pers. obs.)
Atopomyrmex	Myrmicinae	Leptothoracini	3	Afrotropical				TCS	Bolton (1981b)
Atta	Myrmicinae	Attini	15	Neotropical and S Nearctic		Nesting in ground	Fungus cultivators	TCS	Borgmeier (1959)
Aulacopone	Ponerinae	Ectatommini	1	Azerbaijan			Predators	?	Arnol'di (1930); Taylor (1979b)
Axinidris	Dolichoderinae	Dolichoderini	13	Afrotropical		Arboreal		?	Shattuck (1991)
Azteca	Dolichoderinae	Dolichoderini	70	Neotropical			Generalized foragers, visit extrafloral nectaries	DD	Longino (1991) (spp. inhabiting *Cecropia*)
Bajcaridris	Formicinae	Formicini	3	Algeria, Morocco		Epigaeic	Foragers	?	Agosti (1994a)
Baracidris	Myrmicinae	Stenammini	2	Afrotropical	Mesic forest	Forest floor		?C	Bolton (1981b)
Bariamyrma	Myrmicinae	Stenammini	1	Neotropical				?	Lattke (1990)
Basiceros	Myrmicinae	Basicerotini	6	Neotropical	Mesic forest	Forest floor	Predators, esp. of termites	C	Brown and Kempf (1960); Brown (1974a)
Belonopelta	Ponerinae	Ponerini	2	Neotropical	Forest	Litter	Predators, esp. of campodeid diplura	?C	Baroni Urbani (1975b); Brandão (1989) (Brazil)
Blepharidatta	Myrmicinae	Blepharidattini	2	Neotropical	Forest	Nesting in litter, in ground, or under stones	Scavangers	TCS	Kempf (1967c)
Bondroitia	Myrmicinae	Solenopsidini	2	Afrotropical				C	Bolton (1987)
Bothriomyrmex	Dolichoderinae	Dolichoderini	34	S Palearctic, India to Australia				CCS	Shattuck (1992b)
Brachymyrmex	Formicinae	Brachymyrmecini	40	Neotropical, Nearctic, adventive elsewhere		Nesting in seeds, trees, and fallen fruits	Generalized foragers	TCS, ?CS	Santschi (1923a) (out of date)
Bregmatomyrma	Formicinae	Bregmatomyrmini	1	Indomalayan				?	
Calomyrmex	Formicinae	Camponotini	8	Australia, Melanesia	Arid to mesic		Generalized foragers	SC	

Genus	Subfamily	Tribe	No. species	Distribution	Habitat	Nesting	Foraging	Code	References
Calyptomyrmex	Myrmicinae	Stenammini	24	Afrotropical, Indo-Melanesian				?C	Baroni Urbani (1975a) (India); Bolton (1981a) (Africa)
Camponotus	Formicinae	Camponotini	935	Worldwide		Nesting in ground, in dead wood, in and on trees	Generalized foragers	SC	Kusnezov (1951d) (Argentina); Yasumatsu and Brown (1951, 1957) (*herculeanus* complex, E Palearctic); Hashmi (1973) (*Myrmothrix*); Dumpert (1985) (*Karavaievia*); Snelling (1988) (*Myrmentoma*, Nearctic); Wang et al. (1989a, 1989b) (China); Robertson (1990) (*fulvopilosus* group); Dumpert et al. (1995) (*Karavaievia*); McArthur and Adams (1996) (*nigriceps* group, Australia); Radchenko (1996a) (Palearctic Asia); Mackay (1997) (*Myrmostemus*); Mackay and Mackay (1997) (*montivagus* group, *Myrmentoma*)
Cardiocondyla	Myrmicinae	Leptothoracini	35	Warm Old World except Australia, adventive worldwide				O	Bolton (1982) (Afrotropical); Radchenko (1995) (Palearctic)
Carebara	Myrmicinae	Pheidologetonini	18	Afrotropical, Indomalayan, Neotropical				C	Xu (1999)
Carebarella	Myrmicinae	Solenopsidini	3	Neotropical				?C	Kempf (1975b)
Cataglyphis	Formicinae	Formicini	65	S Palearctic S to Ghana, E to N China, India	Steppes and deserts	Nesting in ground	Scavengers	HCS	Agosti (1990, 1994a); Radchenko (1998) (Asia)

Continued on next page

49

Table 5.1 continued

Genus	Subfamily	Tribe	No. Species	Distribution	Habitat	Microhabitat	Biology	Functional Group[b]	Keys
Cataulacus	Myrmicinae	Cataulacini	65	Afrotropical, Madagascar, Indomalayan		Arboreal, nesting in plant cavities		TCS	Bolton (1974a); Snelling (1979a)
Centromyrmex	Ponerinae	Ponerini	6	Neotropical, Afrotropical, Indomalayan		Nesting in termitaria	Cryptic predators of termites	C	Kempf (1967a) (Neotropical); Brown (pers. obs.)
Cephalotes	Myrmicinae	Cephalotini	3	Neotropical		Arboreal, nesting in hollow tree trunks and branches	Some are pollen eaters	TCS	Kempf (1951, 1958a); Andrade and Baroni Urbani (1999)
Cerapachys	Cerapachyinae	Cerapachyini	140	Worldwide in tropics and warm temperate			Army ants, predators of other ants	C, SP[c]	Brown (1975) (world); Radchenko (1993) (Vietnam); Terayama (1996) (Japan)
Chalepoxenus	Myrmicinae	Leptothoracini	8	Palearctic			Parasites on and slavemakers of *Leptothorax*	?CCS	Kutter (1973); Buschinger et al. (1988) (W Palearctic); Radchenko (1989a) (USSR) ; Cagniant and Espadaler (1997a) (Morocco)
Cheliomyrmex	Dorylinae	Ecitonini	4	Neotropical			Army ants	TCS	Borgmeier (1955)
Chelystruma	Myrmicinae	Dacetini	1	Neotropical			Predators		Brown (1950); Kempf (1960c); Bolton (pers. comm.)
Chimaeridris	Myrmicinae	Pheidolini	2	Indomalayan				TCS	Wilson (1989)
Cladarogenus	Myrmicinae	Dacetini	1	Afrotropical			Predators		Brown (1976a, 1976b)
Cladomyrma	Formicinae	Brachymyrmecini	5	Indomalayan, West Malaysia and Borneo		Arboreal, nesting in internodes of trees and vines		?TCS, CS	Agosti (1991); Agosti et al. (1999)
Codiomyrmex	Myrmicinae	Dacetini	1	Neotropical			Predators		Bolton (pers. comm.)
Colobostruma	Myrmicinae	Dacetini	9	Australia, Melanesia			Predators of collembolans	SP	
Concoctio	Ponerinae	Amblyoponini	1	Afrotropical				?C	Brown (1974b, 1974c)

Genus	Subfamily	Tribe	No.	Distribution	Habitat	Nesting	Behavior	Code	References
Creightonidris	Myrmicinae	Basicerotini	1	Neotropical				C	Brown (1949d); Brown and Kempf (1960)
Crematogaster	Myrmicinae	Crematogastrini	427	World tropical and temperate		Arboreal, nest in hollow tree trunks and branches	Generalized foragers	GM	Buren (1968b) (Nearctic); Onoyama (1998) (Japan)
Cryptopone	Ponerinae	Ponerini	8	World tropical and temperate			Cryptic predators	C	Brown (pers. obs.)
Cylindromyrmex	Cerapachyinae	Cylindromyrmecini	10	Neotropical		Nest in ground or logs	Predators of termites	SP	Brown (1975); Andrade (1978)
Cyphoidris	Myrmicinae	Stenammini	4	Afrotropical				TCS	Bolton (1981b)
Cyphomyrmex	Myrmicinae	Attini	37	Neotropical, S Nearctic		Nest in ground	Fungus cultivators	TCS	Kempf (1964, 1965 [*rimosus* group], 1968]; Snelling and Longino (1992) (*rimosus* group)
Dacatria	Myrmicinae	Stenammini	1	Korea	Mesic forest			?	Rigato (1994)
Dacetinops	Myrmicinae	Stenammini	7	Sundaland, Melanesia	Rainforest, savanna			?C	Taylor (1985)
Daceton	Myrmicinae	Dacetini	1	Neotropical		Arboreal, nesting in hollow trees	Predators	?SP	
Decamorium	Myrmicinae	Tetramoriini	2	Afrotropical				?	Bolton (1976)
Dendromyrmex	Formicinae	Camponotini	7	Neotropical		Arboreal	Foragers	SC	Mann (1916)
Diacamma	Ponerinae	Ponerini	33	India to N Australia			Predators	O	Emery (1897) (out of date); Brown (pers. obs.)
Dicroaspis	Myrmicinae	Stenammini	2	Afrotropical				C	Bolton (1981a)
Dilobocondyla	Myrmicinae	Leptothoracini	9	Indo-Melanesian				TCS	Wheeler (1924) (out of date)
Dinoponera	Ponerinae	Ponerini	6	Tropical South America	Forest, savanna	Nesting in ground	Predators, colony size small	SP	Kempf (1971)
Diplomorium	Myrmicinae	Solenopsidini	1	Afrotropical				?	Bolton (1987)
Discothyrea	Ponerinae	Ectatommini	25	World tropical and warm temperate except Palearctic		Nesting in litter	Predators of arthropod eggs	C	Brown (1958)
Doleromyrma	Dolichoderinae	Dolichoderini	1	Australia				O	Shattuck (1992a)
Dolichoderus	Dolichoderinae	Dolichoderini	110	World tropical and temperate except Africa		Arboreal	Generalized foragers	TCS, CCS	Clark (1930) (Australia; out of date); MacKay

Continued on next page

51

Table 5.1 continued

Genus	Subfamily	Tribe	No. Species	Distribution	Habitat	Microhabitat	Biology	Functional Group[b]	Keys
				and Madagascar					(1993) (New World); Xu (1995b) (China)
Dolioponera	Ponerinae	Ponerini	1	Afrotropical			Cryptic predators	?	Brown (1974d, 1974e)
Dorisidris	Myrmicinae	Dacetini	1	Cuba			Predators		Brown (1948)
Doronomyrmex	Myrmicinae	Leptothoracini	4	Palearctic, Nearctic			Parasitic on *Leptothorax*	CCS	Kutter (1945) (Europe); Buschinger (1981)
Dorylus	Dorylinae	Dorylini	60	Afrotropical, Indomalayan			Army ants	TCS	
Dorymyrmex	Dolichoderinae	Dolichoderini	50	Neotropical, Nearctic			Generalized foragers	O (? some DD)	Kusnezov (1951e); (S Neotropical); Snelling and Hunt (1975); (S Neotropical); Snelling (1995a) (Nearctic)
Dysedrognathus	Myrmicinae	Dacetini	1	Indomalayan, W Malaysia			Predators	C	Taylor (1968b)
Echinopla	Formicinae	Camponotini	24	Indomalayan, N Australia		Arboreal	Foragers	SC	
Eciton	Dorylinae	Ecitonini	12	Neotropical, S Nearctic		Epigaeic, form bivouacs	Army ants	TCS	Borgmeier (1955); Watkins (1976, 1982 [Mexico])
Ecphorella	Dolichoderinae	Dolichoderini	1	Afrotropical				?	
Ectatomma	Ponerinae	Ectatommini	14	Neotropical	Forest, savanna	Nesting in ground, hollow trees	Predators, some thieves of other ants' brood, extrafloral nectaries	?O	Kugler and Brown (1982)
Emeryopone	Ponerinae	Ponerini	3	Israel, Indomalayan			Predators	C	Baroni Urbani (1975b) (as Belonopelta)
Epelysidris	Myrmicinae	Solenopsidini	1	Indomalayan				TCS	Bolton (1987)
Epimyrma	Myrmicinae	Leptothoracini	11	Palearctic			Parasites of *Leptothorax*	CCS	Buschinger (1989); Cagniant and Espadaler (1997a) (Morocco)
Epitritus	Myrmicinae	Dacetini	8	Palearctic (S Europe, E Asia), Afrotropical, Indomalayan			Predators	C	Brown (1948, 1949b); Bolton (1972 [world], 1983 [Afrotropical]); Taylor (1968b)

Genus	Subfamily	Tribe	No. species	Distribution	Biome	Nesting	Foraging / prey	Code	References
Epopostruma	Myrmicinae	Dacetini	7	SE, SW Australia			Predators of collembolans	SP	(Malaya); Ogata (1990) (Japan)
Eucryptocerus	Myrmicinae	Cephalotini	3	Neotropical		Arboreal		TCS	Brown (pers. obs.)
Euprenolepis	Formicinae	Lasiini	6	Indomalayan				TCS	Kempf (1951)
Eurhopalothrix	Myrmicinae	Basicerotini	35	Neotropical, S Nearctic, Indomalayan, Australian		Nesting in litter	Predators	C	Brown (1953b); Brown and Kempf (1960); Taylor (1968b, 1980, 1990b) (Indo-Australian)
Eutetramorium	Myrmicinae	?Myrmicini	2	Madagascar				?	Alpert (pers. comm.)
Forelius	Dolichoderinae	Dolichoderini	17	Neotropical, S Nearctic		Epigaeic	Generalized foragers	DD, HCS	Shattuck (1992a)
Forelophilus	Formicinae	Camponotini	1	Indomalayan				SC	Kutter (1931)
Formica	Formicinae	Formicini	160	Nearctic, Palearctic: warm to cold temperate		Epigaeic	Generalized foragers, tend homopterans	CCS, O	Dlussky (1964 [exsecta group in USSR], 1965 [Mongolia, Tibet], 1967 [Palearctic]); Francoeur (1973) (*fusca* group, Nearctic); Dlussky and Pisarski (1971) (Poland); Buren (1968a) (*sanguinea* group in Nearctic); Kupyanskaya (1980) (far eastern Russia); Wu (1990) (China)
Formicoxenus	Myrmicinae	Leptothoracini	7	Nearctic, Palearctic		Xenobiont, nesting in association with other ants		CCS	Francoeur et al. (1985) (world)
Froggattella	Dolichoderinae	Dolichoderini	2	Australia				DD	Shattuck (1992a, 1996b) (Australia)
Gesomyrmex	Formicinae	Gesomyrmecini	5	Indomalayan		Arboreal		?TCS	Cole (1949) (partial)
Gigantiops	Formicinae	Gigantiopini	1	Neotropical	Rainforest, savanna	Nesting in rotten wood on forest floor, foraging on plants		?TCS	Kempf and Lenko (1968)

Continued on next page

Table 5.1 continued

Genus	Subfamily	Tribe	No. Species	Distribution	Habitat	Microhabitat	Biology	Functional Group[b]	Keys
Glamyromyrmex	Myrmicinae	Dacetini	23	Neotropical, Afrotropical, Australia	Forest, savanna	Nesting in litter	Predators	C	Kempf (1960c) (Neotropical); Bolton (1983) (Africa)
Gnamptogenys	Ponerinae	Ectatommini	102	Neotropical. S Nearctic, Oriental, India to Fiji	Forest, savanna	Nesting in the ground and rotten logs	Predators and scavengers	TCS	Lattke (1995) (New World); Brandão and Lattke (1990) (Ecuador); Xu and Zhang (1996) (China)
Goniomma	Myrmicinae	Pheidolini	5	S Palearctic	Arid land		Seed harvesters	HCS	Santschi (1929) (out of date)
Gymnomyrmex	Myrmicinae	Dacetini	7	Neotropical			Predators	C	Kempf (1959, 1960c); Bolton (pers. comm.)
Harpagoxenus	Myrmicinae	Leptothoracini	3	Nearctic, Palearctic			Slavemaking parasites of *Leptothorax*	CCS	
Harpegnathos	Ponerinae	Ponerini	6	India to Philippines, Sundaland			Predators	SP	Bingham (1903) (S. Asia; out of date)
Heteroponera	Ponerinae	Ectatommini	15	Neotropical, Australia and New Zealand	Wet and mesic forests	Nesting in logs	Predators	?CCS	Brown (1958); Kempf (1962)
Huberia	Myrmicinae	Myrmicini	2	New Zealand			Generalized foragers	CCS	Brown (1958)
Hylomyrma	Myrmicinae	Myrmicini	13	Neotropical	Forest, savanna	Nesting in sandy soil, litter	Generalized foragers	TCS	Kempf (1973a)
Hypoponera	Ponerinae	Ponerini	150	Worldwide tropical and warm temperate	Forest, savanna	Nesting in litter	Generalized foragers	C	Brown (pers. obs.)
Indomyrma	Myrmicinae	Stenammini	1	Peninsular India	Mesic forest		Generalized forager	?C	Brown (1985)
Ireneopone	Myrmicinae	Leptothoracini	1	Mauritius	Native forest		Generalized forager	?TCS	Donisthorpe (1946)
Iridomyrmex	Dolichoderinae	Dolichoderini	55	Australia, Indomalayan			Generalized foragers	DD	Shattuck (1992a, 1992b, 1993) (*purpureus* group), (1996a) (*discors* group)
Ishakidris	Myrmicinae	Phalacromyrmecini	1	N Borneo	Rainforest	Nesting in leaf litter	Predator	C	Bolton (1984)

Genus	Subfamily	Tribe	No. of species	Distribution	Habitat	Nesting	Habits	Code	References
Kartidris	Myrmicinae	Pheidolini	5	SE Asia			Cryptic predators	TCS	Bolton (1991); Xu (1999)
Kyidris	Myrmicinae	Dacetini	4	India, China-Japan to New Guinea, Madagascar			Parasites of *Strumigenys*	C	Brown (1949c) (Japan); Wilson and Brown (1956) (New Guinea)
Labidus	Dorylinae	Ecitonini	8	Neotropical to S Nearctic	Forest, savanna	Epigaeic, forms bivouacs	Army ants	TCS	Borgmeier (1955); Watkins (1976)
Lachnomyrmex	Myrmicinae	Stenammini	3	Neotropical	Mesic forest	Nesting in litter		?TCS	Smith (1944); Fernández and Baena (1997) (Colombia)
Lasiophanes	Formicinae	Melophorini	5	Chile and Argentina			Generalized foragers	CCS	Snelling and Hunt (1975)
Lasius	Formicinae	Lasiini	75	Nearctic, Palearctic, temperate Oriental	Epigaeic, arboricolous, subterranean		Generalized foragers, tend homopterans	CCS	Wilson (1955) (world); Seifert (1988a, 1990 [European *Chthonolasius*], 1992 [Palearctic *Lasius* s.s.]); Yamauchi (1978) (Japan)
Lepisiota	Formicinae	Plagiolepidini	65	S Palearctic, Afrotropical, Indomalayan			Generalized foragers	?	Xu (1994a) (China)
Leptanilla	Leptanillinae	Leptanillini	32	S Palearctic, Afrotropical, Indomalayan, Australia			Cryptic mass predators	C	Baroni Urbani (1977)
Leptanilloides	Leptanilloidinae	Leptanilloidini	4	Neotropical, Andes and foothills	Hypogaeic		Predators of centipedes, nomadic	C	Brandão et al. (1999)
Leptogenys	Ponerinae	Ponerini	212	Worldwide in tropics and some subtropics			Predators of isopods and mass-foraging predators, esp. of termites	SP	Bingham (1903) (India); Bolton (1975a) (Africa)
Leptomyrmex	Dolichoderinae	Dolichoderini	16	Australia, New Caledonia, Melanesia W to Maluku		Nesting in cavities in wood		TCS	Wheeler (1934) (out of date)
Leptothorax	Myrmicinae	Leptothoracini	315	Worldwide (?except Australia)	Epigaeic, arboreal	Nesting in ground, under stones, in wood, and in trees	Generalized foragers and parasites	CCS, TCS	Bernard (1956) (W Europe); Kempf (1958c) (*Nesomyrmex*, Neotropical); Baroni Urbani (1978b) *Macromischa*, Neotropical);

Continued on next page

Table 5.1 continued

Genus	Subfamily	Tribe	No. Species	Distribution	Habitat	Microhabitat	Biology	Functional Group[b]	Keys
									Bolton (1982) (Afrotropical); Dlussky and Soyunov (1988) (*Tennothorax*, USSR); Taylor (1989) (Australasian); Radchenko (1994a, 1994b) (C and E Palearctic); Cagniant and Espadaler (1997a) (Morocco); Terayama and Onoyama (1999) (Japan); MacKay (2000) (*Myrafant*)
Linepithema	Dolichoderinae	Dolichoderini	14	Neotropical, adventive worldwide in warm or temperate areas			Generalized foragers	DD	Shattuck (1992a, 1992b)
Liometopum	Dolichoderinae	Dolichoderini	6	Nearctic, Palearctic, Oriental			Generalized foragers	DD	Wheeler (1905); Shattuck (1992b)
Liomyrmex	Myrmicinae	Metaponini	8	Indomalayan		Nesting under bark and in rotten wood		TCS	Ettershank (1966)
Lophomyrmex	Myrmicinae	Pheidolini	4	Indomalayan			Generalized foragers	TCS	Rigato (1994b)
Lordomyrma	Myrmicinae	Stenammini	20	Indo-Melanesian to Japan				TCS	
Loweriella	Dolichoderinae	Dolichoderini	1	Australia			Generalized foragers	TCS	Shattuck (1992b)
Machomyrma	Myrmicinae	Pheidolgetonini	1	Australia				C	
Manica	Myrmicinae	Myrmicini	6	Nearctic, Palearctic, including Orient			Generalized foragers	CCS	Wheeler and Wheeler (1986) (Nearctic)
Mayriella	Myrmicinae	Stenammini	5	Oriental, Australian				?C	
Megalomyrmex	Myrmicinae	Solenopsidini	33	Neotropical	Forest, savanna	Nesting in ground or under leaves or stones	Generalized foragers and commensals in *Attini* nests; tend homopterans	TCS	Brandão (1990); Fernández and Baena (1997)

Genus	Subfamily	Tribe	No.	Distribution	Habitat	Nesting	Biology	Code	References
Melissotarsus	Myrmicinae	Melissotarsini	4	Afrotropical, Madagascar		Nesting in or under tree bark	Tend coccids	?	Bolton (1982)
Melophorus	Formicinae	Melophorini	21	Australia	Mostly xeric habitats		Nest in the ground	HCS	
Meranoplus	Myrmicinae	Meranoplini	55	Afrotropical, Madagascar, Oriental to Melanesia and Australia		Nesting in ground	Seed harvesters and general foragers	HCS	Bolton (1981a) (Afrotropical); Taylor (1990c) (Australasian); Schödl (1998)
Mesostruma	Myrmicinae	Dacetini	6	Australia		Nesting in ground	Predators of collembolans	SP	Taylor (1973)
Messor	Myrmicinae	Pheidolini	106	Nearctic, Palearctic, Afrotropical, Oriental		Nesting in ground	Seed harvesters	HCS	Bernard (1954, 1979); M. Smith (1956a) (Nearctic, as *Veromessor*); Arnol'di (1977); Tohmé and Tohmé (1981); Cagniant and Espadaler (1997b) (Morocco)
Metapone	Myrmicinae	Metaponini	16	Madagascar, Oriental to Melanesia, Australia		Nesting in hollow twigs and other plant cavities	Prey on termites	TCS	
Microdaceton	Myrmicinae	Dacetini	2	Afrotropical			Predators	?C	Bolton (1983)
Monomorium	Myrmicinae	Solenopsidini	296	Worldwide in tropics and warm temperate			Generalized foragers, harvesters	GM, HCS, CCS, TCS	DuBois (1981, 1986) (Nearctic); Bolton (1987) (Afrotropical); Radchenko (1997)
Mycetarotes	Myrmicinae	Attini	2	Neotropical			Cultivators of fungi	TCS	Kempf (1960b); Mahyé-Nunes (1995)
Mycetophylax	Myrmicinae	Attini	6	Neotropical	Coastal dunes	Nesting in ground	Fungus cultivators	TCS	
Mycetosoritis	Myrmicinae	Attini	4	Neotropical, S Nearctic			Fungus cultivators	TCS	
Mycocepurus	Myrmicinae	Attini	4	Neotropical		Nesting in ground	Fungus cultivators	TCS	Kempf (1963)
Myopias	Ponerinae	Ponerini	37	Oriental (N Thailand) to SE Australia			Predators of millipedes, 1 sp. on ants	C	Willey and Brown (1983), Brown (pers. obs.)
Myopopone	Ponerinae	Amblyoponini	1	Indomalayan to N Australia			Predator	C	

Continued on next page

Table 5.1 continued

Genus	Subfamily	Tribe	No. Species	Distribution	Habitat	Microhabitat	Biology	Functional Group[b]	Keys
Myrcidris	Pseudomyrmecinae	Pseudomyrmecini	1	Neotropical		Arboreal		TCS	Ward (1990)
Myrmecia	Myrmeciinae	Myrmeciini	89	Australia, 1 New Caledonia			Generalized predators	SP	Brown (1953c) (partial), (1990); Ogata and Brown (1991)
Myrmecina	Myrmicinae	Myrmecinini	28	Nearctic, Palearctic to			Predators of mites	TCS, CCS	Brown (1949a) (N. America), (1967) (Nearctic); Terayama (1985a) (E Asia)
Myrmecocystus	Formicinae	Formicini	29	W Nearctic, mostly arid habitats			Generalized foragers, store honeydew in repletes	HCS	Snelling (1976, 1982)
Myrmecorhynchus	Formicinae	Melophorini	5	Australia		Mostly arboreal		CCS	
Myrmelachista	Formicinae	?Myrmelachistini	47	Neotropical		Mostly nesting in plant cavities		C	Kusnezov (1951b) (Patagonia)
Myrmica	Myrmicinae	Myrmicini	100	Nearctic, Palearctic, Oriental		Nesting in ground, rotten wood	Generalized foragers	O	Menozzi (1939) (Himalaya, Tibet); Weber (1947, 1948, 1950b) (Nearctic, with synopsis of Palearctic; out of date); Arnol'di (1970 [European USSR], 1976b [central USSR]); Kupyanskaya (1986) (*lobicornis* group of far eastern Russia); Seifert (1988b) (W Palearctic); Radchenko et al. (1997) (Poland); Radchenko et al. (1998) (*ritae* group)

Continued on next page

Genus	Subfamily	Tribe	No. spp.	Distribution	Habitat	Nesting / microhabitat	Biology		Reference
Myrmicaria	Myrmicinae	Myrmicariini	31	Afrotropical, Indomalayan		Many arboreal		TCS	
Myrmicocrypta	Myrmicinae	Attini	24	Neotropical			Cultivators of fungi; general foragers	TCS	Moffett (1985); Agosti (1992)
Myrmoteras	Formicinae	Myrmoteratini	31	India to Sulawesi	Mesic forest	Epigaeic	Predators, mostly on forest floor	SP	
Mystrium	Ponerinae	Amblyoponini	8	W Africa, Madagascar, Indomalayan to NW Australia			Predators, esp. of chilopoda	C	Menozzi (1929)
Neivamyrmex	Ecitoninae	Ecitonini	120	Neotropical, S Nearctic		Hypogaeic, forms bivouacs	Army ants, mainly preying on other ants	TCS	Borgmeier (1955); Watkins (1976, 1982, 1985); Ward (1999b) (Nearctic)
Neoblepharidatta	Myrmicinae	Blepharidattini?	1	India	Forest	Hypogaeic			Sheela and Narendran (1997)
Neostruma	Myrmicinae	Dacetini	6	Neotropical			Predators, esp. of collembolans	C	Brown (1959)
Nomamyrmex	Dorylinae	Ecitoninae	2	Neotropical			Army ants	TCS	Borgmeier (1955); Watkins (1977)
Noonilla	Leptanillinae	Leptanillini	1	Melanesia				C	Petersen (1968)
Nothidris	Myrmicinae	Solenopsidini	3	S Neotropical				CCS	Snelling (1975); Bolton (1987)
Nothomyrmecia	Nothomyrmeciinae	Nothomyrmeciini	1	S Australia	Arid woodland	Nesting in ground, foraging on trees	Nocturnal predators	SP	Taylor (1978a)
Notoncus	Formicinae	Melophorini	7	Australia		Epigaeic	Generalized foragers	CCS	Brown (1955)
Notostigma	Formicinae	Camponotini	3	Australia		Epigaeic	Generalized foragers	SC	
Ochetellus	Dolichoderinae	Dolichoderini	4	Oriental to Australia, adventive in N America		Epigaeic	Generalized foragers	O	Shattuck (1992a)
Ochetomyrmex	Myrmicinae	Ochetomyrmecini	4	Neotropical	Forest	Epigaeic	Generalized foragers	TCS	Kempf (1975b)
Octostruma	Myrmicinae	Basicerotini	10	Neotropical		Nesting hypogaeically, foraging in litter	Predators, some tend homopterans	C	Brown and Kempf (1960); Palacio (1997) (Colombia)
Ocymyrmex	Myrmicinae	Pheidolini	37	Afrotropical	Mostly hot, arid sites	Nesting in ground	Seed harvesters	HCS	Bolton and Marsh (1989)

Table 5.1 continued

Genus	Subfamily	Tribe	No. Species	Distribution	Habitat	Microhabitat	Biology	Functional Group[b]	Keys
Odontomachus	Ponerinae	Ponerini	55	Worldwide in tropical and warm temperate, not W Palearctic		Epigaeic	Predators	O, ?SP	Brown (1976c, 1977b, 1978) (world); Deyrup et al. (1985) (SE United States); Wang (1993) (China)
Odontoponera	Ponerinae	Ponerini	2	Indomalayan		Epigaeic	Predators	SP	Brown (pers. obs.)
Oecophylla	Formicinae	Oecophyllini	1	Afrotropical, Oriental to N Australia		Nesting in arboreal, silk-woven leaf nests	Predator, tend homopterans	TCS	
Oligomyrmex	Myrmicinae	Pheidologetonini	93	Worldwide in tropics, rare in warm temperate			Cryptic foragers, termite thief ants	C	Weber (1950a, 1952) (partial Afrotropical)
Onychomyrmex	Ponerinae	Amblyoponini	8	Australia	Mesic forest		Mass predators	TCS	
Opisthopsis	Formicinae	Camponotini	13	N Australia, Melanesia			Generalized foragers	SC	
Orectognathus	Myrmicinae	Dacetini	29	Australia, Melanesia			Predators	SP	Taylor (1977, 1978b, 1979a, 1980)
Overbeckia	Formicinae	Camponotini	1	Indomalayan				SC	
Oxyepoecus	Myrmicinae	Solenopsidini	11	Neotropical		Nesting in leaf litter	Some spp. in nests of *Pheidole* spp.	TCS	Kempf (1974a)
Oxyopomyrmex	Myrmicinae	Pheidolini	9	S Palearctic			Seed harvesters	HCS	Santschi (1929)
Pachycondyla	Ponerinae	Ponerini	150	Worldwide in tropics and some warm temperate			Predators; 1 species also harvests seeds	SP[d]	Brown (pers. obs.); Xu (1994b) (China)
Paedalgus	Myrmicinae	Pheidologetonini	10	Afrotropical, Oriental				C	Bolton and Belshaw (1993)
Papyrius	Dolichoderinae	Dolichoderini	4	Melanesia, Australia			Generalized foragers	DD	Shattuck (1992a)
Paraponera	Ponerinae	Ectatommini	1	Neotropical		Nesting in ground, foraging arboreally	Predators, tend homopterans, extrafloral nectaries	?	
Paraprionopelta	Ponerinae	Amblyoponini	1	S Neotropical					Kusnezov (1955); Brown (1960)
Paratopula	Myrmicinae	Leptothoracini	9	Indomalayan				TCS	Bolton (1988b)

Genus	Subfamily	Tribe	No.	Distribution	Nesting	Biology	Code	Reference
Paratrechina	Formicinae	Lasiini	107	Worldwide in tropics and temperate		Generalized foragers	O	Trager (1984) (Nearctic)
Pentastruma	Myrmicinae	Dacetini	2	Oriental			C	Brown and Boisvert (1979)
Perissomyrmex	Myrmicinae	Myrmecinini	2	Central America, Oriental		Predators	TCS	Smith (1947) (Guatemala); Baroni Urbani and De Andrade (1993) (Bhutan); Longino and Hartley (1994) (C America)
Peronomyrmex	Myrmicinae	Leptothoracini	1	E Australia			TCS	Taylor (1970a)
Petalomyrmex	Formicinae	Brachymyrmecini	1	Afrotropical			TCS	Snelling (1979b)
Phacota	Myrmicinae	Solenopsidini	1	S Palearctic			?	Bolton (1987)
Phalacromyrmex	Myrmicinae	Phalacro-myrmecini	1	Neotropical			?	Kempf (1960a)
Phasmomyrmex	Formicinae	Camponotini	4	Afrotropical			SC	Wheeler and Wheeler (1930); Petersen (1968)
Phaulomyrma	?Leptanillinae	?Leptanillini	1	Indomalayan			C	
Pheidole	Myrmicinae	Pheidolini	910	Worldwide in tropics and warm temperate	Most nesting in soil, some in rotten wood	Many seed harvesters, many omnivorous	GM	Kusnezov (1951c) (Argentina); Gregg (1958) (Nearctic); Ogata (1982) (Japan); Wilson (forthcoming) (New World); Zhou and Zheng (1999) (China)
Pheidologeton	Myrmicinae	Pheidologetonini	30	Afrotropical, India to Melanesia		Generalized and mass foragers	C	
Philidris	Dolichoderinae	Dolichoderini	7	Indo-Melanesian	Mainly arboreal, most nesting in plants	Foragers	DD	Shattuck (1992a)
Phrynoponera	Ponerinae	Ponerini	3	Afrotropical		Predators	SP	Wheeler (1922a) (out of date)
Pilotrochus	Myrmicinae	Phalacro-myrmecini	1	Madagascar			?	Brown (1977a)

Continued on next page

Table 5.1 continued

Genus	Subfamily	Tribe	No. Species	Distribution	Habitat	Microhabitat	Biology	Functional Group[b]	Keys
Plagiolepis	Formicinae	Plagiolepidini	53	Old World tropics and temperate areas; adventive in New World			Generalized foragers	C	Radchenko (1989b) (USSR); Radchenko (1996b) (central and southern Palearctic)
Platythyrea	Ponerinae	Platythyreini	37	Neotropical, S Nearctic, Old World tropics to S Australia		Arboreal	Predators, many on termites	SP	Brown (1975:4)
Plectroctena	Ponerinae	Ponerini	17	Afrotropical			Predators on millipedes and their eggs	SP	Bolton (1974b)
Podomyrma	Myrmicinae	Leptothoracini	57	Indomalayan, Australia		Mostly arboreal, nesting in plant cavities		TCS, CCS	
Poecilomyrma	Myrmicinae	Leptothoracini	1	Melanesia		Arboreal		TCS	Mann (1921)
Pogonomyrmex	Myrmicinae	Myrmicini	58	Neotropical, Nearctic		Nesting in soil	Generalized foragers and seed harvesters	HCS	Kusnezov (1951a) (S Neotropical); Cole (1968) (Nearctic); Snelling (1981); Shattuck (1987) (*occidentalis* complex); Fernández and Palacio (1997) (Neotropical)
Polyergus	Formicinae	Formicini	4	Nearctic, Palearctic			Parasites and slavemakers of *Formica*	SP	Wheeler (1968) (Nearctic); Agosti (1994a)
Polyrhachis	Formicinae	Camponotini	477	S Palearctic, Afrotropical, Oriental to S Australia		Many arboreal, others nesting on the ground	Generalized foragers	SC	Hung (1967) (subgenera); Hung (1970), (subgenus *Polyrhachis*); Bolton (1973b, 1973c) (Africa); Bolton (1975c) (Africa); Kohout (1988 [*gab*

Continued on next page

group in Australia], 1988 [*sexspinosa* group], 1989 [*relucens* group in Australia]); Kohout and Taylor (1990) (*viehmeyeri* group); Wang and Wu (1991) (China); Dorow and Kohout (1995) (subgenus *Hemioptica*)

Genus	Subfamily	Tribe	No.	Distribution	Habitat	Biology	Code	References
Ponera	Ponerinae	Ponerini	33	Nearctic, Palearctic, Indo-Australian		Predators of small arthropods	C	Taylor (1967) (world); Terayama (1996) (Japan)
Prenolepis	Formicinae	Formicinae	8	Neotropical, Nearctic, Palearctic, Oriental, Indomalayan		Generalized predators	CCS	
Prionopelta	Ponerinae	Amblyoponini	12	Worldwide in tropical and subtropical, except Palearctic		Predators, esp. of small chilopoda	C	Brown (1960) (New World Indo-Australian); Terron (1974) (Afrotropical)
Pristomyrmex	Myrmicinae	Myrmecinini	36	Afrotropical, Mauritius, Oriental to E Australia		Generalized foragers and specialized predators	TCS	Bolton (1981b) (Afrotropical); Xu (1995a) (China)
Proatta	Myrmicinae	Stenammini	1	Indomalayan		Nesting in soil	Scavengers	TCS	Rigato (1994a)
Probolomyrmex	Ponerinae	Platythyreini	13	World tropics except Madagascar	Forest	Nesting in leaf litter	Predators	C	Taylor (1965); Brown (1975) (world); Terayama and Ogata (1988) (Japan); Agosti (1988) (Japan); Agosti (1994b) (Neotropical)
Proceratium	Ponerinae	Ectatommini	29	Worldwide in tropics and temperate		Predators of spider and other arthropod eggs	C	Brown (1958 [world], 1979 [Malagasy]; Terron (1981) (Africa); Ward (1988) (New World)
Procryptocerus	Myrmicinae	Cephalotini	39	Neotropical	Nesting and foraging arboreally	Some are pollen eaters	TCS	Kempf (1951, 1957)

Table 5.1 continued

Genus	Subfamily	Tribe	No. Species	Distribution	Habitat	Microhabitat	Biology	Functional Group[b]	Keys
Proformica	Formicinae	Formicini	24	Palearctic		Epigaeic	Generalized foragers	?	Agosti (1994a) (references to regional keys)
Prolasius	Formicinae	Melophorini	19	Australia, New Zealand in dead wood	Subtropical rainforest	Epigaeic, nesting and soil	Generalized foragers and seed harvesters in forest	CCS	McAreavey (1947) (out of date)
Protalaridris	Myrmicinae	Basicerotini	1	Mountains of Colombia and Venezuela	Mesic forest		Predators	C	Brown (1980a, 1980b)
Protanilla	Leptanillinae	Anomalomyrmini	2	Indomalayan			Predators	C	Taylor (1990a)
Protomognathus	Myrmicinae	Leptothoracini	1	Nearctic			Parasites and slavemakers of *Leptothorax*	CCS	
Psalidomyrmex	Ponerinae	Ponerini	6	Afrotropical			Predators, perhaps of lumbricid worms	C	Bolton (1975b)
Pseudapho-momyrmex	Formicinae	Brachymyrmecini	1	Indomalayan		Arboreal		?TCS	Wheeler (1922b)
Pseudoatta	Myrmicinae	Attini	1	Neotropical		Living in *Acromyrmex* nests	Workerless parasites of *Acromyrmex*	TCS	
Pseudolasius	Formicinae	Lasiini	48	Afrotropical, Oriental to N Australia		Epigaeic	Cryptic foragers	TCS	Weber and Anderson (1950) (Afrotropical); Xu (1997) (China)
Pseudomyrmex	Pseudo-myrmecinae	Pseudo-myrmecini	118	Neotropical and S Nearctic		Mostly arboreal (nesters and foragers), few epigaeic	Generalized predators, visit extrafloral nectaries	TCS	Kempf (1958b [*gracilis* group], 1960c [*tenuis* group], 1961b [groups of *tenuis, oculatus, pallens,* and *latinodus*]); Ward (1985 [Nearctic], 1989 [*oculatus* and *subtilissimus* groups], 1990 [generic revision], 1993 [*Acacia*-inhabiting species], 1999a [*viduus* group])

Genus	Subfamily	Tribe		Distribution				CCS	References
Pseudonotoncus	Formicinae	Melophorini	2	Australia	Woodland		Generalized foragers		
Quadristruma	Myrmicinae	Dacetini	2	Melanesia, ?adventive in Neotropical			Predators	C	Brown (1949b); Bolton (1983)
Recurvidris	Myrmicinae	Pheidologetonini	7	Indomalayan			Generalized foragers	C	Bolton (1992)
Rhopalomastix	Myrmicinae	Melissotarsini	3	Oriental		Nesting and foraging in and under bark		?	Xu (1999) (China)
Rhopalothrix	Myrmicinae	Basicerotini	10	Neotropical, Indo-melanesian, NE Australia			Predators	C	Brown and Kempf (1960) (world); Taylor (1990b) (Indo-Australian)
Rhoptromyrmex	Myrmicinae	Tetramoriini	10	Afrotropical, Oriental to NE Australia			Generalized foragers	?	Bolton (1986)
Rhytidoponera	Ponerinae	Ectatommini	102	Australia and Melanesia W to S Philippines			Generalized predators	O	Clark (1936) (Australia; out of date); Ward (1980 [*impressa* group], 1984 [New Caledonial])
Rogeria	Myrmicinae	?Stenammini	27	Neotropical, S Nearctic, Indo-Melanesian		Nesting in leaf litter	Generalized foragers	TCS	Kugler (1994)
Romblonella	Myrmicinae	Leptothoracini	8	Indo-Melanesian, Australian			Generalized foragers	TCS	Smith (1953a, 1953b, 1956b); Taylor (1990d)
Rossomyrmex	Formicinae	Formicini	2	S Palearctic	Grassland	Nesting in ground	Parasites and slavemakers of *Formica*	?	Agosti (1994a)
Rostromyrmex	Myrmicinae	Solenopsidini?	1	Indomalayan	Lowland rainforest	Litter, nesting in rotten wood		TCS	Rosciszewski (1994) (Malaysia)
Rotastruma	Myrmicinae	Leptothoracini	2	Indomalayan		Probably arboreal		TCS	Bolton (1991)
Santschiella	Formicinae	Santschiellini	1	Afrotropical				?TCS	
Scyphodon	Leptanillinae	Anomalomyrmini	1	Indomalayan				C	Brues (1925); Petersen (1968)
Secostruma	Myrmicinae	Tetramoriini	1	Indomalayan				?	Bolton (1988a)
Sericomyrmex	Myrmicinae	Attini	19	Neotropical		Nesting in soil, some in rotten logs	Fungus cultivators	TCS	
Serrastruma	Myrmicinae	Dacetini	12	Afrotropical			Predators, mainly of collembolans	C	Bolton (1983)

Continued on next page

Table 5.1 continued

Genus	Subfamily	Tribe	No. Species	Distribution	Habitat	Microhabitat	Biology	Functional Group[b]	Keys
Simopelta	Ponerinae	Ponerini	14	Neotropical		Nesting in leaf litter	Mass predators of ants	C	Gotwald and Brown (1966); Brown (pers. obs.)
Simopone	Cerapachyinae	Cerapachyini	16	Afrotropical, Madagascar, Indo-Melanesian			Mass predators of ants	SP	Brown (1975)
Smithistruma		Dacetini	122	Worldwide in tropics and warm temperate		Litter	Predators, mainly of collembolans	C	Brown (1953a, 1964) (world); Bolton (1983) (Afrotropical); Ward (1988) (W Nearctic); Terayama et al. (1995) (Taiwan); Terayama et al. (1996); Ogata and Onoyama (1998) (Japan)
Solenopsis	Myrmicinae	Solenopsidini	180	Worldwide in tropics and warm temperate		Nesting in ground, sand mounds (mainly geminata group), and litter	Generalized foragers and and thief ants	C[e], TCS	Snelling and Hunt (1975) (Chile); Thompson and Johnson (1989) (Florida); Ross and Trager (1990) (*saevissima* complex); Trager (1991) (*geminata* group); Dlussky and Radchenko (1994) (C Palearctic)
Sphinctomyrmex	Cerapachyinae	Cerapachyini	23	Afrotropical, Indomalayan to Australia, SE Brazil			Mass predators of ants	C	Brown (1975)
Stegomyrmex	Myrmicinae	Stegomyrmecini	3	Neotropical	Mesic forest	Nesting in soil	Predator of eggs of millipedes	C	Diniz (1990)
Stenamma	Myrmicinae	Stenammini	42	Nearctic, Palearctic, Oriental			Generalized predators	CCS	Yasumatsu and Murakami (1960) (Japan); Smith (1962) (C America); Snelling (1973)

Continued on next page

Genus	Subfamily	Tribe	No. spp.	Distribution	Ecology	Biology	Code	Reference
Stereomyrmex	Myrmicinae	Leptothoracini	1	Sri Lanka			TCS	(Nearctic); Arnol'di (1975) (USSR); DuBois (1998) (Palearctic and Oriental)
Stigmacros	Formicinae	Plagiolepidini	48	Australia		Generalized foragers	CCS	McAreavey (1957) (out of date)
Streblognathus	Ponerinae	Ponerini	1	South Africa	Arid thorn scrub	Predators of tenebrionid beetles	SP	Brown (pers. obs.)
Strongylognathus	Myrmicinae	Tetramorini	25	Palearctic		Parasites and slavemakers of *Tetramorium*		Pisarski (1966); Baroni Urbani (1969) (*huberi* group, Palearctic); Radchenko (1985, 1991) (USSR)
Strumigenys	Myrmicinae	Dacetini	190	Worldwide in tropics and warm temperate areas, except W Palearctic	Nesting in leaf litter	Predators, esp. of collembolans	C	Brown (1962) (New World); Bolton (1983) (Africa; Lattke and Goitía (1997) (Venezuela)
Talaridris	Myrmicinae	Basicerotini	1	Neotropical	Mesic forest	Predator	C	Brown and Kempf (1960)
Tapinoma	Dolichoderinae	Dolichoderini	60	Worldwide in tropics and temperate regions		Generalized foragers	O, DD	Emery (1925b) (Palearctic; out of date)
Tatuidris	Myrmicinae	Agroeco-myrmecini	1	S Neotropical			?C	Brown and Kempf (1967b)
Technomyrmex	Dolichoderinae	Dolichoderini	60	Old World tropics, 1 sp. adventive worldwide		Generalized foragers	O	
Teleutomyrmex	Myrmicinae	Tetramorini	2	Palearctic		Workerless parasites of *Tetramorium*	?	Kutter (1950); Tinaut (1990)
Terataner	Myrmicinae	Leptothoracini	15	Afrotropical, Madagascar	Mostly arboreal, in plant cavities		TCS	Bolton (1981b)
Teratomyrmex	Formicinae	Lasiini	1	E Australia	Mesic forest	Foraging on leaves of shrubs	TCS	McAreavey (1957)
Tetheamyma	Myrmicinae	Stenammini	1	Borneo	Nesting in forest litter		?	Bolton (1991)

Table 5.1 continued

Genus	Subfamily	Tribe	No. Species	Distribution	Habitat	Microhabitat	Biology	Functional Group[b]	Keys
Tetramorium	Myrmicinae	Tetramoriini	415	Worldwide in tropics and temperate but adventive only in S America			Generalized foragers	O	Bolton (1976 [partial], 1977 [Indo-Australian], 1979 [Madagascar, New World], 1980 [Africa]); Wang et al. (1988) (China); Radchenko and Arakelian (1990) (*ferox* group, Caucasus); Radchenko (1992) (USSR); Cagniant (1997) (Morocco)
Tetraponera	Pseudomyrmecinae	Pseudomyrmecini	78	Old World tropics and warm temperate, except Europe	Mainly forest	Arboreal, nesting in plant cavities		TCS	Ward (1990); Wu and Wang (1990) (China)
Thaumatomyrmex	Ponerinae	Ponerini	6	Neotropical	Forest, savanna	Nesting in litter and bromeliads	Predators of polyxenid millipedes	SP	Kempf (1975a)
Tingimyrmex	Myrmicinae	Dacetini	1	Neotropical		Nesting in litter	Predators	C	Mann (1926)
Trachymyrmex	Myrmicinae	Attini	41	Neotropical, S Nearctic		Nesting in soil	Cultivators of fungi, nests covered with straw, specialized entrances	TCS	
Tranopelta	Myrmicinae	Ochetomyrmecini	4	Neotropical		Nesting and foraging underground or cryptically	Hypogeaic predators of termites	?C	
Trichoscapa	Myrmicinae	Dacetini	150	Worldwide tropical and warm temperate			Predators, mostly of collembolans	C	Brown (1953a); Bolton (1983, unpublished)
Tricytarus	Myrmicinae	Leptothoracini	1	Indomalayan				TCS	Bolton (1994) (male only)
Turneria	Dolichoderinae	Dolichoderini	6	Melanesia, NE Australia		Arboreal, nesting in plant cavities		TCS	Shattuck (1990, 1992b)

Genus	Subfamily	Tribe	No.	Distribution	Forest	Nesting	Diet	Group[b]	Reference
Typhlomyrmex	Ponerinae	Typhlomyrmecini	5	Neotropical		Nesting in rotten logs	Predators	C	Brown (1965)
Vollenhovia	Myrmicinae	Metaponini	50	China to N Australia, Madagascar		Many nesting under bark or in cavities in logs		TCS	
Vombisidris	Myrmicinae	Leptothoracini	12	Indomalayan to E Australia		Mostly arboreal, nesting in plant cavities		TCS	Bolton (1991)
Wasmannia	Myrmicinae	Blepharidattini	8	Neotropical, adventive in W Africa, New Caledonia, Galápagos		Arboreal, nesting in soil or rotten logs	Generalized foragers	TCS	
Willowsiella	Myrmicinae	Leptothoracini	2	Melanesia, E Australia				TCS	Taylor (1990d)
Xenomyrmex	Myrmicinae	Metaponini	3	Neotropical, S Nearctic		Arboreal, nesting in plant cavities		TCs	Creighton (1957)
Yavnella	Leptanillinae	Leptanillini	1	S Palearctic, Oriental				C	Kugler (1986)
Zacryptocerus	Myrmicinae	Cephalotini	71	Neotropical, S Nearctic		Arboreal, nesting in plant cavities	Some are pollen eaters	TCS	Kempf (1951, 1952, 1958a, 1967d, 1973b)

[a]Compiled by the author, with comments added by D. Agosti, A. N. Andersen, C. R. F. Brandão, and X. Espadaler. Functional groups designated to the genera by Andersen (Chapter 3). For further explanation see text.

[b]C, cryptic species; DD, dominant Dolichoderinae; GM, generalized Myrmicinae; H/C/TCS, hot/cold/tropical climate specialists; O, opportunists; SC, subordinate Camponitini; SP, specialist predators.

[c]C applies to subgenus *Cerapachys*; SP to subgenera *Lioponera* and *Phyracaces*.

[d]Most are SP; subgenus *Brachyponera* is TCS.

[e]C applies to subgenus *Diplorhoptrum* only.

species for each genus are largely derived from Bolton (1995a, 1995b), but these numbers have been modified in a few cases where the count is likely to be changed in the near future by the creation of new species, named and unnamed, the existence of which are known to me now. Of course, numbers of species will continue to change in the future. Another drastic change will be a reduction in the number of genera of Dacetonini, which are now deemed excessive by Bolton (1999). Functional groups of the genera have been provided by Anderson (Chapter 3).

Ant Taxonomy: The State of the Art

Characters

Taxonomists (or for that matter most dictionaries) fail in the definition of *character* as it applies to biosystematics. For me, a character is any trait of use in making a comparison. The comparison may be made between different parts of the same organism, but here we are mainly interested in comparisons between whole organisms and populations of them. Thus we accept the distinction between characters and their states, though we may simply use *character* as shorthand for *character state*.

A major change in formicid systematics since 1950 has been the acceleration of the discovery and employment of new characters, a function at least as important as the description of new taxa. In higher-level classification, Barry Bolton's investigations of morphological characters, particularly those of the abdominal segments of the worker caste, leap immediately to mind. He has not neglected other tagmata and other castes, and his sample of the genera and species has been extensive and documented, at least as far as the numbers of species reviewed for each character are concerned. Explicit listing of the species examined for such studies has become a welcome standard.

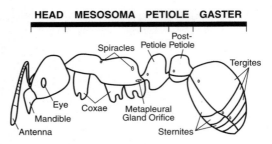

Figure 5.1. Body parts of an ant.

The ancestors of ants were apparently some kind of aculeate wasp, perhaps a vespoid, as claimed by Brothers (1975) and Brothers and Carpenter (1993), but it is probable that the ancestral ant, like most living ants, differed from wasp ancestors by possessing a metapleural gland (Fig. 5.1) far back and down low on each side of the alitrunk (Grimaldi et al. 1997). This relatively obscure feature typically consists of a group of secretory cells that duct separately into an atrium under a more or less obvious bulla having a variously shaped aperture to the exterior on each side of the alitrunk. Its products have disputed functions, one of which may be to protect the insects against microorganisms and fungal spores inhabiting the substrate (Maschwitz 1974). In some ants—notably the carpenter ants and a few others—the metapleural gland has been secondarily lost or reduced.

A much more obvious trait of the family Formicidae, to which all ants belong, is the nodiform or scalelike shape of the second true abdominal somite of the waist, called the petiole (Fig. 5.1), which is separated by a more or less apparent constriction from the following somite (A3), also called the postpetiole, particularly whenever it itself is separated by a constriction from the following somite (A4). The trouble with using the petiole as an ant-diagnostic character is partly that there exist various wasps, often wingless ones, that have developed a nodiform petiole, and some, such as the mutillid Apterogyna, that even have a postpetiole. In such cases, the lack of a metapleural gland and other characters must be used to distinguish them from ants.

To reduce confusion in the naming of successive somites in the region of the waist and beyond, we call them by their sequence as ancestral abdominal units, with the propodeum being the first (actually it is T1, the tergum of the first true abdominal somite, which in Hymenoptera: Apocrita is fused with the thorax to form the alitrunk). Many specialists on apocritan Hymenoptera use *mesosoma* instead of *alitrunk,* after Michener (1944:167). However, *alitrunk* has been in use at least since J. B. Smith's *Glossary* of 1906, and probably much longer. It is used in such well-known works *as The Insects of Australia and New Zealand* by Tillyard (1926) and Torre-Bueno's *Glossary of Entomology* (1937, 1989; Tulloch 1962) as well as several foreign glossaries. When I wrote to Michener to ask why he had avoided use of *alitrunk,* he replied simply that he had not known about it.

The petiole is the second true abdominal segment, and the postpetiole, if present, is the third. The remaining abdominal segments together form the gaster. The somites are of course ancestrally rings, each formed of a dorsal plate, the tergum, and a ventral plate, the sternum; these are often called tergite and sternite (e.g., Gauld and Bolton 1988). A handy convention calls the somites of the true abdomen A1, A2, A3, . . . ; their terga T1, T2, T3, . . . ; and their sterna S1 (although S1 has been lost in the apocritan ancestors of ants), S2, S3, ... Each plate is more or less distinctly divided in the axial direction into regions, particularly an anterior one (the pretergite or presternite, which is the band fitting into the preceding somite and normally at least partly covered by it) and a main exposed region (which I call the tergite or sternite, and Bolton calls the posttergite and poststernite). (I reason by analogy with the mesothoracic sclerites prescutum and scutum,

and because the term *posttergite* or *-sternite* is logically reserved for specialized posterior segmental belts found in many ants.)

Bolton (1990a) has given a special name, *helcium,* to the much-narrowed "presegment," or pretergite with presternite, of the third abdominal segment. The structure of the helcium, especially its sternum, has become important in ant phylogeny and taxonomy (e.g., Agosti 1991).

Other abdominal characters of importance are the fusion or lack of it between the tergum and sternum of each of the second (A2) through fourth (A4) somites, and the structure of their presclerites when present. In Bolton (1990a) and subsequent papers on ant phylogeny by Bolton (1990c, 1990d) and Baroni Urbani, Bolton, and Ward (1992) (hereafter referred to as BBW), the characters are presented in a splendid series of illustrations that should be standard references.

Phylogenies

In the central study of the series, BBW offer a matrix for 68 characters of all adult castes and larvae, with coding binary (0 or 1). (I do suggest that, for making the often complex task of following binary character states easier for the reader, absence of a state might be coded 0 and its presence 1, mnemonic advantage thus outweighing other considerations.) The highlights of the resulting cladogram and classification are presented, with my comments, in the following sections.

THE EXTENDED ARMY ANT CLADE. Aenictinae, Cerapachyinae, Dorylinae, and Ecitoninae, plus Aenictogitoninae and Leptanilloidinae (considered as subfamilies), are placed in a monophyletic cluster, with a more-inclusive cluster being the (Leptanillinae + Ponerinae) and then the next, Apomyrminae. Of these, the first four form the doryline section, while the Aenictogitoninae, Apomyrminae, and Leptanilloidinae are all monogeneric, incompletely known taxa that are probably best considered as tribes of uncertain position.

In his important study of the cerapachyines, Bolton (1990a) made a reasonable case for their separation from the Ponerinae and their inclusion in a "doryline section" also containing subfamilies Dorylinae, Ecitoninae (Brown 1973), and Aenictinae (Bolton 1990c), considered to be monophyletic as based on eight shared derived character states. This relationship is reinforced by the circumstance that most species sufficiently well known are predators with a preference for formicid prey; all follow the army ant lifestyle except *Acanthostichus* and *Cylindromyrmex,* termitotherous specialists; the last may not be nomadic and badly needs study. Of course, we must be aware that the army ant lifestyle could be the basis for a convergent adaptive syndrome of morphological states in some or all four of the subfamilies of the doryline section, a possibility demonstrated by the partial homoplasies of syndromes in species of such disparate ponerine genera as *Leptogenys, Onychomyrmex,* and *Simopelta.* However, the unity of the doryline section, with the cerapachyines near the base of the lineage, is a concept that has been developing at least since Emery (1901), and myrmecologists are probably comfortable with it. In fact, it seems reasonable now to suggest that the "doryline section" is equivalent to one subfamily bearing the prior name Dorylinae.

ANEURETINAE, DOLICHODERINAE, AND FORMICINAE. Aneuretinae, Dolichoderinae, and Formicinae form a second cluster at a distance from the first. These taxa had already received attention from Shattuck (1992c), who found them to be monophyletic in a cladistic analysis. There is substantial accord among ant specialists in regarding dolichoderines and aneuretines as related; the question has become whether they are separate subfamilies (Clark 1951; Wilson et al. 1956) or whether, as treated traditionally,

the Aneuretini are a tribe within the Dolichoderinae. BBW maintained the subfamily status rather hesitantly "in accord with contemporary usage," because they could not list any strong morphological shared derived states to bolster Aneuretinae (BBW, pp. 313, 315–316). Though the sting is present in *Aneuretus* and variously reduced in Dolichoderinae, it seems to me that the ensemble of differences between the two is not strong enough to maintain Aneuretini as more than a tribe of Dolichoderinae.

The placement of the Formicinae with the dolichoderine lineage is a more important matter. Judging from the BBW phylogeny, most of the characters shared by the two subfamilies are probably in the ancestral state. Unique derived (autapomorphic) characters of Formicinae are the presence of the acidopore, or "venom nozzle," and the disarticulation of the sting from its lancets (a stage of sting reduction beyond the most extreme condition in the Dolichoderinae). Associated with this, but not coded, are the production of formic acid uniquely by formicids and what is patently a whole series of characters of the venom production and storage apparatus. On their side, the Dolichoderinae (including Aneuretinae) uniquely have Pavan's gland and also share the lack of an unfused furcula in the sting apparatus. The coding of these complex characters on a simple binary basis is to me one of the unreal aspects of cladistic syntheses, at least as we have them for ants.

Another character, BBW No. 38, is the sclerotized versus flaccid condition of the proventriculus. Coding "according to Eisner (1957) could be an oversimplification" puts it mildly, since the proventriculus of *Dolichoderus* (=*Hypoclinea*) was characterized by Eisner (1957:453, and his Figs. 7, 17–20, and 97) as of the flaccid type, "still conform[ing] to the basic structural plan of *Myrmecia, Pseudomyrmex,* and *Aneuretus,* except that the plicae have become sclerotized toward the base of the bulb." But the subfamilies Dolichoderinae and Formicinae were both coded by BBW as of the sclerotized type. Considering the details of Eisner's review of proventricular morphology, I find it difficult to avoid a conclusion of completely independent evolution of the two morphoclines, regardless of the apparent similarity of some intermediate stages. From all this, I find that close relationship of Formicinae to Dolichoderinae is unlikely even by strict cladistic standards. As is mentioned by BBW, Emery (1925b) provides an interesting commentary on proventricular evolution of formicines versus dolichoderines, but it is worth noting that in his time the function of the organ was not well understood. I would place the Formicinae at an early split in the ant phylogram, mainly on the basis of the complex derived condition of the sting apparatus and the many ancestral states of characters, such as body articulation, antennal and palpal segmentation, and retained worker ocelli. Dolichoderinae seem to me to be a separate lineage, convergently morphoclinal to Formicinae in consonance with their commitment to the adaptive zone of exploiters of sugary fluids.

MYRMECIINAE, PRIONOMYRMECINI, AND NOTHO-MYRMECIINAE. Myrmeciinae, Prionomyrmecini, and Nothomyrmeciinae are placed as terminal taxa on a monophyletic stem, shared at the next lowest node with Pseudomyrmecinae, and at the next lowest with Myrmicinae. Of the fossil *Prionomyrmex,* there is little to argue against in its placement in Myrmeciinae. Concerning Nothomyrmeciinae, things are less certain. Clark (1951:16) started by putting the monotypic *Nothomyrmecia* into a separate subfamily in a key that was rather offhanded, for example, giving subfamily rank to Amblyoponinae, Discothyreinae, Eusphinctinae, and Odontomachinae. Brown (1954b) tended to overemphasize the difference between Nothomyrmeciinae and Myrmeciinae when he split all the living ants into two branches ("myrmecioid complex" and "poneroid complex") and

placed *Myrmecia* in one complex and *Nothomyrmecia* in the other. This was at a time when only the two type specimens of Nothomyrmecia were known but hardly available for study. Due to the efforts of Taylor (1978a) and a series of colleagues, the celebrated *N. macrops* was rediscovered, and it is now one of the best known of all ants. Although its differences from *Myrmecia* are at once apparent, more and more similarities have also been found (e.g., Billen 1990). In addition the BBW opinion that "absence of a postpetiole in *Nothomyrmecia* is undoubtedly primary" need not be accepted unreservedly, because this segment is small and just might represent a reversal of the condition in *Myrmecia.* My original peek did give me the impression that *Nothomyrmecia* was somewhat like lower formicines in habitus, but I would now find it easier to accept placement in a tribe within the Myrmeciinae.

PSEUDOMYRMECINAE. Pseudomyrmecinae were placed by BBW and by Ward (1990) in their cladograms as arising near the *Myrmecia* group of taxa, and for the time being this placement seems reasonable to me, as it did in 1954. Monophyly with the next lowest step, Myrmicinae, is much less appealing.

MYRMICINAE. Myrmicinae, the most speciose of the subfamilies, was placed by BBW and by Ward (1990) next to Pseudomyrmecinae. Both taxa have A3 postpetiolate, that is, reduced in size and pinched off from A4 by a constriction, which has probably long caused them to be considered as allied. Brown (1954b:28 and later papers) has favored a myrmicine derivation from ponerine ancestors, possibly Ectatommini, but the finding by Gotwald (1969), Bolton (1990a), Ward (1990), and BBW that Ponerinae have the tergum and sternum of A4 completely fused seems to weigh heavily against this origin because Myrmicinae do not show complete fusion (but sometimes have the pretergite of A4 fused to its presternite, BBW No. 26). The A4 fusion character is now somewhat diminished by the finding by Ward (1994) of the new Malagasy genus *Adetomyrma,* which, though ostensibly a rather typical amblyoponine ponerine, has the terga and sterna of A3 and A4 unfused. In an honest and earnest probe of the phylogenetic implications of the possibility that tergosternal fusion may be reversed in this (and other) instances, Ward offers Fig. 45, a cladogram assuming reversibility. This schema is fascinating in that it depicts the possibility that Myrmicinae is monophyletic with "remaining Ponerinae," that is, remaining after Amblyoponini is relegated to a neighboring terminal branch on the same stem. It is interesting to compare this tree with one I drew up rather negligently five years ago for a poster exhibit, in which the Myrmicinae are derived from within or near the Ponerinae. The position of the Formicinae is of course entirely different; Ward was not dealing with that issue in 1994. Now a finding by Hashimoto (1996) based on skeletomuscular characters also concludes that Ectatommini, Myrmicinae, and Ponerinae are closely related, perhaps sister groups, and incidentally that the amblyoponine petiole may not be a retention of the typhoid wasp state, but a reversal of a previous formicid condition of petiolar postconstriction.

APOMYRMINI. Apomyrmini, a problematic monogeneric tribe, is probably best considered as an anomalous Leptanillinae for the time being.

Two Lessons

For me, two lessons to be learned from a consideration of the recent history of phylogenetic reconstructions of the ant family are that (1) cladistic techniques are not as robust for ants as had been hoped, perhaps partly owing to problems of coding and discreteness of characters; and (2) characters involving the articulation or fusion, and expansion or reduction, of abdominal segments may be especially subject, even if

only rarely, to reversal. About problem (1), I can only suggest that each character be scanned for complexity, i.e., for whether multiple characters are involved, and for its suitability to be shoehorned into a rigid binary system. For (2), I urge more careful consideration of possible adaptive reasons for particular character states and transitions between them—in short, bolder application of the much-derided "Just-So Stories." Such application might even help to bring phylogeny back into biology. Ward is thinking along these lines when he briefly discusses possible reasons for reversal of tergosternal fusion, though he does not find examples among several taxa with dichthadigyne queens that might be expected to evolve an expandable gaster. It is worth noting, though, that BBW (p. 317) regard the loss of tergosternal fusion in A3 of ecitonine males as secondary!

Be that as it may, reversal of the fusion could conceivably have other adaptive reasons. For example, if a lineage of ants began a behavioral shift toward feeding prominently on sugary fluids, such as nectar or honeydew, flexibility and dissociation of some segments of the abdomen could be at a selective premium. On the other hand, increased commitment to predation might favor strengthening and fusion of integumental elements at the expense of ability to store fluids internally, as seems true of many if not all Ponerinae. Formicinae and Dolichoderinae may be lineages that have taken the fluid-food path and retained or evolved anew separated terga and sterna, while Ponerinae, the dorylines with Cerapachyinae, and the minor predatory subfamilies (usually) have fusion at least in A3.

At the Front with Revisionary Taxonomy

Taxonomy in the "old days" was largely accomplished on a faunal basis, especially when tropical colonialism was in flower. Collections made by missionaries, official travelers, and others were sent back to a specialist in the home country—a Frederick Smith, a Gustav Mayr, a Carlo Emery, an Auguste Forel—who then duly produced a paper on "New and Little-Known Ants Collected by Mr. Whatsisname in Southern Whereverland," replete with descriptions of new species and varieties, and, fairly often, new genera. This system, practiced by specialists who communicated and exchanged specimens all too rarely, produced such a confused welter of taxa that it was a wonder that Emery could make as much sense of it as he managed to do with his fascicles in the *Genera Insectorum* (1910, 1911, 1913, 1921, 1922, 1925a). Even so, the synonymy and other confusion resulting from the compartmental colonial-faunal approach left a mess that thwarted most efforts at effective identification or inventory of ant species, particularly in the larger genera. I need only mention the examples of *Camponotus, Crematogaster, Pheidole,* and *Solenopsis* to make this point about identifiability, even to this very day.

For a half century this state of affairs has been changing. The Era of Revision has happily arrived with a vengeance. By keeping descriptions of new species largely within the context of world- or at least continentwide revision, genus by genus, we have been able to avoid most of the duplicate description and consequent synonymy of the old days. The process of revision as it is understood today means to gather large collections in the field, compare them with the type specimens in the classical collections to fix their names, and generally sort out all the species. The output is monographs of whole genera, tribes, or even subfamilies, with keys to genera and species, illustrations, and other identification aids by now familiar to all. These monographs allow others to make accurate identifications for the first time, so that one result of the appearance of a revision is a subsequent flock of descriptions of new species. If these descriptions are carefully carried out with-

in the context of the revision itself, that is all to the good. The best result of a revision is the encouragement of new revisions.

Some Cases in Point. Subfamily Ponerinae has been revised in a series of steps, primarily by Brown since 1952, all listed in Bolton's *Catalogue.* A major part of this study dealing with the subtribe Poneriti, which has been long delayed, deals with the expanded version of *Pachycondyla,* consisting of nearly 150 valid species after extensive synonym pruning. This and allied genera contain many ground-dwelling species. Other parts deal with the large genera *Leptogenys* and *Hypoponera*—the latter a genus of great interest to this manual because of its abundance in leaf litter samples. The African *Leptogenys* have already been completed for the known species by Bolton (1975a). *Hypoponera,* separated from the smaller genus *Ponera* by Taylor (1960), is a real challenge because of the great uniformity of the many species, and because ergatoid queens commonly resemble the workers of different species. In addition some adventives are common and widespread in both hemispheres.

MacKay revised *Acanthostichus* (1996) and added several species to this New World genus of termitotheres.

Following Bolton's (1990b, 1990d) works on the Cerapachyinae and Leptanillinae, nothing has been done on the latter, though Taylor has had excellent descriptions of *Anomalomyrma* and *Protanilla,* with fine illustrations, in manuscript for two decades. Their descriptions are ascribed to Taylor in Bolton's paper, but the real authorship is confused. *Scyphodon,* known only from the tiny male, has been found again in Borneo, and its peculiar mandibles suggest that it is the male corresponding to the worker of *Protanilla.*

Revisionary work on the Pseudomyrmecinae by Ward (1985, 1989, 1990) following useful partial revisions by Kempf (1958–1967) has been completed for the Neotropical species.

Probably the largest bloc of unrevised species is genus *Pheidole,* of which the Nearctic species have been partly revised by Gregg (1959). Wilson has revised the New World species of this genus, which is principally tropical in the New World, and after synonymic debridement has found 335 new species, or approximately one-tenth of the known New World ant fauna! This revision is essentially complete, with illustrations, and will be published in 2000 by Harvard University Press. *Pheidole* is of course another of the prevalent ground-dwelling taxa; in most tropical forests, it is the dominant one in terms of individuals (Chapter 8). In Africa, *Pheidole* has many species, but fewer than it does in the Americas; in Asia, Melanesia, and Australia there are more species, but it is not yet known how the number compares with that in tropical America.

Charles Kugler's (1994) revision of *Rogeria* is another first full examination of a long-neglected myrmicine genus, and his study of the sting apparatus of Myrmicinae (1978) plus later papers on the sting promise to be of importance as more taxa are examined.

The tribe Dacetonini, exceedingly common in tropical litter samples, has also been under stepwise revision by Brown (preliminary revision in 1948, plus many later papers) and by Bolton for the Afrotropical species. Bolton is now engaged in a major revision of the classification of the tribe, and it is clear that many of the smaller genera, particularly among the short-mandibulate forms that once seemed so distinct, are now compromised by the more recent discovery of intermediate species. This tribe is astonishingly diverse in tropical forests, especially the largest genus *Strumigenys,* with hundreds of species worldwide, mostly in leaf litter and rotten wood. There are probably fifty endemic species of this genus on Madagascar, only one of which has been described and named (B. Fisher, pers. comm.). I have revised the New World species and have a large, incom-plete manuscript covering the known Indo-Australian species.

The genera of the Basicerotini were included in the Dacetonini in the Emery-Wheeler classifications, and the tribe was set up by Brown (1949d) and revised by Brown and Kempf (1960) for the world. A supplement to the revision of *Basiceros* was produced by Brown (1974g), who also described the genus and species *Protalaridris armata* from the northern Andes (Brown 1980a, 1980b). Scattered Neotropical species were published from time to time by various authors, and major groups of basicerotine species were studied for the Indo-Australian area by Taylor (1968a, 1968b, 1970b, 1980, 1990b), bringing the total of Paleotropical species to 26.

Excellent recent revisions of Dolichoderinae are those by Shattuck: genera *Turneria* (1990), *Axinidris* (1991), and *Iridomyrmex* (1992a); a generic revision of Dolichoderinae (1992b); and two Australian species groups of *Iridomyrmex.* References are contained in his catalogue (Shattuck 1994). This work is exemplary in its treatment of a long-confused subfamily; the author recognizes 912 currently valid species, including Aneuretinae and a few fossils.

In Formicini, the tribe Myrmoteratini, with the single genus *Myrmoteras,* has attracted successive revisers, perhaps because of its easily recognizable and distinctive species, originally few in number. After limited pre-1980 efforts, reviews were published by Moffett (1985) and then Agosti (1992), primarily consisting of diagnoses of new species. The species count has reached the astounding number of 31. Agosti (1994a) has extended research to the classification of the tribe Formicini and to the species-level taxonomy of *Cataglyphis* (1990) and *Cladomyrma* (1991).

Paratrechina remains to be revised; the only large regional revision is by Trager for North America (1984), but this is a frequently encountered ground-dwelling genus in most tropical forests.

Other large genera known to be rich in ground-dwelling species are *Tetramorium, Monomorium,* and *Solenopsis* (Chapter 8). The first two of these are predominantly Old World in distribution; they have been revised by Bolton (1986–1988) in a very useful series of contributions. *Solenopsis* (including *Diplorhoptrum* sensu Baroni Urbani 1968), is a very common and speciose genus in the New World tropics, but much less so in the Old World. Creighton's review of the tropical species really dealt only with the larger fire ants of the *saevissima* group. Trager (1991) revised the same group, and his revision has largely been superseded since, but the many species of the small-sized groups, composing the thief ants (subgenus *Diplorhoptrum*), are without modern revision. Treatment of these requires massive collections of nest series containing queens, and if possible males, because the workers are often devoid of striking differences.

Then there are the Attini, fungus cultivators of the Americas, predominantly the tropics. Their diversity is being studied by Schultz and colleagues (Chapela et al. 1994; Hinkle et al. 1994; Schultz and Meier 1995; Mueller et al. 1998; Schultz 1998; Wetterer et al. 1998), who are also determining the mode of co-evolution of the ants with their fungi, venturing into comparative studies of the ant larvae and also of DNA.

Other taxa, such as *Camponotus, Crematogaster,* and *Pseudomyrmecinae,* are certainly common and ubiquitous in the tropics, but they are less closely tied to the leaf litter and tend to be arboreal.

Army ants, although often common and conspicuous in tropical forest, usually occur unevenly in samples, but they and other less densely distributed genera often are indicators of particular conditions or habitats, so they clearly should and will receive attention in surveys. The New World Ecitoninae were revised by Borgmeier (1955), with additional taxonomic studies by Watkins (1976, 1982, 1985).

Much more could be said about the systematics of particular groups and particular taxonomists, but many of these studies are based on work performed in study areas outside the forest litter zone and are often published in languages and journals not readily accessible to most workers. China now has a number of specialists on ant systematics, but as far as I have seen they have not focused on revisions, but rather on random detection and description of new species from the Chinese fauna. In Japan the local fauna is also emphasized, but some workers are now commencing revisionary work in the Asian tropics and elsewhere, and experts such as Imai and Kubota are the leaders in the field of ant karyology. The world revisionary initiative has been strong in South America and Australia, with revisers of the caliber of Brandão, Lattke, Shattuck, and Taylor, but there are still not very many of them.

Systematic Infrastructure

The year 1950 was a turning point for world ant taxonomy; in that year Creighton published *The Ants of North America,* the first major work to apply the Modern Synthesis (Mayr 1942) to the family. This meant that the "quadrinomial" system of the last century, congealed in Emery's great *Genera Insectorum* fascicles of 1910–1929 and kept relatively unchanged in the works of W. M. Wheeler and his contemporaries, was abandoned—we hope forever! The quadrinomial (four-name) system in the Formicidae—almost unique in the persistence of its application to species-group taxa of names for genus, species, subspecies, and variety—was really a pentanomial (five-name) system, because in practice it routinely used a fifth infrageneric category, the subgenus. Thus we suffered such monstrosities as *Lasius* (*Chthono-*

lasius) *umbratus mixtus aphidicola;* today this species is represented (in North America) by *Lasius umbratus.* Although many pentanomials and quadrinominals are still listed in catalogues and regional lists, almost everyone agrees that they should be eliminated. Under the International Code of Zoological Nomenclature, changes in the status of the variety have helped mightily in this direction, and in my opinion they could help still further by removing the subspecies category from Linnaean nomenclature (Wilson and Brown 1953).

The elimination of the subgenus as a formally named category is a worthy goal in systematic biology, and one that is well on the way to realization in ant systematics (Brown 1973). For nomenclatorial purposes the International Rules recognize the subgenus as equivalent to a genus. The confusion that results when generic names are synonymized or found to be homonyms reveals subgenera to be dangerous as well as merely inconvenient and burdensome: witness the recurrent case of *Cryptocerus* and its subgenera. Most myrmecologists seem to be moving in the direction of using informal species groups instead of subgenera, and groups serve every purpose except to satisfy the cravings of authorship. In practice, formicid taxonomy is fast becoming binomial again, after two centuries.

Another nomenclatorial issue is the ending for subtribal names, which in the beetles and other groups has unfortunately settled on *-ina,* this neither unambiguously plural nor uncommon as an ending of generic names, and identical in the vernacular of most European languages with the already overburdened *-ine* (e.g., Ponerinae, ponerine; Ponerini, ponerine; Ponerina, ponerine). I have discussed the problem (Brown 1958) and suggested instead the subtribal ending *-iti,* which does not have the disadvantages of *-ina* and enjoys much better classical credentials. Bolton feels that the subtribal category may be unnecessary in ant

systematics. I think he is wrong in this, and I believe that, in a taxonomy increasingly based on phylogeny, fine subdivisions such as the subtribe will be needed more and more.

One more practice needs attention: the superfluous citation of authors with the ant name, either with the species binomial or just the genus. I used to teach this practice in taxonomy courses until at last I asked myself: What is its purpose? The author's name has almost become part of the true scientific name of the organism for many who publish research papers, in or outside systematics (true even for many journal editors!), but the Rules state that the author is not part of the scientific name. Author citation has scant reference value in practice, though it is often put forward as a rationale. Abandonment or radical deemphasis of author citation would save much time, effort, and page space; would cut down on the clutter of titles, lists, and captions that slavish citations now engender; and would facilitate computer searching through bibliographies. The rule I follow in this case is to give the author's name only when the instance normally calls for a full bibliographic citation of author, date, volume, and page.

Looking Ahead

We can discern some of the taxonomic needs expected to arise in the future, and the likely responses to them; indeed this manual defines some of them. Requirements of litter ecology overlap, but there is much more to ant systematics than the litter inhabitants. Genera such as *Camponotus, Crematogaster, Paratrechina,* and *Solenopsis* are all there waiting, but it cannot be said that they make very attractive subjects for Ph.D. theses. The hope is that a Bolton or a Shattuck will soon have the courage (and time and money) to take them on. There is also the matter of phylogeny and major taxon relationships. More characters will have to be found and studied. Some of these will be molecular genetic, but their study will not be all that simple, as Crozier (1990) hints in his review of the prospects for mitochondrial DNA research on phylogeny and populations. If these dual tasks of defining species and developing new charac-ter systems continue apace, we may look forward to dramatic advances in ant systematics and phylogenetics that will favorably impact diverse biological disciplines, including the field of biodiversity assessment that is the focus of this book.

Chapter 6

Ants as Indicators of Diversity

Leeanne E. Alonso

With the increasing loss of habitats and biodiversity around the world, there is an urgent need for biodiversity assessments to be carried out during the conservation planning process. Since time, money, and limited available taxonomic expertise prohibit a complete survey of all taxonomic groups, several "rapid" strategies for measuring biodiversity have been developed and implemented (e.g., Schulenberg and Awbrey 1997; Mack 1998). One approach to is to focus on selected taxonomic groups, referred to variously as indicator taxa (Lawton et al. 1998), priority taxa (New 1987), surrogate taxa (Oliver and Beattie 1996a), predictor sets (Kitching 1993), focal groups (Di Castri et al. 1992), or target taxa (Kremen 1992). Measurements of the species richness or diversity of such indicator groups have been proposed as a representative

measure of the species richness or diversity of other taxa, and therefore as an indicator of the overall diversity of an area.

Indicator taxa are also used to detect environmental change. The ecological responses of selected taxa sensitive to habitat modification have been used as indicators of responses in other taxa (Landres et al. 1988; Noss 1990; Pearson and Cassola 1992; Spellerberg 1992; see Chapter 7 for a discussion of the responses of ants to environmental change).

To be a successful indicator of species richness or diversity, selected taxa should meet four basic criteria. They should (1) be easily sampled and monitored, (2) represent fairly diverse groups and/or groups of biological importance in the ecosystem under study, (3) have known relationships to the diversity of other taxa, and

(4) respond to environmental change in ways similar to other taxa (Oliver and Beattie 1996a).

Most groups selected as potential indicator taxa meet the first two criteria. Biodiversity surveys generally focus on vascular plants and vertebrates (e.g., mammals, birds, reptiles, and amphibians; Landres et al. 1988). These groups are, for the most part, readily sampled using standardized techniques (Heyer et al. 1994; Wilson et al. 1996) and fairly easily identified through keys and collections. Ecologically, many of these groups have been proposed (although not often verified) to have broad requirements that encapsulate those of other species, thus serving as "umbrella species" (Noss 1990; Launer and Murphy 1994). Plants and their associated habitat types are often considered to be reliable predictors of the overall diversity of a habitat, since all organisms directly or indirectly rely on plants for food or shelter (e.g., Lesica 1993). Many invertebrate groups have been considered as potential indicator taxa owing to their high diversity and ecological importance (Kremen 1992; Kremen et al. 1994).

The usefulness of indicator taxa basically rests on the third and fourth criteria: that the species richness or diversity of the selected taxa and their responses to habitat change overlap with those of other organisms. Conservation decisions based on these taxa are assumed to be appropriate for other organisms living in the area. However, few data have been collected on the relationships between groups of organisms. For those groups for which we do have information, few correlations have been found between groups (e.g., Wilcox et al. 1986; Prendergast et al. 1993; Lawton et al. 1998; Pharo et al. 1999).

Why Consider Ants as an Indicator Taxon?

Ants have been used as bioindicators in Australia for many years and have been considered for use in other areas of the world as well. They appear to be an ideal candidate for use as an indicator group because they are diverse (approximately 9000 described species), found abundantly in almost every terrestrial habitat in the world, and easily collected (Majer 1983).

Ants are particularly appropriate for inventory and monitoring programs because most species have stationary, perennial nests with fairly restricted foraging ranges (ranging from less than a meter to a few hundred meters). Therefore—in contrast to other insects that move frequently between habitats in search of food, mates, or nesting sites—ants are a more constant presence at a site and can thus be more reliably sampled and monitored. Ants are important ecologically because they function at many levels in an ecosystem—as predators and prey, as detritivores, mutualists, and herbivores.

Correlations between the Species Richness or Diversity of Ants and That of Other Organisms

If ants are to be used as an indicator of the diversity of other organisms, the relationship between the species richness or diversity of ants and that of other target organisms must be understood. In recent years, several studies have investigated this relationship.

The most comprehensive study was conducted by Lawton et al. (1998), who investigated nine taxa, including canopy ants and ground-dwelling ants, in a semideciduous humid forest in southern Cameroon, Africa. Species richness of these taxa was compared across a gradient of habitat types of increasing intensity and frequency of disturbance. They found few correlations between taxa in change in species richness across the disturbance gradient (Table 6.1). Of all the groups, canopy ants were positively correlated with the most other taxa, including butterflies, canopy beetles, and ground-dwelling ants (Table 6.1).

Table 6.1 Correlations between Ants and Other Taxa in Changes in Species Richness across Plots along a Disturbance Gradient[a]

Other Taxa	Canopy Ants	Ground-Dwelling Ants
Birds	5,—	4, 0.47, P = 0.53
	3, 0.78, P = 0.12	
Butterflies	3, –0.49, P = 0.67	5, 0.025, P = 0.69
	*4, 0.97, P = **0.03***	
Flying beetles		
Malaise traps	2,—	4, 0.43, P = 0.57
	5, –0.75, P = 0.15	
Interception traps	2,—	4, 0.21, P = 0.80
	3, 0.30, P = 0.81	
Canopy beetles	3, 0.86, P = 0.33	5, 0.67, P = 0.22
	*4, 0.97, P = **0.03***	
Termites	2,—	4, 0.84, P = 0.16
	3, 0.46, P = 0.70	
Nematodes	3, 0.04, P = 0.98	5, –0.21, P = 0.73
	4, 0.70, P = 0.30	
Ground-dwelling ants	*3, 0.99, P = **0.01***	

[a]Data presented are number of plots compared, Pearson's *r*, and associated probability, *P*. Dashes indicate sample sizes too small for correlations to be calculated. Data in *italics* were calculated using an assumed species richness of zero for canopy-inhabiting taxa in the absence of a canopy. Data in **boldface** indicate statistically significant values. Data from Lawton et al. (1998).

Most other studies have been carried out in Australia. Results from seven studies from Australia and Tasmania are shown in Table 6.2. These results indicate that, in general, ant species richness does not correlate positively with many taxa and that correlations are not consistent between different habitat types. The species richness of ants correlates positively only with that of vascular plants and a few invertebrate groups (Table 6.2).

These seven studies differed in terms of the taxa and habitat types studied, sampling methods, and data analysis methods. Two studies, Majer (1983) and Andersen et al. (1996), found significant positive correlations between ants and several other taxa. Majer (1983) compared the species richness of ants to the species richness and abundance of plants and several invertebrate taxa at several Western Australian sites. He found significant positive associations between the species richness of ants and plants in rehabilitated bauxite minesites of differing ages and rehabilitation treatments (see Majer et al. 1984 for details). He also found a significant correlation between the species richness of ants and that of collembolans and termites in pitfall traps (Table 6.2). The species richness of ants in sweep and beat samples was also significantly positively correlated with the total species richness and abundance of all the invertebrate taxa sampled.

Andersen et al. (1996) also found several positive associations between ant species composition and that of seven other taxa in the Kakadu region of Australia's Northern Territory. Ants and several other invertebrate groups were collected in pitfall traps in (1) natural *Eucalyptus* woodland, (2) disturbed sites on rehabilitated bauxite minesites (woodlands or shrublands dominated by either *Acacia* spp.

Table 6.2 Comparisons between the Species Richness of Ants and the Species Richness of Other Taxa in Australia[a]

Taxon Compared	Habitat	Variance, Probability	Reference
Positive association between ants and:			
Plants	Rehabilitated bauxite minesites	$r^2 = 0.24$, $P < 0.05$	Majer (1983)
		$r^2 = 0.35$, $P < 0.001$	Andersen et al. (1996)
	Eucalyptus woodland	$r^2 = 0.22$, $P < 0.01$	Abensperg-Traun et al. (1996)
	Wet sclerophyll forest, dry eucalypt forest, heathland, and swamp	—	Cranston and Trueman (1997)
Invertebrates			
Beetles	Rehabilitated bauxite minesites	$r^2 = 0.21$, $P < 0.001$	Andersen et al. (1996)
Collembola	Rehabilitated bauxite minesites	—, $P < 0.05$	Majer (1983)
Scorpions	*Eucalyptus* woodland	$r^2 = 0.25$, $P < 0.01$	Abensperg-Traun et al. (1996)
Termites	Rehabilitated bauxite minesites	$r^2 = 0.30$, $P < .001$	Majer (1983)
		$r^2 = 0.30$, $P < 0.05$	Andersen et al. (1996)
	Tropical hummock-grassland	—, $P < 0.05$	Majer (1983)
	Eucalyptus woodland	$r^2 = 0.17$, $P < 0.05$	Abensperg-Traun et al. (1996)
Ground invertebrates	Rehabilitated bauxite minesites	$r^2 = 0.21$, $P < 0.001$	Andersen et al. (1996)
Vegetation invertebrates	Rehabilitated bauxite minesites	$r^2 = 0.32$, $P < 0.001$	Andersen et al. (1996)
Soil invertebrates	Rehabilitated bauxite minesites	$r^2 = 0.07$, $P < 0.05$	Andersen et al. (1996)
Total invertebrates	Rehabilitated bauxite minesites	—, $P < 0.05$	Majer (1983)
Negative association between ants and:			
Birds	Forest (unlogged)	$r^2 = -0.14$, $P < 0.05$	Oliver et al. (1998)
Beetles	Grassland, dry and moist forests	$r^2 = -0.75$, $P < 0.001$	Oliver and Beattie (1996b)
Termites	Shrubland	$r^2 = -0.18$, $P < 0.05$	Abensperg-Traun et al. (1996)
No association between ants and:			
Plants	Forest	$r^2 = 0.00$, $P > 0.05$	Oliver et al. (1998)
	Shrubland	—, $P > 0.05$	Abensperg-Traun et al. (1996)
Vertebrates			
Birds	Woodland, heath, plantations	$r^2 = 0.09$, $P > 0.05$	Burbridge et al. (1992)
	Logged forest	$r^2 = 0.16$, $P > 0.05$	Oliver et al. (1998)
Mammals	Woodland, heath, plantations	$r^2 = -0.001$, $P > 0.05$	Burbridge et al. (1992)
	Forest (logged and unlogged)	$r^2 = 0.008$, $P > 0.05$	Oliver et al. (1998)
Reptiles	Woodland, heath, plantations	$r^2 = 0.03$, $P > 0.05$	Burbridge et al. (1992)
	Eucalyptus woodland	—, $P > 0.05$	Abensperg-Traun et al. (1996)
	Forest (logged and unlogged)	$r^2 = 0.06$, $P > 0.05$	Oliver et al. (1998)
Amphibians	Forest (logged and unlogged)	$r^2 = 0.13$, $P > 0.05$	Oliver et al. (1998)
Invertebrates			
Beetles	*Eucalyptus* woodland	—, $P > 0.05$	Abensperg-Traun et al. (1996)
	Shrubland	—, $P > 0.05$	Abensperg-Traun et al. (1996)
	Forest (logged and unlogged)	$r^2 = 0.12$, $P > 0.05$	Oliver et al. (1998)
	Wet sclerophyll forest, dry eucalypt forest, heathland, and swamp	[b]	Cranston and Trueman (1997)
Butterflies	Shrubland	—, $P > 0.05$	Abensperg-Traun et al. (1996)

Continued on next page

Table 6.2 continued

Taxon Compared	Habitat	Variance, Probability	Reference
Centipedes	Wet sclerophyll forest, dry eucalypt forest, heathland, and swamp	[b]	Cranston and Trueman (1997)
Cockroaches	*Eucalyptus* woodland	—, $P > 0.05$	Abensperg-Traun et al. (1996)
Collembola	Hummock-grassland	—, $P > 0.05$	Majer (1983)
	Wet sclerophyll forest, dry eucalypt forest, heathland, and swamp	[b]	Cranston and Trueman (1997)
Earwigs	*Eucalyptus* woodland	—, $P > 0.05$	Abensperg-Traun et al. (1996)
Grasshoppers	Rehabilitated bauxite minesites	$r^2 = 0.05$, $P > 0.05$	Andersen et al. (1996)
True bugs	*Eucalyptus* woodland	—, $P > 0.05$	Abensperg-Traun et al. (1996)
Hymenoptera (non-ant)	Wet sclerophyll forest, dry eucalypt forest, heathland, and swamp	[b]	Cranston and Trueman (1997)
Isopods	*Eucalyptus* woodland	—, $P > 0.05$	Abensperg-Traun et al. (1996)
Millipedes	Wet sclerophyll forest, dry eucalypt forest, heathland, and swamp	[b]	Cranston and Trueman (1997)
Spiders	Grassland, dry and moist forests	—, $P > 0.05$	Oliver and Beattie (1996b)
	Wet sclerophyll forest, dry eucalypt forest, heathland, and swamp	[b]	Cranston and Trueman (1997)
Thrips	Wet sclerophyll forest, dry eucalypt forest, heathland, and swamp	[b]	Cranston and Trueman (1997)

[a]Statistical tests included correlation analysis (Abensperg-Traun et al. 1996), Mantel tests (Andersen et al. 1996), simple regression analysis (Majer 1983), Spearman rank correlation analysis (Burbridge et al. 1992), and Pearson product-moment correlation analysis (Oliver and Beattie 1996a, 1996b; Oliver et al. 1998). Dashes indicate that values were not given in the published report.

[b]Cranston and Trueman (1997) used site rankings, which did not provide variance and probability values.

or *Eucalytus tetrodonta*), and (3) waste rock sites, which consisted of *Acacia* or mixed shrubland. Vegetation invertebrates and grasshoppers were collected with sweep nets. Significant positive correlations were found between ant species richness and that of plants; ground, vegetation, and soil invertebrate assemblages; beetles; and termites (Table 6.2). No correlation was found between ants and grasshoppers.

Abensperg-Traun et al. (1996) sampled ants and several other invertebrate groups in pitfall traps and censused plants in 20×20-m plots in *Eucalyptus* woodlands and shrublands of west-

ern Australia. Termites were collected by examining soil and dead wood and butterflies were collected using hand nets. The study found a positive correlation between ant species richness and that of scorpions in woodlands, termites in woodlands, and vascular plants in *Eucalyptus* woodlands (trees, shrubs, herbs, and grasses) but not in shrublands (Table 6.2). In contrast, ant species richness was significantly negatively correlated with termite species richness in shrublands. No associations were found between ants and several other invertebrate groups, including isopods, cockroaches, beetles, and earwigs in woodlands, and hemipterans, beetles, and butterflies in shrublands (Table 6.2). Furthermore, no associations were found between ants and lizards (collected by hand searching) in woodlands.

The study by Cranston and Trueman (1997) is the only other study to find a positive relationship between the species richness of ants and that of another taxon, in this case plants. They compared the species richness of ants with that of several other invertebrate groups (collected in pitfall traps, in yellow-pan traps, and by leaf litter extraction in Tullgren funnels) and with the species richness of plants in five sites in northeastern Tasmania. Since they had only one site in each habitat, correlation analysis was not possible. Instead, the five sites were ranked independently according to the species richness of each taxon, and then the rank order was compared between taxa. It was identical only to the site ranking based on plant species richness (Table 6.2). The results of this study should be viewed with caution because the small sample size and superficial analysis do not provide a sound comparison of the patterns of species richness between taxa.

Burbridge et al. (1992) sampled ants in undisturbed woodlands and heaths in Western Australia using pitfall traps, hand collecting, and leaf litter extraction using Winkler sacks. They compared the species richness of ants to the number of vertebrate species sampled by a separate study conducted in the same area (A. H. Burbridge and J. Rolfe, unpubl. data). Reptiles and small mammals were collected in a pit line. They found no significant correlations between ant species richness and the species richness of reptiles, birds, or mammals (Table 6.2).

Oliver and Beattie (1996a, 1996b) studied the relationship between the species richness of ants and that of beetles or spiders collected in pitfall traps in New South Wales, Australia, in four habitat types along a transect that represented a transition from dry soils and a fairly

open canopy to higher soil moisture and denser canopy cover. When the four forest types were ranked in order of species richness for each taxon, site rankings for ants, beetles, and spiders were all different. They found a significant negative correlation between ant and beetle species richness in each forest type (Oliver and Beattie 1996b; Table 6.2). In addition, ordination analysis of species turnover between the four forest types revealed that ants and beetles had similar levels of turnover but that spiders showed lower levels.

Finally, Oliver et al. (1998) investigated the relationship between species richness and turnover of ants and other groups between logged and unlogged forests in northeastern New South Wales, Australia. They conducted plant and bird point surveys at 100-m intervals along transects through each site. Small mammals were captured with Elliot traps at each of these points; reptiles and amphibians were collected by timed visual searches and in pitfall traps; and invertebrates were sampled using pitfall traps. Only ants and three families of beetles were sorted and identified from the invertebrate samples. No significant positive correlations were found between ants and any other group in unlogged or logged forest sites (Table 6.2). A significant negative correlation was found between ants and birds in unlogged forest but not in logged forest (Table 6.2). Species turnover between sites was in the order plants > invertebrates > vertebrates, indicating that these three groups do not display similar patterns of response to environmental change.

Limitations to the Indicator Approach

The finding that there are few strong positive correlations between ant species richness and that of other taxa is not surprising. There is no strong a priori reason why the diversity of a selected indicator taxon (or taxa) should correlate with the diversity of other taxa (Goldstein 1999). Every species has a unique evolutionary history that influences its distribution. Higher taxonomic levels, such as genera and families, may be influenced by factors that are not necessarily the same as those that affect other genera and families, even within the same habitat. Different organisms have distinct ecological requirements and are unlikely to respond to environmental change in similar ways (Lawton et al. 1998).

Studies of the relationships between the species richness of a variety of taxa have found similar patterns. Kremen (1992) found butterflies to be poor predictors of plant species richness in Madagascar; Wilcox et al. (1986) found marginal correlations between the richness of butterflies, birds, and mammals in the western United States; Prendergast et al. (1993) found low overlap in hotspots of species richness for butterflies, dragonflies, liverworts, aquatic plants, and breeding birds in England; and Pharo et al. (1999) found no correlation between vascular plant diversity and bryophyte and lichen diversity in Australia. Of the studies reported here, Abensperg-Traun et al. (1996), Oliver and Beattie (1996a), Cranston and Trueman (1997), and Oliver et al. (1998) did not find any correlations between the richness of other taxonomic groups.

Knowledge of the biology of the species under investigation is an essential aspect of biodiversity studies if a biologically meaningful interpretation of the data and understanding of the relationships between taxa are to be developed (see Chapters 2 and 8). Such knowledge is not possible unless the species names are known. Furthermore, conservation and management plans should not be made based solely on the number of species present in an area, but also on the identity and biology of the species present (Goldstein 1999).

Potential Use of the Indicator Approach

It is possible that the species richness of different taxa may be related in some areas. In undisturbed habitats, historical evolutionary factors may have produced similar levels of species richness in unrelated taxa (Cranston and Trueman 1997). This is possible in areas of high stability, such as refugia, where high speciation rates, low extinction rates, and close co-evolved mutualistic associations could occur. In areas that have been subject to some type of disturbance, whether it be natural or anthropogenic, corresponding levels of species richness may be expected for taxa that have similar habitat or microclimate needs. For example, soil-dwelling arthropods and reptiles may both increase in species richness in an area as the plant species richness—and thus the availability of nesting sites and more appropriate soil moisture conditions—changes. Taxa with similar dispersal or colonizing abilities may also display correlated patterns of species richness in disturbed areas. In addition, partners involved in tight mutualistic interactions may display similar patterns of species richness.

Potential Use of Ants as Indicators

Patterns in ant species richness and diversity might be correlated with patterns in taxa that have similar nesting or feeding needs, taxa that are affected by similar environmental factors, or taxa with which ants have significant interactions. Such taxa could include invertebrates that also live in the litter or soil, such as spiders, collembolans, or mites; invertebrates that have similar but restricted diets; and organisms that may serve as specialized prey items for some ant species. Although ants are regarded as generalist feeders and nesters, many species have very specific diets and microclimate needs. Similar patterns of species richness may also be found in ants and particular plant, homopteran, and beetle species with which ants have obligate mutualistic interactions.

Tables 6.1 and 6.2 present some evidence in support of the hypothesis that the species richness of ants may correlate more closely with that of taxa that have similar microhabitat requirements. The species richness of canopy ants was positively correlated with that of other taxa that occur in the canopy (birds, butterflies, and canopy beetles) as well as with the richness of another ant group, the ground-dwelling ants (Table 6.1; Lawton et al. 1998). Positive associations were also found between the species richness of ants and that of plants, beetles, scorpions, termites, ground-foraging invertebrates, and low-vegetation-dwelling invertebrates (Table 6.1). All the insect taxa that were found to correlate positively with ants were collected with pitfall traps, as were the ants (with the exception of the total invertebrate fauna of Majer 1983). This suggests that these taxa live and operate in a microhabitat similar to that of ants and may therefore have similar habitat requirements.

Plant species richness would be expected to correlate with that of ants if a diverse community of ants required a variety of plants to provide nesting sites or food, or to regulate the microclimate they needed. This would certainly be the case in relatively disturbed and harsh habitats, such as the rehabilitated bauxite minesites (Majer 1983; Andersen et al. 1996). The ant species richness of any habitat, including the *Eucalyptus* woodlands (Abensperg-Traun et al. 1996) and other natural Australian habitats (Cranston and Trueman 1997), would be predicted to increase with increasing plant species richness as microhabitats and microclimate became available for specialist species with specific requirements.

Future Directions

Most scientists now agree that individual taxa or restricted groups of taxa are not sufficient for use as indicators of overall biodiversity (Noss 1990; Kremen et al. 1994). Oliver et al. (1998) concluded that "the evaluation of sites for conservation based on the species richness of a few better known taxonomic groups does not adequately represent the biodiversity of other groups." Similarly, using changes in the species richness of one or a limited number of indicator taxa to predict changes in the richness of other groups does not provide an accurate picture of overall change (Lawton et al. 1998).

A better approach is the combined use of a number of diverse indicator taxa, including taxa with diverse ecological requirements, such as plants, vertebrates, and invertebrates (Noss 1990; Kremen et al. 1992; Lawton et al. 1998). This multispecies approach theoretically provides a better assessment of the overall diversity of an area, more accurately reflects changes in diversity caused by habitat modification, and provides more complete information for proper management of habitats for diversity (Lambeck 1997).

More studies comparing the relationships between the species richness and diversity of a range of taxa, both invertebrate and vertebrate, in a variety of habitat types are needed. Furthermore, more basic information on the ecology and habitat requirements of potential indicator groups should be collected so that the patterns of species richness or diversity for selected groups can be properly interpreted.

Ants have great potential for use as an indicator taxon. Their high abundance, ease of sampling, relatively good resources for taxonomic identification, and ecological importance make them ideal candidates. Ground-living ant species, in particular, make useful indicators, since standardized, quantitative methods for sampling them have been developed (see Chapter 14). Used in conjunction with other taxa that have different ecological requirements, ants can provide valuable information on an area's overall species richness or diversity.

ACKNOWLEDGMENTS

I thank Alfonso Alonso, Michael Kaspari, Jonathan D. Majer, and two anonymous reviewers for helpful comments on this chapter.

Chapter 7

Using Ants to Monitor Environmental Change

Michael Kaspari and Jonathan D. Majer

Ecological assemblages are in a constant state of flux. Individuals reproduce and die. Populations cycle and are buffeted by factors ranging from random to predictable. Species are introduced and go extinct. Parts of the landscape are disturbed and recover. The role of human perturbation in this dynamic is not new, nor is it restricted to advanced industrial societies. But the availability of cheap energy combined with human population pressures has produced agriculture, urbanization, and resource extraction on a grand scale. As a result, habitats are increasingly changed and fragmented. Introduced species invade these disturbed areas and infiltrate pristine habitats. Waste products that result from the production and use of this "cheap" energy accumulate in the soil, water, and atmosphere.

The Challenge

Society calls upon the ecologist and resource manager primarily to do two things. First, we are asked to monitor extant, pristine environments and warn society of looming change. But ecosystems naturally vary in almost every property. We require a detailed "baseline" so as not to cry wolf every time a population dips or becomes locally extinct. In other words, ecologists need an *expectation of normalcy* and a protocol to achieve it. Second, ecologists and resource managers are presented with degraded ecosystems in various stages of recovery and asked to evaluate the reconstruction of these ecosystems.

Central to both endeavors is an understanding of the variability inherent in ecosystems. But ecosystems are complex, dynamic things, with countless taxa exhibiting an array of interactions between organisms and the abiotic environment. As described in Chapter 6, selected taxa are often used as indicators of the diversity or ecological responses of other taxa and sometimes even as representatives of an entire ecosystem. In monitoring environmental change, taxa that are hypersensitive to perturbations (Kareiva et al. 1993) are most often chosen. Spellerberg (1991) suggests a set of criteria for including taxa in a monitoring program.

Ants are an ideal indicator group for inclusion in such a program. Many ant species have narrow tolerances and thus respond quickly to environmental change. Ants' small size and reliance on relatively high temperatures make them especially sensitive to climate and microclimate change. In addition, some ant colonies are long lived and have permanent nests that can be marked and revisited. Long-lived species thus allow us to monitor the health of a colony as the environment changes around it. In contrast, short-lived ant species may show high turnover and immediate responses to a stressor. Ant assemblages, therefore, allow a monitoring program that is sensitive to change on a number of temporal scales. (See Chapter 6 for other attributes of ants as an indicator taxon.)

Here we explore the potential role for ground-dwelling ant assemblages in programs aimed at monitoring environmental change and evaluating remediation (recovery) efforts. We first review the evidence for long-term stability in ant assemblages. We then review how ant assemblages have been used to study remediation efforts after perturbation. We conclude with modest suggestions—based on our best evidence and a fair bit of conjecture—of the properties of ant assemblages that would be most valuable for a successful environmental monitoring program.

Baseline Change: Variable Ant Populations in a Changing World

The assumption of any monitoring program is that a pristine ecosystem is sensitive to changing conditions. Remediation programs attempt to reconstruct ecosystems that behave, to some degree of accuracy, like pristine systems. The point of baseline monitoring studies, as controls for remediation or as worthy endeavors in themselves, is to determine the degree to which ecosystem properties (e.g., productivity, biomass, species composition and richness) vary naturally. Even if populations fluctuate wildly, such a "fuzzy target" becomes our expectation of normalcy and should temper our interpretation of any remediation effort.

To assess this variability, we need long-term data sets, ideally greater than the average lifespan of our longest-lived organism (Connell and Sousa 1983). Such data sets are rare. This section summarizes a few studies that followed ant populations or assemblages for at least four years. In each, we look for evidence of stasis in population trends and community dominance. We find, instead, dynamic populations and assemblages. At least one study points to long-term trends in climate as a potential cause for these ecosystem changes.

A Neotropical Ant Guard Assemblage

In a Neotropical second-growth rainforest, ants were monitored attending *Calathea ovandensis,* an understory herb (Horvitz and Schemske 1990). The flowers of *C. ovandensis* produce a sugary solution that attracts ants. These ants, in turn, remove herbivores from *Calathea.* Four plots, from 25 to 64 m^2 in area and from 80 to 250 m apart, were monitored every two weeks from 1983 through 1986. Inflorescences were surveyed for ants. In effect, Horvitz and Schemske report data from a four-year bait study. Their results are sobering.

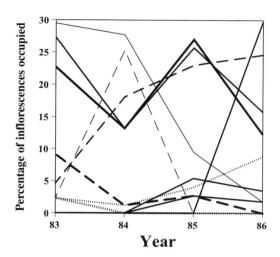

Figure 7.1. Abundance (measured as percentage of surveyed inflorescenses occupied) of 11 species of ant guard on *Calathea ovandensis* in a Neotropical rainforest (for species names, see Horvitz and Schemske 1990). Flowers act as "ant baits," and this bait study suggests great interannual variation in population densities, colony sizes, bait attractiveness, or some combination of the three factors.

In three of four sites, the numerically dominant ant at *C. ovandensis* changed over the four years. In each plot, ant numbers were highly dynamic, with no apparent trends in shifting species composition. Figure 7.1 summarizes data from one of their plots. For example, *Pheidole gouldi,* the dominant ant in this plot from 1983 to 1984, was found on fewer than 5% of the flowers by 1986. *Pheidole* "sp. A" alternated between dominance and rarity. The causes of these fluctuations are difficult to pinpoint without further study. Apart from changes in the successional stage of the forest, changes in ant numbers could also reflect changes in, for example, population densities, colony sizes, or the availability of alternate food types and hence bait attractiveness. Bait studies, one would conclude, should be interpreted with caution.

Such fluctuations in insect numbers are common (Andrewartha and Birch 1954). For example, in a 14-year light trap sample on Barro Colorado Island, Panama, one in five Homoptera species showed a 10% change in numbers (Wolda 1992). Do these changes represent normal variation around an equilibrium (hence "baseline" variation)? Or are the Homoptera "indicating" subtle changes in the forest? This is, as we shall see, a basic problem in interpreting monitoring data. Interestingly, even as individual species waxed and waned, Wolda found that two measures of species richness were rather constant.

Two Northern Harvester Ants

The remaining studies all come from arid North America. All resulted from counts of large, soil-nesting species that build conspicuous nest mounds. Two population studies come from northern desert-grasslands, two from southern desert-grasslands. The former two feature the genus *Pogonomyrmex,* harvester ants that construct nest disks and mounds of fine gravel. These harvester ants store large quantities of seeds in underground middens.

A population of *Pogonomyrmex salinus* was monitored in a Great Basin sagebrush habitat in Idaho (Porter and Jorgensen 1988). Mounds were censused on three plots (two of 0.25 ha and one of 2.72 ha) three times over 9 years. The populations varied little from 1977 to 1986, although there was considerable population turnover (Fig. 7.2).

In shortgrass prairie of Nebraska a population of *Pogonomyrmex occidentalis* was monitored in a 1-ha plot (Keeler 1993). Mounds were marked and censused yearly for 15 years. In contrast to those of *P. salinus, P. occidentalis* densities at this site increased 41% from 1977 to 1991, during a period of no apparent change in grazing pressure or site characteristics but higher than average rainfall (K. Keeler, pers. comm.). The causes of this increase are unknown.

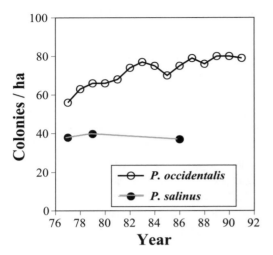

Figure 7.2. Abundance of two harvester ants, *Pogonomyrmex occidentalis* and *P. salinus,* over 15 and 9 years, respectively, in North America. *P. salinus* shows no discernible trend, but *P. occidentalis* appears to have increased in numbers. Data from Porter and Jorgensen (1988) and Keeler (1993).

Two Studies from the Chihuahuan Desert

Two more long-term studies reflect how different investigators focusing on different species may achieve complementary results. Chew and De Vita (1980) studied three species (*Aphaenogaster cockerelli, Myrmecocystus depilis,* and *M. mexicanus*) in Chihuahua desert scrub. A 9.3-ha cattle exclosure was censused six times over 23 years. One species, the diurnal *M. depilis,* varied in density about 50%, while the numbers of its congener *M. mexicanus* increased 11-fold (Fig. 7.3a). *A. cockerelli,* in contrast, was locally extirpated over the same time period. The increase of *M. mexicanus,* given its negative association in space with the other two species, suggested competitive release from *A. cockerelli,* but there was little to suggest why the assemblage had changed.

A second experiment nearby found evidence for major reorganization of species composition over 18 years (Brown et al. 1997). Brown and colleagues followed the responses of plants, rodents, ants, and birds to various experimental treatments on a set of 0.25-ha plots. On the site's control plots, the numbers of two harvester ants, *Pogonomyrmex rugosus* and *P. desertorum,* decreased over the 18 years (Fig. 7.3b). *P. rugosus,* like *A. cockerelli* in the previous study, went locally extinct. A third species, *Pheidole xerophila,* though showing threefold variation in density, had no downward trend. These changes in ant composition occurred at the same time as a threefold increase in shrub cover and shifting abundance of a number of dominant rodent species. Changes in ant and rodent densities appeared to ramify throughout the desert community, affecting horned lizards (which prey on *P. rugosus*) and burrowing owls (which nest in rodent burrows).

Brown et al. (1997) link these community changes to an increase in winter rainfall from four El Niño years. The increased winter rainfall favors shrubs at the expense of desert grasses, and it may wet and ruin the stored seeds of harvester ants such as *Pogonomyrmex* and *Aphaenogaster.* As in Wolda's study of Panama light traps, such profound changes in species composition in this ecosystem yielded little change in species richness (Valone and Brown 1995). In both cases, the loss of some species may be compensated by the arrival of others.

Recovery from Perturbation: Inertia, Resilience, and Nonlinearities

We now turn to studies of ecosystems recovering from stressors. For our purposes, a stressor is anything that alters the ecosystem properties of a site relative to a control site. Stressors create a perturbation; the site recovers to some degree once the stressor is removed. Given the inherent variability of ecosystems (as we have seen previously), monitoring site recovery

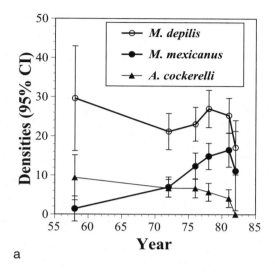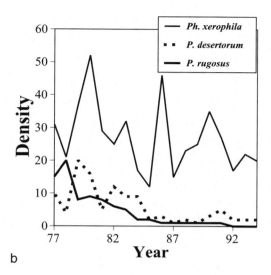

Figure 7.3. Changes in two Chihuahuan desert ant assemblages as reported by (a) Chew and De Vita (1980) and (b) Brown et al. (1997). In both systems, over roughly the same time period, populations of seed-harvesting ants decreased.

requires the simultaneous monitoring of multiple control sites. Over time, one accumulates data on the trajectories of both control and perturbed sites with the goal of determining when those trajectories have converged.

The trajectory taken by each disturbed ecosystem is by definition unique. However, all have a number of features that can be quantified by a monitoring program and measured against control sites. To discuss these features, we use the metaphor of a spring stretched and allowed to recoil. The resulting terminology has been elegantly set forth by Westman (1986), and we develop some of his terminology in the following sections (Fig. 7.4).

Inertia

Inertia reflects the ability of an ecosystem to retain its properties in the face of a stressor. Some ecosystem properties are highly sensitive to certain stressors. For example, in response to chemical pollution, a lake's species richness is likely to change more rapidly than its produc-

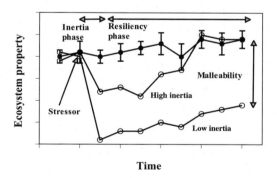

Figure 7.4. Model illustrating the response of an ecosystem to perturbation using the terminology of Westman (1986). A number of control assemblages (with error bars) are compared with two disturbed assemblages, before and after the stressor. One assemblage (labeled High inertia) has high inertia and resilience—responding less to the stressor and quickly attaining properties of the control assemblages. The other (labeled Low inertia) has low inertia and resilience and fails to recover control conditions completely. It is considered more malleable and may have reached an alternate equilibrium point.

tivity (Schindler 1990). In this case, species richness has low inertia and may be an important property to monitor, as it is often the harbinger of more profound and irreversible ecosystem changes.

Properties of ant assemblages show different degrees of inertia relative to similar stressors. For example, an Australian mallee ant assemblage retained its entire species complement following a hot woodland fire (Andersen and Yen 1985). An ant assemblage in an English heathland, in contrast, was dramatically changed by fire (Brian et al. 1976). Likewise, logging will likely have a greater impact on insect diversity in a Neotropical forest, where the canopy is rich with insects (including ants), than in a temperate pine woodland, where insect diversity may be concentrated on the ground (Jeanne 1979; Erwin 1986; Blackburn et al. 1990). Thus one of the first steps in a remediation program is to determine the actual impact of the stressor.

Resilience

Resilience reflects an ecosystem's ability to recover the properties of matched control sites. Resilient ecosystems recover quickly and converge on original ecosystem properties. What properties of ant assemblages yield high resilience?

One important factor appears to be rainfall (Fig. 7.5). Species richness on six 3-year-old mine sites increased most rapidly in sites with the highest amount of rainfall (Majer 1992). In tropical rainforests, wetter sites with higher productivity and higher levels of ant activity recovered ant density and diversity on 1-m^2 plots more quickly than drier sites (Kaspari 1996a). Ant density and diversity in rainforests may be quite resilient to drought. A severe El Niño drought in a seasonal Panama rainforest decreased ant densities to their lowest recorded values (as measured by Berlese funnels), yet the drought's signature had disappeared only a few weeks into the wet season (Wheeler and

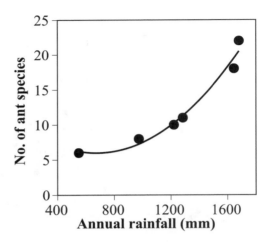

Figure 7.5. Fitted line for the relationship between the number of ant species in 3-year-old rehabilitated mines and annual rainfall for a range of sites throughout Australia. Adapted from Majer (1992).

Levings 1988). Recovery of drier sites may be much slower without extra remediation efforts.

A second factor that must enhance resilience is proximity of the disturbed area to sources of new immigrants, or "propagules" (MacArthur and Wilson 1967). Large-scale perturbations should recover species richness more slowly than small-scale perturbations embedded within pristine habitat. Species richness of ants in recovering bauxite mines decreased with increasing distances from the forest border (Majer 1980). The processes by which species richness—and other properties such as productivity and biomass—may recover from perturbation deserve further investigation.

Malleability

Some disturbed ecosystems may never recover to control levels. Instead, they may reach a different, stable set of properties. Malleability is the difference between the disturbed ecosystem's final properties and those of the control plots. The greater the difference, the more malleable the ecosystem (Westman 1986).

Malleability is a function of the stressor and the ecosystem. In one study, ant colonization was followed over 30 rehabilitated bauxite mines (Majer et al. 1984). One site had been accidentally cleared and revegetated with pines, with no mining having taken place. Ten years after restoration the species richness of ants in this plot was high relative to that of mined plots of similar age that had also been planted with pines (means of 12 and 10.5 species per transect respectively). Preservation of the original soil profile may have reduced that site's malleability.

Yet who is to say that temporary periods of stasis will not give way to further convergence of disturbed sites on control sites? Data cited in the next section should make one view short-term dynamics with caution.

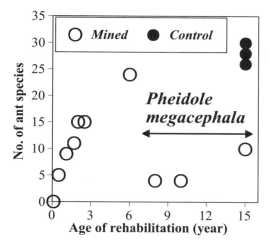

Figure 7.6. Pattern of recolonization of ants in rehabilitated sand-mined areas on North Stradbroke Island, Queensland. The arrival of the tramp ant *Pheidole megacephala* appears to have dramatically restructured the system, lowering diversity relative to three controls. Adapted from Majer (1985).

Oscillations and Other Nonlinear Behavior

Just as a perturbed spring may oscillate before reaching an equilibrium, so may an ecosystem's properties fluctuate following a perturbation. Species richness in particular may be highest at intermediate time periods following perturbation (Connell 1978). If so, then remediation projects that use species richness alone to gauge success may end prematurely.

Ant assemblages in postmining ecosystems commonly show sharp fluctuations in species richness, as dominant species are lost and re-placed. In the next two examples, species richness on recovering mining sites shows opposite yet symmetric relationships between age and species diversity.

One site, a dune system in Queensland, Australia, increased in species richness from the cessation of mining up to year 8 (Fig. 7.6). In that year there was an abrupt increase in the population of the multiple-queen tramp ant *Pheidole megacephala*. Like other introduced species (see Chapters 2 and 4), *P. megacephala* frequents disturbed ecosystems and can have major effects on the local ant assemblage. The

arrival of *P. megacephala* at this site was accompanied by an equally abrupt decrease in species richness and the introduction of new species into the newly depauperate assemblage (Majer 1985).

In another dune-heath system in New South Wales, Australia, Fox and Fox (1982) found a gradual decrease in species diversity after the cessation of mining. The cause was another dominant ant, a territorial *Iridomyrmex* that gradually increased in abundance for 8 years. Then, in year 9, this species was replaced by another *Iridomyrmex* species. This dramatic switch was accompanied by an increase in species diversity. In both cases, species diversity seemed to hinge on the identity of the dominant species. Succession in ant assemblages, then, may not always represent the gradual accumulation of species (Haskins and Haskins 1988).

Oscillations and nonlinearities in ecosystem properties create problems for the design of remediation studies. In addition to oscillations in species richness, the assemblage can exhibit nonlinear trends in species composition during succession. A study of ants in rehabilitated mine sites at Richard's Bay, South Africa, illustrates this phenomenon. Here a mosaic of cleared and pristine coastal dune forest is being mined for mineral sands. The company is attempting to rehabilitate much of the area to coastal dune forest by planting forest species and species that follow shifting cultivation-type farming. Ordination—which collapses species lists into fewer, descriptive, variables—was used to study succession in plots ranging from 0.3 to 13 years old (plus three forest controls). The first part of the succession was not "directed" toward the original forest assemblage. Rather, only in the older plots does the ant assemblage start to resemble that of the original forest. Again, introduced species played a role. Early stages were dominated by *P. megacephala,* which progressively attains massive densities in the youngest rehabilitation. As the tramp ant declines to negligible levels from years 6 to 13, species composition approaches that of the control plots (Majer and de Kock 1992).

Lessons from Baseline and Perturbation Studies

The emerging picture from long-term studies of ant assemblages in both disturbed and pristine ecosystems is one of flux. A close study of Fig. 7.6 should give any ecologist pause—a progress report written in year 6 would look dramatically different from one written two years later.

Some tentative lessons from these case histories include the following:

1. Populations and assemblages are dynamic entities and may be highly sensitive to the way in which they are constructed. Since

we are far from understanding ecosystem dynamics, a series of control plots is vital to creating a realistic target for remediation. Such control plots, aside from allowing the monitoring of a remediation program, will also provide society with much-needed long-term baseline ecological data (Palmer et al. 1997).

2. Stressors come in many spatial scales. Apart from local perturbation, such as timber harvesting or mining, large-scale changes in the environment, such as the accumulation of greenhouse gases, may be changing the abiotic environment. Hence control plots will also likely change, albeit at a slower rate.

3. A key strategy for enhancing remediation may be the control of introduced ant species such as *Pheidole megacephala* and *Solenopsis wagneri* (=*S. invicta*). On the other hand, if those hardy species are able to modify the environment (e.g., through soil preparation), they may be valuable remediation tools.

4. Arid ecosystems may take longer to reconstruct.

What to Measure? A Cautious Assessment

One admonition in conservation biology is to "save all the parts." That is, in something as complex as an ecosystem the loss of any part may have unintended consequences. A conservative approach would therefore be to try not to lose any species.

Ecologists monitoring an ecosystem face a similar problem. We are asked to define an expectation of normalcy for an ecosystem when the critical elements that hold an ecosystem together—if indeed there is such a class of elements—are still poorly known. Little wonder that Spellerberg's (1991) first rule of ecosystem monitoring is "Any variable or process which can be readily measured and dated may be valuable in detecting changes in ecosystems." A corollary to "save all the parts" thus seems to be "monitor everything you can." However, all monitoring programs are limited by time, money, and taxonomists.

Given the sensitivity of ant assemblages outlined in this chapter, we argue that ants would be an ideal animal group to monitor in an ecosystem. But what aspects of an ant assemblage should be monitored? We suggest a set of three parameters—those of individuals, populations, and diversity.

Individual-Based Changes in Ant Assemblages

Individual organisms can be collected and assessed for physiological responses to environmental change. In long-lived colonies (e.g., the harvester ant *Pogonomyrmex*) the same colony can be monitored over many years. This might be especially useful if early warnings of stressors are first reflected at the individual level.

Thus far such studies in ant ecology have been rare. However, there are a number of properties of individual colonies that might be monitored. One is colony activity. If it is sampled at the same time every year, in similar weather, a concerted decline in the numbers of foragers recorded outside a colony may suggest pathology long before the colonies die.

A second possibility is deformation of individual workers. A promising new field in conservation biology is the study of fluctuating asymmetry (Palmer and Strobeck 1986; Leary and Allendorf 1989). Organisms tend to develop symmetrically (i.e., their right and left sides are mirror images of each other). Environmental stressors can disturb this development and cause asymmetry. Yearly samples of large, long-lived species may thus detect changes in symmetry in colonies exposed to stressors compared to controls. Working on social insects has

the added benefit of holding the genotype constant as the environment changes.

Population-Based Changes in Ant Assemblages

Within any diverse assemblage there are likely to be species that are acutely sensitive to a variety of stressors (Carpenter et al. 1993; Tilman 1996). A monitoring program focusing on the population responses of these species stands a good chance of picking up the effects of stressors on ecosystem function long before permanent damage is done (Schindler 1990).

Measurements of colony density—based on quadrat sampling, coupled with hand, Winkler, or Berlese extraction (Chapter 9)—probably remain the best foundation on which to build an ant population monitoring program, because of the quadrat's lack of ambiguity. The more indirect and relative the estimate of abundance (e.g., that arrived at using baits or pitfall traps), the harder it is to interpret changes in numbers (see Fig. 7.1).

Large-colony species like the harvester ants may be relatively easy to detect and should be part of any monitoring program. Small-colony species, such as those that nest in the litter, are also readily sampled using a quadrat-based method. Combined, the two groups of species would monitor for change on a variety of spatial and temporal scales.

Certain species may characterize pristine ecosystems (Chapter 6). If so, the abundance of those species should be an index of an ecosystem's recovery from perturbation. Such an index (percentage recovery of target species) may be used to gauge recovery. Ordination methods, which can detect trends in (or add weighting to) target species, might be used to detect such trends (see Chapter 13).

Diversity-Based Changes in Ant Assemblages

Monitoring has often focused on some measure of diversity, be it the number of species (species richness), the makeup of the species in the assemblage (species composition), or some index reflecting a combination of these two values.

Species richness (the number of species in a given area and time), as we have seen, is a tricky measure upon which to base a monitoring program for at least two reasons. First, species richness is often a nonlinear function of time and is expected to overshoot the control plots during the course of recovery. As a result, remediation programs may be halted prematurely when control species richness equals that of the recovering plots. For example, recovering bauxite and manganese mines may yield ant assemblages whose species richness approaches that of control sites after 7.5 years. But these sites can have quite different species (Majer 1984). Second, species richness may remain constant even while the assemblage undergoes major restructuring (Brown et al. 1997).

The use of species richness has its advantages, including its relative lack of ambiguity and its already wide use. We suggest that it be incorporated, but not relied upon solely, as one index among many in a monitoring program.

Andersen (Chapter 3; 1997b) advocates ant functional groups (collections of species based on an amalgam of phylogeny, habitat, and microclimate) as another potential index.

Conclusion

The study of ecology, although it has made great strides in the past hundred years, is still coming to grips with the complexity of ecosystems—a complexity manifest in the intricate dynamics we have reviewed in this chapter. We feel that the three sampling approaches just highlighted—based on individuals, populations, and diversity—will likely capture much of the phenomena required to describe and reconstruct ecosystem structure and function. We also foresee that attempts at ecosystem remediation, and the monitoring they require, will only *increase* our understanding of these dynamics and point to relationships yet unknown. Many, if not most, of the profound advances in ecosystem and community ecology will come from these tentative applications of our current understanding (Palmer et al. 1997).

Broad-Scale Patterns of Diversity in Leaf Litter Ant Communities

Philip S. Ward

Ants are an important biological component of the leaf litter stratum in temperate and especially tropical forests. A number of regional studies have documented the predominance and diversity of soil and leaf litter ants (Kempf 1961a; Levings 1983; Andersen and Majer 1991; Olson 1991, 1994; Agosti et al. 1994; Belshaw and Bolton 1994a; Longino 1994; Delabie and Fowler 1995; Fisher 1996a, 1997, 1998; Kaspari 1996a; Majer et al. 1997; see also Chapter 15), but there have been few attempts at interregional comparisons. Emery (1920), Brown (1973), Hölldobler and Wilson (1990), Bolton (1995a), and Fisher (1997) provide valuable summaries of the global distributions of ant genera, but without focusing on particular habitats. Wilson (1976) presented data on the most prevalent arboreal and ground-dwelling ant genera at a series of tropical sites, but he did not specifically analyze the leaf litter fauna.

This chapter characterizes, in rather general terms, diversity patterns among assemblages of leaf litter ant species inhabiting forest and woodland habitats in different biogeographic regions. It is based primarily on a set of leaf litter ant collections made by the author over a 12-year period. The survey is somewhat uneven in its geographic coverage, but it does reveal some robust patterns: the number of ant species at a given site (alpha diversity) is strongly negatively correlated with latitude and altitude; there is a slight secondary decline in species richness at the lowest elevations in tropical (but not temperate) forests; and, at higher taxonomic levels (genus, subfamily), there is substantial faunal turnover (beta diversity) among biogeographic regions.

The study also highlights certain glaring knowledge gaps: without much greater effort devoted to alpha taxonomy (i.e., the description and discrimination of ant species), we will be unable to obtain complete measures of beta diversity at the species level. We are also ignorant of the ecological habits of most leaf litter ants and their effects on other cohabiting organisms. Thus, although we can point to biogeographic patterns—such as the predominance of *Stenamma* in Nearctic and montane Central American forests or the profusion of *Tetramorium* species in the Old World tropics—we have little idea about the ecological significance of these observations.

Data Sources and Methods

The primary database for this analysis is a set of 110 Winkler litter samples collected by the author from various tropical and temperate localities (Table 8.1; Fig. 8.1). The sites ranged in latitude from 47°43′N to 35°34′S and in elevation from 10 to 2700 m. Defining biogeographic realms broadly, the provenance of these samples is as follows: Nearctic (23 samples), Neotropical (49), Malagasy (19), and Indo-Australian (19). The last includes a collection each from Singapore and peninsular Malaysia that have been grouped with the Australian and New Guinea samples for the purposes of comparing taxonomic diversity across broadly defined regions. When analyzing the distribution of leaf litter ant genera I have included data from additional sources and used a finer division of biogeographic regions, as discussed later in the chapter.

Most of the Winkler samples were taken from closed canopy forest (see under "Habitat" in Table 8.1), and all were taken under conditions in which the leaf litter was moist from precipitation. None of the collections was made during or immediately after heavy rain, however, because experience suggests that Winkler ex-

tractions from water-saturated litter underestimate the diversity of the ant fauna. At each site handfuls of moist leaf litter and rotten wood were gathered haphazardly over an area of about 1 ha or less and sifted through a sieve (8 cm diameter) until a total of 6 liters of sifted litter had been acquired. An attempt was made to sample each area broadly, avoiding undue concentration on a few localized accumulations of litter. The 6 liters of litter were placed in three mesh bags, which were hung in a cloth Winkler sack ("Gesiebeautomat") for passive arthropod extraction (for further description of this method, see Besuchet et al. 1987; Ward 1987; Fisher 1998; Chapter 9). Extraction was usually carried out near the field site, under ambient temperatures, typically in a sheltered field camp or in a local hotel. The total extraction time varied depending upon the circumstances (10–72 hours; mean 32.3 hours). Extraction time was treated as an independent variable in multiple regression analyses of several measures of taxonomic diversity (species richness, genus richness, number of subfamilies represented), but it was found not to have a significant effect on these measurements. This is probably because representatives of most of the ant species in a sample fall out of the mesh bags within a few hours.

The ants from each sample were sorted to subfamily, genus, and morphospecies, with higher classification following Bolton (1995b). The approach taken toward species identification was as follows. All specimens in a sample were first rough-sorted in alcohol. Several workers of each putative morphospecies were then point-mounted for examination (as were all uniques in a sample). For difficult genera, such as *Pheidole* and *Solenopsis,* this approach often revealed additional species masquerading as a single form when first examined in alcohol.

The alcohol residue was then reexamined more carefully and additional specimens point-mounted and checked. This process was

Table 8.1 List of Winkler Leaf Litter Collection Sites

Accession No.	Region	Country[a]	Locality[b]	Habitat	Latitude	Longitude	Altitude (m)	Date	Extraction Time (hours)	Number of Workers	Number of Species	Number of Genera
9522	NEA	CAN	NS: 6 km N Greenfield	Deciduous forest	44°19'N	64°51'W	80	17-Sep-88	36	32	3	3
9520	NEA	CAN	NS: 8 km N Greenfield	Pine-hemlock–northern hardwood	44°21'N	64°51'W	80	17-Sep-88	24	64	8	8
9534	NEA	CAN	NS: Hemlock Ravine	Mixed coniferous forest	44°41'N	63°40'W	20	20-Sep-88	48	2	1	1
7050	NEA	CAN	NS: Aldershot	Mixed coniferous forest	45°06'N	64°31'W	15	09-Sep-84	72	47	8	1
6371	NEA	CAN	ONT: Lake Opinicon	Deciduous forest	44°33'N	76°22'W	150	22-Oct-83	24	4	3	3
7080	NEA	UNI	VA: 25 km NNW Madison	Deciduous forest	38°36'N	78°22'W	1030	15-Sep-84	72	15	1	1
5936	NEA	UNI	WA: Seattle	*Pseudotsuga-Tsuga-Acer* forest	47°43'N	122°22'W	10	02-May-83	72	2	2	2
7531	NEA	UNI	OR: 10 km ESE Oakridge	*Pseudotsuga-Tsuga-Acer* forest	42°43'N	122°20'W	500	20-May-85	72	17	4	3
7523	NEA	UNI	OR: 10 km NW LaPine	*Pinus ponderosa* forest	43°43'N	121°36'W	1400	20-May-85	36	4	3	3
7533	NEA	UNI	OR: W. L. Finley NWR	Riparian woodland	44°25'N	123°19'W	75	01-Jan-85	72	0	0	0
6984	NEA	UNI	NV: 14 km WSW Carson City	Pine-fir forest	39°08'N	119°55'W	2080	19-Jul-84	72	9	2	2
8463	NEA	UNI	NV: 3 km SSE Mt. Rose	Pine-hemlock forest	39°19'N	119°54'W	2640	20-Jul-86	72	1	1	1
6986	NEA	UNI	NV: 9 km SSE Incline Village	Mixed coniferous forest	39°10'N	119°55'W	1950	19-Jul-84	72	5	1	1
11621	NEA	UNI	CA: 8 km NW Quincy	Mixed coniferous forest	40°00'N	120°59'W	1030	26-Jun-92	24	5	2	2
6733	NEA	UNI	CA: Lang Crossing	Oak woodland	39°19'N	120°39'W	1425	12-May-84	72	36	4	4
10663	NEA	UNI	CA: 4 km NNW Guinda	Riparian woodland	38°52'N	122°12'W	120	07-Apr-90	72	49	6	5
8222	NEA	UNI	CA: 4 km E Yolo	Oak woodland	38°44'N	121°46'W	15	01-Feb-86	72	99	4	4
8254	NEA	UNI	CA: Cold Canyon	Oak woodland	38°30'N	122°06W	120	29-Mar-86	72	46	7	5
9377	NEA	UNI	CA: Gold Creek Road	Oak woodland	34°19'N	118°20'W	550	07-Mar-88	72	25	4	3
12033	NEA	UNI	CA: Centinela, Santa Cruz Island	Coastal pine forest	34°01'N	119°48'W	435	26-Jun-93	72	28	6	4
9372	NEA	UNI	CA: Banner	Mixed coniferous forest	33°04'N	116°33W	820	29-Feb-88	72	14	3	2
7213	NEA	UNI	TX: 8 km E George West	Riparian woodland	28°20'N	98°02'W	40	16-Dec-84	72	91	14	9
7149	NEA	UNI	TX: Santa Ana NWR	Subtropical dry forest	26°05'N	98°08'W	30	13-Dec-84	42	245	8	7

Continued on next page

Table 8.1 continued

Accession No.	Region	Country[a]	Locality[b]	Habitat	Latitude	Longitude	Altitude (m)	Date	Extraction Time (hours)	Number of Workers	Number of Species	Number of Genera
12826	NEO	ARG	Tuc.: 11 km N Tafi Viejo	Tropical dry forest	26°38'S	65°14'W	820	01-Feb-95	60	882	21	8
9085	NEO	BOL	Beni: 42 km E San Borja	Tropical dry forest	14°48'S	66°23'W	210	05-Sep-87	12	193	20	12
12314	NEO	BOL	SC: 10 km NW Terevinto	Tropical moist forest	17°40'S	63°27'W	380	09-Dec-93	24	1289	52	19
12199	NEO	BOL	SC: 35 km SSE Flor de Oro	Rainforest	13°50'S	60°52'W	450	29-Nov-93	24	965	65	22
12174	NEO	BOL	SC: Aserradero Moira	Rainforest	14°34'S	61°12'W	180	27-Nov-93	12	463	45	20
12438	NEO	BOL	SC: Buena Vista	Second-growth rainforest	17°27'S	63°40'W	350	18-Dec-93	20	1116	73	26
12266	NEO	BOL	SC: Las Gamas	Rainforest	14°48'S	60°23'W	700	03-Dec-93	30	259	39	17
12285	NEO	BOL	SC: Las Gamas	Rainforest	14°48'S	60°23'W	700	04-Dec-93	12	219	42	19
9159	NEO	BRA	AM: 80 km NNE Manaus	Rainforest	2°25'S	59°46'W	80	15-Sep-87	24	361	37	16
7912	NEO	COL	Magd.: 4 km N San Pedro	Rainforest	10°57'N	74°03'W	550	14-Aug-85	24	350	32	18
7858	NEO	COL	Magd.: Cañaveral	Tropical dry forest	11°19'N	73°56'W	300	11-Aug-85	48	383	31	16
7891	NEO	COL	Magd.: El Campano	Montane rainforest	11°07'N	74°06'W	1300	13-Aug-85	24	233	23	14
6468	NEO	COS	Gste.: Parque Nacional Santa Rosa	Tropical dry forest	10°48'N	85°41'W	10	16-Dec-83	12	34	6	4
6423	NEO	COS	Gste.: Parque Nacional Santa Rosa	Tropical dry forest	10°51'N	85°37'W	290	14-Dec-83	24	276	15	9
6530	NEO	COS	Limón: 3 km SSE Cahuita	Rainforest	9°43'N	82°50'W	70	24-Dec-83	24	734	39	18
7771	NEO	COS	Pts.: 3 km N Ciudad Neily	Second-growth rainforest	8°41'N	82°57'W	210	31-Jul-85	36	342	29	17
7692	NEO	COS	Pts.: Parque Nacional Manuel Antonio	Second-growth rainforest	9°23'N	84°09'W	10	27-Jul-85	36	202	28	14
7650	NEO	COS	Pts.: Reserva Biológica Carara	Rainforest	9°47'N	84°36'W	500	25-Jul-85	72	257	36	18
7832	NEO	COS	San José: 1 km N La Ese	Montane rainforest	9°27'N	83°43'W	1400	05-Aug-85	12	157	27	13
7811	NEO	COS	San José: near San Gerardo	Montane rainforest	9°28'N	83°35'W	1600	04-Aug-85	12	109	18	13
11726	NEO	DOM	16 km ENE Pedernales	Montane rainforest	18°07'N	71°37'W	800	09-Sep-92	24	593	23	14
11751	NEO	DOM	16 km ENE Pedernales	Tropical-temperate mesic forest	18°07'N	71°37'W	800	10-Sep-92	16	198	16	11
11770	NEO	DOM	4 km NNW Villa Altagracia	Rainforest	18°42'N	70°11'W	200	12-Sep-92	12	724	17	11
11418	NEO	ECU	19 km ENE La Mana	Second-growth rainforest	0°53'S	79°03'W	1100	10-Aug-91	24	514	23	10

					Lat	Long	Elev	Date				
11364	NEO	ECU	Jatun Sacha	Rainforest	1°04'S	77°37'W	400	05-Aug-91	24	653	75	27
11503	NEO	ECU	Maquipucuna	Second-growth rainforest	0°07'N	78°38'W	1500	17-Aug-91	24	358	20	15
11581	NEO	MEX	Chis.: 5 km E Rayon	Second-growth rainforest	17°13'N	92°58'W	1700	23-Dec-91	12	235	11	8
11570	NEO	MEX	Chis.: Lago Pojoj	Tropical-temperate mesic forest	16°06'N	91°40'W	1500	21-Dec-91	24	23	7	6
9283	NEO	MEX	Col.: 19 km NNE Comala	Tropical-temperate mesic forest	19°29'N	103°41'W	1650	25-Dec-87	12	9	4	4
9273	NEO	MEX	Jal.: 10 km S Autlan	Tropical-temperate mesic forest	19°41'N	104°23'W	1600	20-Dec-87	24	55	3	2
9326	NEO	MEX	Jal.: 14 km SSW Puerto Vallarta	Rainforest	20°30'N	105°18'W	130	31-Dec-87	24	290	13	10
9280	NEO	MEX	Jal.: 16 km SW Ciudad Guzman	Mixed coniferous forest	19°35'N	103°34'W	2700	24-Dec-87	24	0	0	0
9327	NEO	MEX	Jal.: 6 km NE El Tuito	Tropical-temperate mesic forest	20°22'N	105°19'W	730	31-Dec-87	24	149	11	10
9255	NEO	MEX	Jal.: Estación Biológica Chamela	Tropical dry forest	19°30'N	105°02'W	100	18-Dec-87	36	285	13	9
7369	NEO	MEX	Ver.: Los Tuxtlas	Rainforest	18°35'N	95°05'W	500	21-Mar-85	60	626	42	19
7333	NEO	MEX	Ver.: Los Tuxtlas	Rainforest	18°35'N	95°05'W	200	20-Mar-85	72	1632	53	24
7314	NEO	MEX	Ver.: 10 km S Orizaba	Cloud forest	18°45'N	97°05'W	1500	19-Mar-85	72	73	12	9
7414	NEO	MEX	Ver.: 11 km N San Andres Tuxtlas	Cloud forest	18°33'N	95°12'W	1400	23-Mar-85	12	80	18	12
7415	NEO	MEX	Ver.: 11 km N San Andres Tuxtlas	Cloud forest	18°33'N	95°12'W	1600	23-Mar-85	12	44	11	7
6391	NEO	PAN	CZ: 3 km NW Gamboa	Rainforest	9°08'N	79°43'W	40	10-Dec-83	24	566	38	18
8701	NEO	PER	15 km WSW Yurimaguas	Rainforest	5°59'S	76°13'W	200	22-Aug-86	24	269	38	21
8684	NEO	PER	30 km NNE Tarapoto	Rainforest	6°15'S	76°15'W	220	21-Aug-86	24	222	25	15
9011	NEO	VEN	17 km SSW Ciudad Bolivia	Second-growth rainforest	8°02'N	70°46'W	240	28-Aug-87	36	1071	32	19
8511	NEO	VEN	49 km ENE Tumeremo	Second-growth rainforest	7°28'N	61°06'W	200	09-Aug-86	17	146	22	9
8927	NEO	VEN	5 km SW Guarico	Rainforest	9°36'N	69°50'W	1350	23-Aug-87	24	206	16	9
8537	NEO	VEN	66 km ESE El Dorado	Rainforest	6°09'N	61°30'W	250	11-Aug-86	24	652	42	20
8920	NEO	VEN	9 km SE Barbacoas	Montane rainforest	9°46'N	70°00'W	2000	27-Aug-87	12	12	4	4
8572	NEO	VEN	Campamento: Rio Grande	Rainforest	8°07'N	61°42'W	250	14-Aug-86	24	290	23	14
8548	NEO	VEN	km 114 El Dorado–Santa Elena	Rainforest	6°01'N	61°24'W	1000	12-Aug-86	18	93	23	11

Continued on next page

Table 8.1 continued

Accession No.	Region	Country[a]	Locality[b]	Habitat	Latitude	Longitude	Altitude (m)	Date	Extraction Time (hours)	Number of Workers	Number of Species	Number of Genera
11862	MAL	MAD	Berenty Reserve	Tropical dry forest	25°01'S	46°18'E	25	09-Feb-93	18	192	13	9
11831	MAL	MAD	3 km E Mahamavo	Montane rainforest	24°45'S	46°45'E	1050	05-Feb-93	12	137	17	6
11820	MAL	MAD	6 km SSW Eminiminy	Rainforest	24°44'S	46°48'E	330	04-Feb-93	12	230	30	13
11935	MAL	MAD	15 km E Sakaraha	Tropical dry forest	22°54'S	44°41'E	760	15-Feb-93	24	87	16	8
10413	MAL	MAD	3 km W Ranomafana	Rainforest	21°15'S	47°25'E	950	27-Apr-89	24	699	32	13
10435	MAL	MAD	28 km SSW Ambositra	Montane rainforest	20°46'S	47°11'E	1660	29-Apr-89	12	178	11	4
11971	MAL	MAD	Manjakatompo	Montane rainforest	19°21'S	47°19'E	1600	20-Feb-93	18	168	10	6
10966	MAL	MAD	16 km S Moramanga	Rainforest	19°05'S	48°14'E	950	18-Nov-90	24	210	22	9
10956	MAL	MAD	6 km ESE Perinet	Rainforest	18°57'S	48°28'E	900	17-Nov-90	14	287	30	10
11146	MAL	MAD	1 km SSW Andasibe	Rainforest	18°56'S	48°25'E	920	12-Dec-90	12	109	14	9
11086	MAL	MAD	25 km NNE Ankazobe	Rainforest	18°06'S	47°11'E	1500	05-Dec-90	24	508	18	8
10358	MAL	MAD	Nosy Mangabe	Rainforest	15°30'S	49°46'E	150	21-Apr-89	24	409	22	9
10320	MAL	MAD	Nosy Mangabe	Rainforest	15°30'S	49°46'E	300	18-Apr-89	24	393	27	12
10340	MAL	MAD	Nosy Mangabe	Rainforest	15°30'S	49°46'E	20	20-Apr-89	24	151	21	9
10379	MAL	MAD	19 km ESE Maroantsetra	Rainforest	15°29'S	49°54'E	350	22-Apr-89	20	1162	40	11
10471	MAL	MAD	4 km ESE Hellville	Rainforest	13°25'S	48°18'E	200	02-May-89	24	375	23	8
11010	MAL	MAD	Reserve Ankarana	Rainforest	12°54'S	49°07'E	150	28-Nov-90	48	180	16	10
10264	MAL	MAU	Le Pouce	Low closed forest	20°12'S	57°31'E	700	09-Apr-89	24	30	6	6
10503	MAL	MAU	Bassin Blanc	Disturbed rainforest	20°27'S	57°28'E	500	06-May-89	24	303	8	8
8140	AUS	AUS	WAust.: 4 km E Walpole	Open tall eucalypt forest	34°59'S	116°47'E	150	08-Dec-85	36	65	5	4
9685	AUS	AUS	NSW Kioloa State Forest	Closed eucalypt forest	35°34'S	150°18'E	35	17-Dec-88	24	314	21	13
8214	AUS	AUS	NSW Royal National Park	Rainforest	34°09'S	151°01'E	50	15-Dec-85	24	332	21	15
8217	AUS	AUS	NSW Royal National Park	Closed eucalypt forest	34°09'S	151°01'E	50	15-Dec-85	24	254	15	13
9770	AUS	AUS	NSW Mt. Kaputar National Park	Closed eucalypt forest	30°17'S	150°08'E	1180	26-Dec-88	10	186	11	10
9833	AUS	AUS	QLD 6 km SSW North Tamborine	Rainforest	27°56'S	153°11'E	500	31-Dec-88	12	106	18	12
6246	AUS	AUS	QLD 10 km SE Kenilworth	Closed eucalypt forest	26°40'S	152°47'E	340	25-Aug-83	12	82	14	12
6279	AUS	AUS	QLD 10 km SE Kenilworth	Rainforest	26°40'S	152°47'E	340	27-Aug-83	12	174	21	12

6202	AUS	AUS	QLD 1 km SW Eungella	Rainforest	21°09'S	148°29'E	840	23-Aug-83	12	67	5	5
6013	AUS	AUS	QLD 27 km NNE Coen	Rainforest	13°44'S	143°20'E	530	06-Aug-83	24	190	17	10
6028	AUS	AUS	QLD 27 km NNE Coen	Rainforest	13°44'S	143°20'E	530	06-Aug-83	19	93	19	11
6073	AUS	AUS	QLD 12 km WNW Lockhart River	Rainforest	12°44'S	143°14'E	30	09-Aug-83	24	243	22	11
10050	AUS	PNG	Varirata National Park	Rainforest	9°27'S	147°21'E	800	26-Jan-89	12	250	30	17
10121	AUS	PNG	24 km N Madang	Rainforest	5°01'S	145°46'E	80	02-Feb-89	24	285	41	19
10127	AUS	PNG	38 km N Madang	Rainforest	4°53'S	145°45'E	40	03-Feb-89	72	588	49	21
10142	AUS	PNG	5 km SW Mt. Uluman	Rainforest	4°41'S	145°57'E	800	05-Feb-89	24	478	34	23
10186	AUS	PNG	Ambunti	Rainforest	4°13'S	142°49'E	150	12-Feb-89	20	418	56	25
9576	ORI	SIN	Bukit Timah Nature Reserve	Rainforest	1°21'N	103°47'E	100	20-Nov-88	48	116	22	16
9586	ORI	MAL	Kota Tinggi Falls	Rainforest	1°50'N	103°50'E	100	22-Nov-88	24	161	24	13

[a]Abbreviations: ARG, Argentina; AUS, Australia; BOL, Bolivia; BRA, Brazil; CAN, Canada; COL, Colombia; COS, Costa Rica; DOM, Dominican Republic; ECU, Ecuador; MAD, Madagascar; MAL, Malaysia; MAU, Mauritius; MEX, Mexico; PAN, Panama; PER, Peru; PNG, Papua New Guinea; SIN, Singapore; UNI, United States; VEN, Venezuela.

[b]Abbreviations: AM, Amazonas; CA, California; CZ, Canal Zone; Gste., Guanacaste; Jal., Jalisco; Magd., Magdalena; NS, Nova Scotia; NSW, New South Wales; NV, Nevada; ONT, Ontario; OR, Oregon; Pts., Puntarenas; QLD, Queensland; SC, Santa Cruz; Tuc., Tucumán; TX, Texas; VA, Virginia; Ver., Veracruz; WA, Washington; WAust., Western Australia.

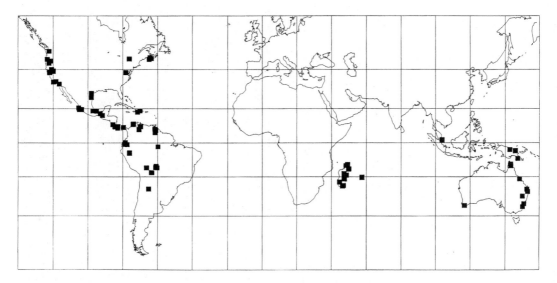

Figure 8.1. Locations of the 110 leaf litter collection sites. Grid lines are shown for every 20° of latitude (between 60°N and 60°S) and for every 20° of longitude.

continued until no further distinctions could be made. All point-mounted specimens were retained as vouchers. I drew upon my experience as an ant taxonomist to make judgments about the kinds of morphological discontinuities that would indicate the presence of two biological species. The morphospecies so designated can be thought of as working hypotheses about species identities, which can be independently assessed in the future by examination of the voucher specimens. Specific names were assigned where feasible (i.e., for taxonomically well-understood genera), but in many instances it was necessary to develop a system of code names for the species in a given geographical area (e.g., *Pheidole* BOL-32 for one of about 40 *Pheidole* species from eastern Bolivia). Such code numbers have local applicability only, and the task of reconciling the specific identities of code-named taxa from different geographic regions has not yet been completed.

Indeed, although sorting species from samples taken within the same geographical area is challenging and time consuming, such difficul-

ties pale in comparison to those that attend any attempt to resolve species identities over large geographical scales. As a result, this study is primarily about patterns of alpha diversity (i.e., geographical variation in *within-site* species richness) and about regional variation in faunal composition at higher taxonomic levels (genera, subfamilies). Large-scale measurements of species beta diversity (species turnover) remain constrained by insufficient taxonomic knowledge.

All the results reported here are based on workers only, although the presence of other castes was noted. For each sample the number of individual workers of each morphospecies was recorded. Variables of interest for each sample include the total number of workers and the numbers of species, genera, and subfamilies represented. Among the independent variables recorded for each sample were biogeographic region, habitat, latitude, longitude, and altitude (see Table 8.1). Latitude was converted to absolute decimal latitude for all statistical analyses.

Voucher specimens were deposited in the Museum of Comparative Zoology, Harvard University (MCZC) and in the P. S. Ward collection at the University of California at Davis (PSWC). In addition, duplicate specimens from Australia, Papua New Guinea, Madagascar, Mauritius, Argentina, Bolivia, Brazil, Colombia, Ecuador, and Venezuela were returned to the following host institutions, respectively: Australian National Insect Collection, Canberra (ANIC); Entomology Collection, University of Papua New Guinea (UPNG); Parc Botanique et Zoologique de Tsimbazaza, Antananarivo (PBZT); Mauritius Sugar Industry Research Institute (MSIR); Instituto Miguel Lillo, Tucumán (IMLA); Museo de Historia Natural "Noel Kempff Mercado," Santa Cruz (UASC); Instituto Nacional de Pesquisas da Amazônia, Manaus (INPA); Museo de Historia Natural, Universidad Nacional de Colombia, Bogotá (UNCB); Museo Ecuatoriano de Ciencias Naturales, Quito (MECN); and Instituto de Zoología Agrícola, Universidad Central de Venezuela, Maracay (IZAV).

The primary database was supplemented by information on Winkler leaf litter collections from Malaysia (data kindly provided by Annette Malsch) and West Africa (Belshaw and Bolton 1994a). The 11 Malaysian samples are all from Pasoh Forest Reserve, and each consists of 9 m² of rainforest leaf litter sifted to yield about 6 liters of concentrated litter. The West African samples are from 34 sites (20 localities) in Ghana; at each site ten 1-m² quadrats were randomly placed in an area of approximately 1000 m², and all the leaf litter in these quadrats was collected, sifted, and extracted. Because the sampling methods used in the Malaysian and Ghanaian studies differ from those described here, I have not used these data for the analysis of diversity patterns. They have been employed mainly to broaden the geographic base for a genus-level comparison of faunal composition (Tables 8.11 and 8.12).

Content of the Winkler Leaf Litter Samples

The 110 Winkler samples yielded a total of 29,942 worker ants from 6 subfamilies, 103 genera, and approximately 911 species (Table 8.2). Because of taxonomic uncertainties the cumulative tally of the number of species should be considered provisional. I estimate that it could be higher or lower by as much as 10%. Site richness (alpha diversity) per sample ranged from 0 to 75 species (mean 20.3 ± 15.9 s.d.), from 0 to 27 genera (mean 10.6 ± 6.4 s.d.), and from 0 to 5 subfamilies (mean 2.8 ± 1.0 s.d.). The number of worker ants per Winkler

Table 8.2 Summary of Winkler Leaf Litter Samples: Taxonomic Content

Subfamily	Number of Genera	Number of Species (% of Total)	Number of Workers (% of Total)
Cerapachyinae	2 (1.9)	4 (0.4)	45 (0.2)
Dolichoderinae	6 (5.8)	10 (1.1)	137 (0.5)
Ecitoninae	2 (1.9)	3 (0.3)	107 (0.4)
Formicinae	14 (13.6)	ca. 97 (10.6)	3,873 (12.9)
Myrmicinae	57 (55.3)	ca. 594 (65.2)	22,067 (73.7)
Ponerinae	22 (21.4)	ca. 203 (22.2)	3,713 (12.4)
Total	103	ca. 911	29,942

Table 8.3 Forty Ant Genera Most Frequently Encountered in the Survey

Genus	Number (%) of Winkler Samples Occupied	Mean Number of Species per Sample Belonging to Genus	Mean Proportion of Species per Sample Belonging to Genus
Hypoponera	83 (75.5)	2.27	0.099
Pheidole	83 (75.5)	3.82	0.148
Strumigenys	75 (68.2)	1.55	0.061
Solenopsis	64 (58.2)	1.63	0.069
Paratrechina	59 (53.6)	0.84	0.034
Pachycondyla	48 (43.6)	0.70	0.028
Oligomyrmex	42 (38.2)	0.57	0.020
Cyphomyrmex	33 (30.0)	0.46	0.018
Rogeria	30 (27.3)	0.45	0.019
Anochetus	28 (25.5)	0.31	0.012
Brachymyrmex	28 (25.5)	0.40	0.017
Monomorium	28 (25.5)	0.54	0.025
Tetramorium	28 (25.5)	0.71	0.034
Stenamma	27 (24.5)	0.46	0.086
Wasmannia	27 (24.5)	0.25	0.009
Gnamptogenys	25 (22.7)	0.35	0.012
Crematogaster	23 (20.9)	0.31	0.011
Prionopelta	23 (20.9)	0.28	0.008
Octostruma	22 (20.0)	0.27	0.010
Smithistruma	20 (18.2)	0.21	0.009
Acropyga	17 (15.5)	0.21	0.007
Neostruma	17 (15.5)	0.16	0.006
Odontomachus	17 (15.5)	0.16	0.005
Discothyrea	15 (13.7)	0.15	0.008
Adelomyrmex	14 (12.7)	0.22	0.009
Leptothorax	14 (12.7)	0.14	0.036
Ponera	13 (11.8)	0.16	0.009
Aphaenogaster	12 (10.9)	0.11	0.018
Cryptopone	12 (10.9)	0.15	0.007
Apterostigma	11 (10.0)	0.12	0.003
Myrmecina	11 (10.0)	0.11	0.007
Lasius	10 (9.1)	0.10	0.042
Myrmicocrypta	10 (9.1)	0.10	0.002
Hylomyrma	9 (8.2)	0.09	0.002
Rhytidoponera	9 (8.2)	0.11	0.005
Eurhopalothrix	8 (7.3)	0.08	0.002
Heteroponera	7 (6.4)	0.06	0.005
Proceratium	7 (6.4)	0.06	0.004
Camponotus	6 (5.5)	0.05	0.002
Leptogenys	6 (5.5)	0.05	0.002

sample varied from 0 to 1632 (mean 272.2 ± 301.4 s.d.).

Over the entire collection of Winkler samples, the predominant subfamily is the Myrmicinae (57 genera, about 594 species), followed by the Ponerinae (22 genera, about 203 species), Formicinae (14 genera, about 97 species), Dolichoderinae (6 genera, 10 species), Cerapachyinae (2 genera, 4 species), and Ecitoninae (2 genera, 3 species). The predominance of the Myrmicinae is greater, and that of the Ponerinae is less, when numbers of individual workers rather than numbers of species are considered (Table 8.2). The most frequent and species-rich genera are listed in Table 8.3. The six most frequent genera are *Pheidole* (represented in 83 out of 110 samples), *Hypoponera* (83/110), *Strumigenys* (75/110), *Solenopsis* (64/110), *Paratrechina* (59/110), and *Pachycondyla* (48/110). The six most species-rich genera are an overlapping but not identical set: *Pheidole* (mean number of species per sample: 3.82), *Hypoponera* (2.27), *Solenopsis* (1.63), *Strumigenys* (1.55), *Paratrechina* (0.84), and *Tetramorium* (0.71). Based on the proportion of species in any sample that belong to a particular genus, the most species-predominant genera are *Pheidole* (mean proportion: 0.15), *Hypoponera* (0.10), *Stenamma* (0.09), *Solenopsis* (0.07), and *Strumigenys* (0.06). The mean proportion for *Stenamma* is accompanied by a high variance and does not indicate large numbers of species because this genus is largely confined to species-poor samples taken from temperate Nearctic localities.

Latitudinal and Altitudinal Patterns of Leaf Litter Ant Diversity

Results of a multiple regression analysis of species richness (number of species in a sample) on latitude, altitude, and extraction time are

Table 8.4 Multiple Regression Analysis of Sample Species Richness[a]

Variable	Coefficient	Standard Error	*P*
Constant	40.412	2.642	0.000
Latitude	−0.803	0.103	0.000
Altitude	−0.008	0.002	0.000
Extraction time	0.033	0.061	0.593

[a]Independent variables are latitude (absolute), altitude (m), and extraction time (hours). Multiple R^2 = 0.471, n = 110.

shown in Table 8.4. Latitude and altitude alone account for about 47% of the variance in species richness, and extraction time has no significant effect (P = 0.593). Plots of species richness as a function of latitude and altitude are shown in Figs. 8.2 and 8.3. The latitudinal species gradient is steeper for low-altitude than for high-altitude sites, and the decline in ant species richness with altitude is less pronounced at high latitudes. There is also an indication that species richness shows a slight decline below 500 m in tropical regions (Figure 8.3a). In fact, for tropical and subtropical sites (latitude < 30°) at elevations below 500 m there is a significant *positive* correlation between species richness and altitude (r = 0.453, n = 42, P = 0.003), whereas for sites at or above 500 m the relationship is negative (r = −0.561, n = 42, P = 0.000). No such midelevation peak in species richness was detected among the high-latitude (>30°) sites, although the sample size is admittedly small (n = 26).

A latitudinal gradient in the species richness of leaf litter ants is hardly surprising and accords with the pattern seen generally in ants (Kusnezov 1957; Jeanne 1979) and in numerous other taxa (Stevens 1989). A sharp attenuation of the ant fauna at higher elevations in tropical forests has also been well documented (e.g., Weber 1943a; Brown 1973; Janzen et al. 1976; Olson 1994; Fisher 1996a, 1998). Darlington

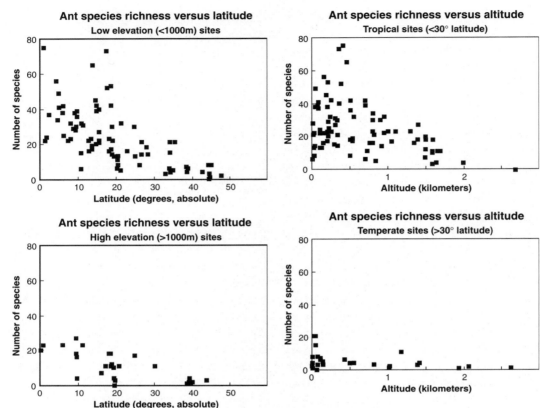

Figure 8.2. Species richness of ant leaf litter samples as a function of latitude. Values are plotted separately for low-elevation and high-elevation sites. Total number of sites: 110.

Figure 8.3. Species richness of ant leaf litter samples as a function of altitude. Values are plotted separately for tropical (<30° latitude) and temperate sites.

(1971) and Olson (1994) have discussed the possible consequences of this decline in ant diversity for other ground-dwelling arthropods. In Olson's (1994) study of leaf litter invertebrates in Panama, carabid beetles and weevils showed peaks of abundance and diversity at the highest elevations, where ants were relatively scarce, suggesting ecological release or replacement.

A midelevation peak in species richness of leaf litter ants was documented by Olson (1994) for Panama and by Fisher (1998) for Madagascar. Sampson et al. (1997) described a similar pattern for ground-dwelling and arboreal ants in the Philippines. The present results

extend the generality of these findings. Midelevation diversity peaks have also been reported for other taxa and on larger spatial scales (e.g., McCoy 1990; Colwell and Hurtt 1994; Rahbek 1995, 1997; cf. Stevens 1992). Various nonexclusive hypotheses have been put forward to explain this phenomenon, including the coincidence of midelevation sites with regions of either intermediate (Rosenzweig 1995) or maximum (Janzen et al. 1976) productivity, coupled with the respective assumptions that species richness bears a unimodal or monotonic relationship to productivity. It has also been argued that topographical constraints such as bounded ranges (Colwell and Hurtt 1994) and narrower midelevation zone widths (Rahbek 1997) produce greater overlap of faunas at midelevation (see also Fisher 1998). The apparent absence of a midelevation ant diversity peak in temperate regions (Fig. 8.3b; see also Cole 1940:11; Gregg 1963:201) argues against a unifying explanation for elevational gradients in ant diversity or suggests that patterns of covariation between elevation and other environmental variables differ between temperate and tropical areas.

Regional Differences in Taxonomic Diversity of Leaf Litter Ants

The mean numbers of ant species, genera, subfamilies, and individual workers, per Winkler sample, are given in Table 8.5 for the different biogeographic regions. All of these variables show significant regional heterogeneity (ANOVA, $P = 0.000$, for all four comparisons). This is due largely to the relatively low ant diversity and abundance in the Nearctic region,

however, and an analysis of covariance of sample species richness, with region as the grouping variable and latitude and altitude as covariates, reveals no significant effect of biogeographic region ($P = 0.448$; Table 8.6). Thus in this study no intrinsic differences between regions in species-level alpha diversity were detected, other than those that could be attributed to differences in latitude and altitude. By contrast, a comparable analysis of covariance of genus richness shows significant regional variation ($P = 0.007$; Table 8.7), independent of latitude and altitude. Relatively low genus-level alpha diversity occurs not only in the Nearctic region, where it is expected on the basis of high latitudes, but also in samples from the Malagasy region.

The finding that sites in the Malagasy region are depauperate in ant genera but have levels of species richness comparable to other regions of the same latitude and altitude suggests the existence of a climatically influenced asymptote to local species richness, which can be achieved even in an island fauna with a limited

Table 8.5 Data on Winkler Samples from Different Biogeographic Regions[a]

Region	Species	Genera	Subfamilies	Workers
Nearctic	4.1 ± 3.3	3.5 ± 2.4	1.87 ± 1.01	36.5 ± 53.5
($n = 23$)	(0–14)	(0–9)	(0–4)	(0–245)
Neotropical	26.8 ± 17.3	13.5 ± 6.1	2.90 ± 0.85	385.6 ± 360.8
($n = 49$)	(0–75)	(0–27)	(0–4)	(0–1632)
Malagasy	19.8 ± 9.0	8.8 ± 2.4	3.05 ± 0.52	305.7 ± 263.1
($n = 19$)	(6–40)	(4–13)	(2–4)	(30–1162)
Indo-Australian	23.4 ± 13.5	13.8 ± 5.5	3.32 ± 0.75	231.7 ± 145.5
($n = 19$)	(5–56)	(4–25)	(2–5)	(65–588)
ANOVA	$F = 15.062$	$F = 25.050$	$F = 13.032$	$F = 8.721$
($n = 110$)	$P = 0.000$	$P = 0.000$	$P = 0.000$	$P = 0.000$
ANOVA, excluding Nearctic	$F = 1.538$	$F = 5.720$	$F = 2.044$	$F = 1.836$
($n = 87$)	$P = 0.221$	$P = 0.005$	$P = 0.136$	$P = 0.166$

[a]Means, standard deviations, and ranges of the numbers of species, genera, subfamilies, and individual workers are given.

Table 8.6 Analysis of Covariance of Sample Species Richness[a]

Source	Sum of Squares	Degrees of Freedom	Mean Square	F-ratio	P
Region	368.169	3	122.723	0.892	0.448
Latitude	1,910.767	1	1,910.767	13.890	0.000
Altitude	2,676.789	1	2,676.789	19.458	0.000
Error	14,306.663	104	137.564		

[a]Grouping variable is biogeographic region; covariates are latitude and altitude. Multiple R^2 = 0.48, n = 110.

source pool, given sufficient time for autochthonous speciation.

Subfamilies

There are significant differences among biogeographic realms in the prevalence of different ant subfamilies (Tables 8.8–8.10). The relative scarcity of the Cerapachyinae in the New World is reflected in their absence from the Nearctic and Neotropical litter samples. They are present in 10% and 16% of Malagasy and Indo-Australian litter collections, respectively. Conversely, army ants of the subfamily Ecitoninae, absent from the Old World, are found in 4% of Nearctic samples and 10% of Neotropical samples. Old World army ants (Aenictinae and Dorylinae) were not recovered from the leaf litter samples that were taken during this survey, although they were observed at some of the Indo-Australian localities.

Ponerinae are most frequent in the Indo-Australian region (making up about one-third of the species in Winkler samples and present in 100% of samples), poorly represented in the Nearctic samples (3.9% of species, on average, per sample; present in 22% of samples), and of intermediate occurrence in the Neotropical and Malagasy regions (Tables 8.8 and 8.10). The subfamily Formicinae, although present in a higher percentage of tropical than temperate (Nearctic) samples (Table 8.8), shows a greater proportional representation of species in the Nearctic sites (Table 8.10).

Table 8.7 Analysis of Covariance of Sample Genus Richness[a]

Source	Sum of Squares	Degrees of Freedom	Mean Square	F-ratio	P
Region	211.199	3	70.40	4.236	0.007
Latitude	295.267	1	295.267	17.766	0.000
Altitude	470.126	1	470.126	28.287	0.000
Error	1728.465	104	16.620		

[a]Grouping variable is biogeographic region; covariates are latitude and altitude. Multiple R^2 = 0.61, n = 110.

Table 8.8 Percentage of Winkler Samples in Which One or More Species of a Given Subfamily Were Present

Subfamily	Nearctic	Neo-tropical	Mala-gasy	Indo-Australian
Cerapachyinae	0.0	0.0	10.5	15.8
Dolichoderinae	8.7	12.2	5.3	26.3
Ecitoninae	4.4	10.2	0.0	0.0
Formicinae	65.2	75.5	89.5	89.5
Myrmicinae	87.0	98.0	100.0	100.0
Ponerinae	21.7	93.9	100.0	100.0

Table 8.9 Mean Number of Species per Subfamily per Winkler Sample

Subfamily	Nearctic	Neo-tropical	Mala-gasy	Indo-Australian
Cerapachyinae	0.00	0.00	0.11	0.16
Dolichoderinae	0.09	0.12	0.05	0.32
Ecitoninae	0.04	0.12	0.00	0.00
Formicinae	0.91	2.06	1.84	2.21
Myrmicinae	2.83	18.22	12.63	13.11
Ponerinae	0.26	6.27	5.16	7.63
All sub-families	4.13	26.80	19.79	23.47

The mean proportions of species, per sample, belonging to a given subfamily vary significantly among regions for the Cerapachyinae, Formicinae, Myrmicinae, and Ponerinae (Kruskal-Wallis tests, $P = 0.017$, 0.011, 0.000, and 0.000, respectively; Table 8.10). These differences are sustained even if one excludes the Nearctic region. Thus even on the coarse scale of ant subfamilies there is significant regional heterogeneity in taxonomic composition. This finding cautions against the use of "indicator taxa" (i.e., the use of a small subset of tribes or subfamilies) when making interregional comparisons of leaf litter ant communities.

Genera

For the analysis of genus-level differences in faunal composition among leaf litter ant assemblages I have used a finer subdivision of biogeographic regions. The 2 "standard" Winkler samples from Singapore and Malaysia have been combined with the 11 additional leaf litter samples from Pasoh Forest, Malaysia (collected by A. Malsch), to provide an assessment of the

Table 8.10 Mean Proportion of Species per Subfamily per Winkler Sample[a]

Subfamily	Nearctic	Neotropical	Malagasy	Indo-Australian	P[b]
Cerapachyinae	0.000	0.000	0.003	0.007	0.017
Dolichoderinae	0.012	0.006	0.003	0.018	ns
Ecitoninae	0.003	0.005	0.000	0.000	ns
Formicinae	0.278	0.066	0.102	0.112	0.011
Myrmicinae	0.667	0.695	0.624	0.520	0.000
Ponerinae	0.039	0.228	0.268	0.343	0.000

[a]Two samples (one Nearctic and one Neotropical) that yielded no ants are excluded.

[b]Kruskal-Wallis test. ns, Not significant.

Table 8.11 Most Frequent Genera in Each Biogeographic Region[a]

Neotropical (n = 49)		Nearctic (n = 23)		Australian (n = 17)		Oriental (n = 13)		Malagasy (n = 19)	
Solenopsis	91.8	Stenamma	69.6	Hypoponera	100.0	Strumigenys	100.0	Hypoponera	100.0
Pheidole	89.8	Leptothorax	56.5	Pheidole	94.1	Tetramorium	100.0	Pheidole	100.0
Hypoponera	83.7	Lasius	43.5	Strumigenys	94.1	Monomorium	92.3	Strumigenys	89.5
Strumigenys	79.6	Aphaenogaster	30.4	Solenopsis	76.5	Oligomyrmex	92.3	Tetramorium	89.5
Cyphomyrmex	67.3			Oligomyrmex	70.6	Odontoponera	84.6	Monomorium	84.2
Paratrechina	61.2			Paratrechina	70.6	Pheidole	84.6	Paratrechina	78.9
Pachycondyla	57.1			Ponera	58.8	Myrmecina	76.9	Oligomyrmex	57.9
Wasmannia	55.1			Monomorium	52.9	Odontomachus	69.2	Pachycondyla	57.9
Rogeria	53.1			Rhytidoponera	52.9	Hypoponera	61.5	Anochetus	36.8
Gnamptogenys	51.0			Tetramorium	47.1	Lophomyrmex	53.8	Prionopelta	31.6
Brachymyrmex	49.0			Heteroponera	41.2	Vollenhovia	46.2		
Octostruma	44.9			Pachycondyla	41.2	Cerapachys	38.5		
Anochetus	40.8			Myrmecina	35.3	Crematogaster	38.5		
Oligomyrmex	36.7			Acropyga	29.4	Pachycondyla	38.5		
Neostruma	34.7			Cryptopone	29.4	Pristomyrmex	38.5		
Crematogaster	32.7			Lordomyrma	29.4	Pseudolasius	38.5		
Odontomachus	30.6			Prionopelta	29.4	Anochetus	30.8		
Smithistruma	30.6			Pristomyrmex	29.4	Cryptopone	30.8		
Adelomyrmex	26.5			Prolasius	29.4	Myrmoteras	30.8		
						Pheidologeton	30.8		
						Rostromyrmex	30.8		
						Smithistruma	30.8		
						Solenopsis	30.8		

[a]Genera are those present in 25% or more of Winkler samples in each biogeographic region. Figures refer to the percentage of Winkler samples in which a given genus was represented. n, Number of samples.

Oriental region. The 17 remaining Indo-Australian samples, from Australia and Papua New Guinea, can be considered representative of the Australian biogeographic region.

Table 8.11 summarizes the distribution and prevalence of the most common leaf litter ant genera in the five biogeographic regions considered here; the full data set is given in Table 8.12. A few genera, including *Hypoponera, Pachycondyla, Pheidole,* and *Strumigenys,* are predominant in all four tropical realms. For most other genera, however, there are striking regional differences in the frequency of occurrence. Here I focus on the most frequently encountered ant genera. Rare ant taxa, including those endemic to a region or otherwise of biogeographic inter-

est (e.g., *Kyidris, Mystrium, Perissomyrmex*), are largely ignored since they appear to contribute in only a minor way to the composition of the leaf litter community.

The Nearctic region stands out as distinctly different from the others and low in genus-level diversity. The most frequent genera in the leaf litter samples are, in decreasing order of importance, *Stenamma, Leptothorax, Lasius, Aphaenogaster, Hypoponera, Solenopsis, Prenolepis, Formica,* and *Myrmecina. Stenamma* has a primarily Holarctic distribution, being absent from the Old World tropics and of sharply diminished importance at Neotropical sites with increasing distance from the Nearctic region. The same is true of most other Nearctic leaf litter ants; *Hypoponera* and *Solenopsis* are obvious exceptions.

The Neotropical samples reveal a generic diversity comparable to that of the Oriental and Australian regions. Common leaf litter genera that are largely or entirely confined to the Neotropics include *Adelomyrmex, Brachymyrmex, Neostruma, Octostruma, Rogeria, Wasmannia,* and all the attine genera (of which *Cyphomyrmex* appears most frequently in Winkler samples). *Solenopsis,* represented mostly by small to minute species, reaches its zenith in the New World tropics. Despite its morphological homogeneity, Neotropical *Solenopsis* is species rich (mean number of species per Winkler sample: 3.14; range: 0–9; mean proportion of species per sample: 0.12) and undoubtedly of considerable ecological importance. In mean species richness per sample it is exceeded in the Neotropics only by *Pheidole* (mean: 5.45 species; range: 0–16; mean proportion of species per sample: 0.19). Another distinctive hallmark of the New World leaf litter fauna is the virtual absence of species of *Monomorium* and *Tetramorium.*

The wet forests of the Australian region are characterized by high frequencies of *Cryptopone, Discothyrea, Heteroponera, Ponera,* and *Rhytidoponera,* ponerine genera that tend to be uncommon or absent elsewhere. Other distinctive and common elements include the myrmicine genera *Lordomyrma, Myrmecina, Pristomyrmex,* and *Tetramorium,* and the endemic formicine genus *Prolasius.* The solenopsidine genera *Monomorium, Oligomyrmex,* and *Solenopsis* are also prevalent leaf litter ants in this region.

The Oriental samples are from a geographically restricted area in peninsular Malaysia and Singapore. Insofar as they are representative of the Oriental region as a whole, they indicate a leaf litter ant fauna with a distinctive complexion. Some genera are prominent both here and in the Australian region (*Cryptopone, Monomorium, Myrmecina, Oligomyrmex, Pristomyrmex, Tetramorium*), but others are more common in, and in some instances unique to, the Oriental region: *Acanthomyrmex, Lophomyrmex, Myrmoteras, Odontoponera, Pheidologeton, Pseudolasius, Rostromyrmex,* and *Vollenhovia.* The genus *Cerapachys* also appears in a high percentage of the Oriental leaf litter samples; more extensive geographic sampling is needed to confirm the generality of this result.

Barry Bolton (pers. comm.) has recently analyzed a larger set of leaf litter ant samples from Pasoh Forest, Malaysia, with results that are consistent with the foregoing generalizations, although *Cerapachys* shows lesser prominence.

The Malagasy region is depauperate in genera, as befits an island fauna, but species-rich in certain groups, such as *Hypoponera, Monomorium, Pheidole,* and *Tetramorium.* The genus *Solenopsis,* a very significant component of litter and soil faunas in the Australian and Neotropical regions, is completely lacking in Madagascar, its presence in the Malagasy samples being due to a single Mauritian species. In Madagascar the ecological counterparts of the small species of *Solenopsis* appear to be drawn from the genus *Monomorium.* Another significant absence from Madagascar and adjacent islands is that of Old World army ants (Aenictinae, Dorylinae), with the possible consequence that there is a relatively rich cerapachyine ant fauna (Fisher 1997).

Winkler leaf litter samples, collected using the same methods as previously described, are unavailable from the Ethiopian region (mainland Africa), but Belshaw and Bolton's (1994a) detailed census of the leaf litter ant fauna in Ghana provides useful and approximately comparable information. In that study the most widespread and species-rich genera included *Tetramorium* (27 species), *Monomorium* (16 species), *Oligomyrmex* (12 species), *Smithistruma* (12 species), *Pheidole* (11 species), *Pachycondyla* (8 species), *Strumigenys* (7 species), *Anochetus* (6 species), *Hypoponera* (6 species), and *Technomyrmex* (5 species).

Table 8.12 Distribution and Prevalence of Leaf Litter Ant Genera in Different Biogeographic Regions

Genus	Nearctic		Neotropical		Malagasy		Australian		Oriental	
	Percentage	No. spp.	Percentage	No. spp.	Percentage	No. spp.	Percentage	No. spp.	Percentage	No. spp.
Acanthognathus	—	—	2.04	0.02	—	—	—	—	23.08	0.23
Acanthomyrmex	—	—	—	—	—	—	—	—	—	—
Acromyrmex	—	—	2.04	0.02	5.26	0.05	**29.41**	0.41	7.69	0.23
Adelomyrmex	—	—	20.41	0.24	—	—	5.88	0.06	—	0.08
Amblyopone	—	—	**26.53**	0.47	10.53	0.11	17.65	0.18	7.69	0.08
Anochetus	—	—	**40.82**	0.45	**36.84**	0.58	5.88	0.06	**30.77**	0.31
Anonychomyrma	—	—	—	—	—	—	17.65	0.18	—	—
Anoplolepis	—	—	—	—	—	—	—	—	15.38	0.15
Aphaenogaster	**30.43**	0.30	8.16	0.08	—	—	5.88	0.06	—	—
Apterostigma	—	—	22.45	0.27	—	—	—	—	—	—
Azteca	—	—	2.04	0.02	—	—	—	—	—	—
Basiceros	—	—	6.12	0.08	—	—	—	—	—	—
Belonopelta	—	—	2.04	0.02	—	—	—	—	—	—
Brachymyrmex	8.70	0.09	**48.98**	0.82	10.53	0.11	—	—	—	—
Camponotus	4.35	0.04	6.12	0.06	10.53	0.11	—	—	15.38	0.15
Cardiocondyla	—	—	—	—	—	—	—	—	23.08	0.23
Cerapachys	—	—	—	—	10.53	0.11	5.88	0.06	**38.46**	0.46
Colobostruma	—	—	—	—	—	—	17.65	0.18	—	—
Crematogaster	8.70	0.09	**32.65**	0.53	5.26	0.05	17.65	0.24	**38.46**	0.38
Cryptopone	—	—	10.20	0.10	—	—	**29.41**	0.59	**30.77**	0.31
Cyphomyrmex	—	—	**67.35**	1.04	—	—	—	—	—	—
Dacetini gen. indet.	—	—	2.04	0.02	—	—	—	—	—	—
Dacetinops	—	—	—	—	—	—	5.88	0.06	—	—
Discothyrea	—	—	20.41	0.20	—	—	23.53	0.29	15.38	0.15
Dolichoderus	—	—	2.04	0.02	—	—	—	—	—	—
Echinopla	—	—	—	—	—	—	—	—	7.69	0.08
Ectatomma	—	—	2.04	0.02	—	—	—	—	—	—
Eurhopalothrix	—	—	8.16	0.08	—	—	23.53	0.29	—	—

Continued on next page

Formica	13.04	0.13	—	—	—	—	—	—	—	—
Gen. nov. (Madagascar)	—	—	—	—	10.53	0.11	—	—	—	—
Glamyromyrmex	—	—	6.12	0.06	—	—	—	—	—	—
Gnamptogenys	—	—	**51.02**	0.80	—	—	—	—	15.38	0.31
Gymnomyrmex	—	—	2.04	0.02	—	—	—	—	—	—
Heteroponera	—	—	—	—	—	—	**41.18**	0.41	—	—
Hylomyrma	17.39	—	18.37	0.20	—	—	—	—	—	—
Hypoponera	—	0.17	**83.67**	2.73	**100.00**	2.95	**100.00**	3.12	**61.54**	0.85
Kyidris	—	0.04	—	—	15.79	0.16	—	—	7.69	0.08
Labidus	4.35	—	8.16	0.08	—	—	—	—	—	—
Lachnomyrmex	—	—	8.16	0.08	—	—	—	—	—	—
Lasius	**43.48**	0.48	—	—	—	—	—	—	—	—
Lenomyrmex	—	—	2.04	0.02	—	—	—	—	—	—
Leptogenys	—	—	2.04	0.02	15.79	0.16	5.88	0.06	7.69	0.08
Leptothorax	**56.52**	0.61	2.04	0.02	—	—	—	—	—	—
Linepithema	—	—	8.16	0.08	—	—	—	—	—	—
Liomyrmex	—	—	—	—	—	—	5.88	0.06	—	—
Lophomyrmex	—	—	—	—	—	—	**29.41**	0.41	**53.85**	0.54
Lordomyrma	—	—	—	—	—	—	11.76	0.12	7.69	0.08
Mayriella	—	—	—	—	—	—	—	—	—	—
Megalomyrmex	—	—	6.12	0.06	—	—	—	—	—	—
Meranoplus	—	—	—	—	—	—	5.88	0.06	15.38	0.15
Monomorium	4.35	0.04	—	—	**84.21**	2.11	**52.94**	0.82	**92.31**	1.77
Mycocepurus	—	—	4.08	0.04	—	—	—	—	—	—
Myopias	—	—	—	—	—	—	17.65	0.24	—	—
Myopopone	—	—	—	—	—	—	5.88	0.06	—	—
Myrmecina	13.04	0.13	2.04	0.02	—	—	**35.29**	0.41	**76.92**	1.00
Myrmicocrypta	—	—	20.41	0.22	—	—	—	—	—	—
Myrmoteras	—	—	—	—	—	—	—	—	**30.77**	0.31
Mystrium	—	—	—	—	5.26	0.05	5.88	0.06	—	—
Neivamyrmex	—	—	4.08	0.04	—	—	—	—	—	—
Neostruma	—	—	**34.69**	0.37	—	—	—	—	—	—
Notoncus	—	—	—	—	—	—	11.76	0.12	—	—
Ochetomyrmex	—	—	10.20	0.10	—	—	5.88	—	—	—

Table 8.12 continued

Genus	Nearctic		Neotropical		Malagasy		Australian		Oriental	
	Percentage	No. spp.	Percentage	No. spp.	Percentage	No. spp.	Percentage	No. spp.	Percentage	No. spp.
Octostruma	—	—	**44.90**	0.61	—	—	—	—	—	—
Odontomachus	—	—	30.61	0.33	—	—	11.76	0.12	69.23	0.69
Odontoponera	—	—	—	—	—	—	—	—	84.62	0.85
Oligomyrmex	—	—	**36.73**	0.51	57.89	0.68	**70.59**	1.35	92.31	1.54
Orectognathus	—	—	2.04	0.02	—	—	11.76	0.12	—	—
Oxyepoecus	—	—	2.04	—	—	—	—	—	—	—
Pachycondyla	4.35	0.04	**57.14**	0.94	57.89	0.74	**41.18**	0.82	38.46	0.54
Paratrechina	4.35	0.04	**61.22**	0.94	78.95	1.47	**70.59**	0.88	15.38	0.31
Perissomyrmex	—	—	2.04	0.02	—	—	—	—	—	—
Pheidole	8.70	0.30	**89.80**	5.45	**100.00**	4.21	**94.12**	3.41	**84.62**	2.23
Pheidologeton	—	—	—	—	—	—	5.88	0.06	**30.77**	0.31
Plagiolepis	—	—	—	—	10.53	0.11	—	—	—	—
Polyrhachis	—	—	—	—	—	—	5.88	0.06	15.38	0.15
Ponera	4.35	0.04	2.04	0.02	5.26	0.05	58.82	0.88	15.38	0.23
Prenolepis	13.04	0.13	—	—	—	—	—	—	—	—
Prionopelta	—	—	24.49	0.35	**31.58**	0.47	**29.41**	0.29	—	—
Pristomyrmex	—	—	—	—	—	—	**29.41**	0.41	**38.46**	0.38
Probolomyrmex	—	—	—	—	—	—	—	—	7.69	0.08
Proceratium	—	—	10.20	0.10	5.26	0.05	5.88	0.06	—	—
Prolasius	—	—	—	—	—	—	**29.41**	0.41	—	—
Protalaridris	—	—	2.04	0.02	—	—	—	—	—	—
Pseudolasius	—	—	—	—	—	—	17.65	0.18	**38.46**	0.38
Rhopalothrix	—	—	2.04	0.02	—	—	—	—	—	—
Rhytidoponera	—	—	—	—	—	—	**52.94**	0.71	—	—
Rogeria	8.70	0.09	**53.06**	0.94	—	—	11.76	0.12	**30.77**	0.31
Rostromyrmex	—	—	—	—	—	—	—	—	—	—
Sericomyrmex	—	—	10.20	0.14	—	—	—	—	—	—
Serrastruma	—	—	—	—	10.53	0.11	—	—	—	—
Smithistruma	—	—	**30.61**	0.35	15.79	0.21	11.76	0.12	**30.77**	0.31

	Percentage	No. spp.	Percentage	No. spp.	Percentage	No. spp.	Percentage	No. spp.	Percentage	No. spp.
Solenopsis	17.39	0.22	**91.84**	3.14	5.26	0.05	**76.47**	1.06	**30.77**	0.31
Sphinctomyrmex	—	—	—	—	—	—	11.76	0.12	—	—
Stegomyrmex	—	—	2.04	0.02	—	—	—	—	—	—
Stenamma	**69.57**	1.00	22.45	0.57	—	—	—	—	—	—
Stigmacros	—	—	—	—	—	—	5.88	0.06	—	—
Strumigenys	4.35	0.04	**79.59**	1.82	**89.47**	1.84	**94.12**	2.53	**100.00**	2.46
Tapinoma	8.70	0.09	—	—	—	—	11.76	0.12	—	—
Tatuidris	—	—	4.08	0.04	—	—	—	—	—	—
Technomyrmex	—	—	—	—	5.26	0.05	5.88	0.06	23.08	0.23
Tetramorium	—	—	2.04	0.02	**89.47**	3.11	**47.06**	0.76	**100.00**	2.54
Thaumatomyrmex	—	—	2.04	0.02	—	—	—	—	—	—
Trachymyrmex	—	—	10.20	0.10	—	—	—	—	—	—
Typhlomyrmex	—	—	10.20	0.16	—	—	—	—	—	—
Vollenhovia	—	—	—	—	—	—	17.65	0.18	**46.15**	0.54
Wasmannia	—	—	**55.10**	0.57	—	—	—	—	—	—
Total no. samples	23		49		19		17		13	

[a] Values are based on the Winkler leaf litter survey discussed in this chapter. "Percentage" is percentage of samples occupied, with values greater than 25% in **boldface.** "No. spp." is mean number of species per sample.

Leaf Litter Ant Diversity and Composition: Ecological Trends

Thus far we have considered differences in ant diversity and faunal composition without reference to the ecological roles of the constituent taxa. In fact, for leaf litter ants, which are mostly small in size and cryptic in habits, this is largely terra incognita. There does appear to be substantial variation in functional roles, from host-specific predators (e.g., Cerapachyinae, some Ponerinae, and myrmicine tribes such as Basicerotini, Dacetonini, and Myrmecinini) to generalist predators (many Ponerinae), seed harvesters (some *Pheidole* and *Acantho-myrmex*), and omnivores or scavengers (many myrmicines and formicines). For many regionally prominent leaf litter ants (e.g., species of *Brachymyrmex, Rogeria, Stenamma, Tetra-morium, Vollenhovia*) and even members of such cosmopolitan genera as *Oligomyrmex* and *Pheidole,* we know little about their feeding habits and ecological effects. It may be possible to assign a functional group label of "cryptic species" or "tropical climate specialist" to such leaf litter ants (cf. Andersen 1995), but this reveals little about their biology.

Some ecological variation in leaf litter ant communities appears to have a strong geographical component. For example, the relative prevalence of species of Ponerinae, a group of mostly predacious ants, is strongly negatively correlated with latitude ($r = -0.698$, $P = 0.000$; proportions arcsine-transformed) (Fig. 8.4). Ponerine species are also overrepresented in leaf litter samples from the Indo-Australian region compared to other tropical continents, as noted previously. Other ecologically well-defined ant taxa (e.g., leaf litter species of the fungus-growing tribe Attini; army ants of the subfamily Ecitoninae; mite-catching *Myrme-cina;* some of the collembolan-hunting dacetine genera) also have geographically restricted distributions. All of this hints at complex geo-

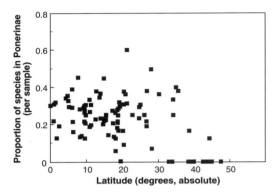

Figure 8.4. Proportion of ant species in a sample belonging to the subfamily Ponerinae as a function of latitude.

graphical variation in leaf litter community structure that we have hardly begun to investigate.

Latitudinal gradients have been reported in ant worker body size (Cushman et al. 1993) and in ant colony size (Kaspari and Vargo 1995). These studies were not specifically concerned with the leaf litter habitat, but many of the ant species from which data were taken, especially in the colony size study, are leaf litter inhabitants. In both studies the authors favored the hypothesis that abiotic factors select for larger body or colony sizes at higher latitudes, as a buffer against starvation. It would be interesting to have comparable data for elevation gradients.

Caveats and Concluding Remarks

This study is an attempt to characterize large-scale geographical variation in ant leaf litter communities. The analysis is based primarily on a series of Winkler litter samples, and the results should be considered provisional. The samples cover a broad but by no means comprehensive set of geographical locations. All samples in this study come from moist leaf litter in woodland and forest habitats. Xeric environments, to which the Winkler method is poorly suited, have been largely ignored (only 6 of the

110 Winkler samples are from sites that could be characterized as tropical dry forest). Questions can also be raised about the appropriateness of the Winkler method for exhaustive sampling of the leaf litter ant fauna, although this procedure appears to work better than any other single method (Olson 1991; Fisher 1996a, 1998). Some taxa are nevertheless undersampled, especially the "army ant" groups (Aenictinae, Dorylinae, Ecitoninae). These ants are nomadic and hence episodic in their occurrence at any given site. Yet there is evidence that they have a potent impact on the leaf litter ant fauna (Franks and Bossert 1983; Gotwald 1995).

Habitat-based differences in ant abundance and diversity (independent of biogeographic region) certainly exist but have not been explored in detail here. Future studies focusing on the effects of habitat on the diversity and composition of leaf litter ant communities would benefit from the adoption of a standard classification of forest communities.

A comprehensive assessment of species turnover (beta diversity) has not been attempted. Knowledge of beta diversity patterns is essential for a better understanding of biological diversity and for intelligent conservation planning, but measurement of species-level beta diversity in groups such as leaf litter ants is frustrated by the "taxonomic impediment" (Taylor 1983) imposed by the lack of high-quality descriptive taxonomy. Many of the most prevalent and species-rich leaf litter ant genera (Tables 8.4 and 8.11) have never had the benefit of a modern taxonomic revision. Three genera alone—*Hypoponera, Pheidole,* and *Solenopsis*—constitute, on average, 32% of the species in a given leaf litter sample (this figure increases to 40% for the Neotropics), and all of them are in a state of taxonomic anarchy. These three genera, and others such as *Pachycondyla* and *Paratrechina,* should be high on the list of priorities for revisionary studies.

Some results of the present study appear to be robust and unaffected by taxonomic constraints. One is the finding of significant heterogeneity in faunal composition at the level of ant genera and subfamilies across major biogeographic regions. This observation argues against the use of subsets of "indicator taxa" to assess overall diversity patterns over large geographical scales. A second finding is that the taxonomic diversity of leaf litter ant communities is strongly affected by latitude and altitude (Figures 8.2 and 8.3), with a general trend toward increasing diversity at lower latitudes and altitudes. In tropical regions, however, species richness appears to reach a maximum at about 500 m and then decline slightly at lower elevations. Finally, whereas species alpha diversity shows no differences among biogeographic regions other than those predicted by variation in latitude and elevation, local genus richness is lower in the insular Malagasy region than in comparable continental areas. This finding implicates historical constraints on genus richness, which nevertheless do not prevent the achievement of climatically characteristic levels of local species richness.

ACKNOWLEDGMENTS

I thank Alan Andersen, Brian Fisher, and Mike Kaspari for valuable discussions; Barry Bolton and Annette Malsch for sharing data; and Donat Agosti for his invitation to participate in the leaf litter ant conference. In addition, I extend my gratitude to the many persons, too numerous to list, who have facilitated my field work on ants in various parts of the world.

Chapter 9

Field Techniques for the Study of Ground-Dwelling Ants

An Overview, Description, and Evaluation

Brandon T. Bestelmeyer, Donat Agosti, Leeanne E. Alonso,
C. Roberto F. Brandão, William L. Brown Jr.,
Jacques H. C. Delabie, and Rogerio Silvestre

The precise nature of the methods used to estimate the abundance and composition of organisms in biodiversity assessment is of critical importance. Owing to the inevitable limitations of field methods, these estimates are often biased; that is, some species in a given habitat are either over- or underrepresented relative to their true abundances. The estimates obtained from different sampling techniques or from variations in the execution of a particular technique may bias the data in different ways. This fact, in conjunction with differences in sampling design or analytical procedures between studies of a particular system, impedes the direct comparison and integration of data. Integrated data sets based on sound and repeatable methodologies are essential for long-term ecological monitoring as well as for developing a general understanding of patterns of biodiversity.

In this chapter we describe seven field techniques that are used to study ground-dwelling ants and recommend a set of protocols for the use and execution of these techniques. We discuss the special considerations required for each method, and the utility and limitations of each technique for different kinds of research questions and habitats. Our goal is to provide a set of standard, repeatable methods that (1) can be adapted to different research programs and logistical situations, (2) will provide data that are as accurate as possible, and (3) will produce results that are comparable between studies and researchers.

An Overview of Ant Sampling: Challenges and Opportunities

In spite of the abundance and ease of collection of ants in most ecosystems (see Chapter 1), several features of ant biology complicate their sampling. First, ants are variably and non-randomly distributed on several spatial scales. Individual ants are aggregated into colonies on small scales, and colonies are often regularly dispersed across the landscape owing to competition (Wiernasz and Cole 1995; Crist and Wiens 1996). Thus caution should be exercised with sampling designs and statistical procedures that require the assumption that the subjects are randomly distributed. Second, ants may be studied and sampled both as populations of individual foragers (ignoring colony membership) and as populations of colonies. Forager-based studies often emphasize ecological or functional relationships to the environment (Greenslade 1973; Andersen 1991a), whereas colony-based studies emphasize population or genetic structure (e.g., Herbers and Grieco 1994).

The relationship between the activity and abundance of foragers and colony abundance and distribution varies greatly between species, so that forager- and colony-based comparisons of communities may not be equivalent. For example, given equal colony density, the foragers of highly active ant species with large foraging distances from the colony will be sampled more frequently than those of sedentary species that forage near the colony (Andersen 1991b). Finally, the diversity of behavior and habitat selection exhibited in ants results in different sampling probabilities between species and methods. An obvious example is that arboreal ants are seldom found in leaf litter samples.

The first of the challenges just outlined is of special relevance to the sampling design and analytical procedures used in studies of ant communities and is discussed in Chapter 10. The remaining points also have implications for the field techniques used to census ants. Different methods should be used if one's focus is on ant colonies rather than ant foragers. Different methods of ant collection are required to sample in the different habitats that ants occupy. Furthermore, the several methods that may be employed for a given research question in a particular habitat each have biases owing to practical limitations and differences in species behavior. These biases must be recognized in order to interpret and compare field data correctly.

Two broad categories of questions are commonly addressed in biodiversity assessment: those related to evaluating differences in communities between habitats or sites (e.g., to assess environmental degradation or restoration) and those concerned with species inventory within sites (Heyer et al. 1994; Chapter 13). For purposes of monitoring or comparing ant communities, several features can be examined that respond to environmental variation: richness, species composition, forager abundance, foraging behavior, and colony density. Furthermore, different kinds of ants and ants occupying different microhabitats (litter-dwelling, ground-dwelling, and arboreal) will reflect this variation in unique ways (e.g., Ward 1987). Thus, many different techniques can be used to compare ant communities as long as they provide the desired data, are applied consistently between the sites, and are logistically feasible. On the other hand, the primary goal of a species inventory is to record as many of the species present at a site as possible. For a given habitat, a certain set of complementary techniques will best achieve this objective (Majer, 1996).

See Chapter 2 for more information on how the biology of the ant species under study can influence the choice of sampling methods.

Field Techniques

In the sections that follow, we provide detailed descriptions and evaluations of seven field methods that are commonly used to study ant communities. Some techniques—such as colony, intensive, and direct sampling—overlap slightly with respect to methodology and the kinds of data collected, but we separate the techniques here because of their distinct objectives. Some of the suggestions we make regarding the choice or execution of certain techniques should be evaluated and modified in the course of pilot studies in a given ant community. For example, the abundance and activity of ants will influence the duration of pitfall trapping or quadrat sampling. If too few ants are sampled the research question will remain unanswered, and if too many ants are sampled the investigator may be unable to sort and identify all of the specimens.

The techniques we describe may be divided into two broad classes: passive and active sampling. Passive sampling methods—including pitfall trapping, baiting, and quadrat sampling—are easy to replicate and rely on ant activity at sample stations to obtain data. Active sampling methods—such as direct sampling, colony counts, and intensive sampling—require that investigators seek out ants over the study area and are difficult to replicate precisely between investigators. In general, passive techniques suffer from biases because of differences in the behavior of different ant species in different habitats or alterations in the natural foraging behaviors of the ants. Each technique will systematically miss some ant species. Active sampling techniques introduce bias through differences in the effectiveness of researchers in different habitats, through differences in the detectability of ant species, and through variations in the execution of the sampling techniques. The resulting lack of comparability of samples will hinder spatial or temporal comparisons, such as those in long-term monitoring. Litter extraction is subject to both active and passive sampling biases because the techniques used by investigators vary widely and also depend upon the reactions of ants to behavioral stimuli.

In addition to these broad characteristics, each technique has a particular set of challenges and advantages that investigators should keep in mind when choosing their methods. Here we outline the types of questions that are best addressed by each technique and summarize important criticisms of the data generated by them.

General Materials and Methods

Materials

General preparation for all techniques will require several items, including vials or plastic twirl bags (small plastic bags with a wire-twist closure), ethanol solution, card stock or paper tags, pencils, field notebook, and materials for setting up study plots, including meter tapes, a compass, a random number table or calculator, flagging, stakes, and field tags. See Appendix 1 for a complete list of materials and sources.

Methods

In all of the techniques, ants are collected in the field or afterwards into vials or twirl bags filled with alcohol. Vials should have tight-fitting caps to retard the evaporation of the alcohol. The vials should be filled with at least 75% ethanol, and preferably 90% for long-term storage. Every sample receptacle (e.g., cup, twirl bag, vial) should be clearly labeled with either a temporary or a permanent label. The proper label is a thick paper or card stock tag with the sample code written in pencil. The label should be placed inside the receptacle whenever possible. The sample code should also be recorded in a field book with appropriate information about the identity of the sample (e.g., location, date and time, habitat; see below).

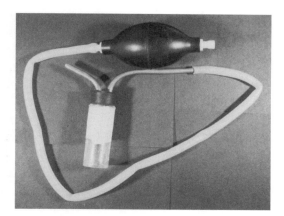

Figure 9.1. An aspirator. Note that the black gas-collecting bulb can be removed and that a vacuum can be created with the mouth. Photo by Brandon Bestelmeyer.

For general ant collecting, forceps and aspirators are required (see Appendix 1). Feather-weight forceps are preferred in order to minimize damage to delicate ants. Aspirators are very useful for collecting small or fast-moving species (Fig. 9.1). When air is drawn through the aspirator and the tip of the aspirator is held close to an ant, the ant is drawn into the collecting vial. A small screen prevents ants from being sucked back out of the vial. Air can be sucked through the aspirator with the mouth, but volatile compounds released from several formicine and dolichoderine ants may be irritating to the lungs or cause formicosis. The use of a gas-collecting bulb (see Fig. 9.1) will eliminate this hazard. Studies should be conducted along transects or within plots of standardized area. Metal tags and sturdy metal or wooden stakes may be called for in long-term studies.

General Collection Data

Regardless of collection method, ant specimens are most valuable when accompanied by the fullest possible collection information. Data documentation involves several levels: regional, local, and sample.

REGIONAL. At the regional level one must note the country and lesser political subdivisions, such as state, county, district, or national park. Additional regional information may include geographic features, such as a watershed, peninsula, mountain range, or valley.

LOCAL. Local information includes type of habitat or vegetation in which the collection is being carried out, for example, lowland humid forest, dry forest or scrubland, and altitude. Given the varied names of habitats and ecological communities, it is best to choose a system of nomenclature already in print for ease of comparison. Use the most specific vegetation classification that is available.

At a finer level, microgeographic characteristics of the collection site can be described, including slope, aspect, the presence of gullies or bluffs, soil type, and so forth (see Appendix 2). This information can be especially useful in characterizing the ecological preferences of ant species.

SAMPLE. Each sample—be it from a leaf litter sack, pitfall trap, or individual nest collection—carries its own individual record. It receives a unique collection number that goes into the field notebook. The sample code is the only certain means by which multiple specimens may be recognized as coming from the same colony, or by which trap sample specimens may be linked to a specific sample and to data entries. Codes may be created to reflect the hierarchical structure of a sampling design; for example, SBE2-9 could indicate a pitfall trap sample from site "S," habitat "BE," transect 2, and point 9.

Other collection data at this level can include a brief description of the microhabitat at the sample location, such as a rotten log, under a stone, in a bromeliad, beneath the bark of tree (specify type)—specific data that may help determine microhabitat preferences of ant species. All of the field data we describe should be recorded as soon as possible; the sooner it is recorded the less chance there is of forgetting a detail that may later prove important, or of mixing up information. Errors should never be erased but crossed out.

Materials

Maps should be of geological survey or cartographic quality, with geographic or Universal Transverse Mercator (UTM) coordinates and elevation contour lines. Road maps are frequently used in lieu of higher-quality maps, but their standards are poor and they are of no use for precisely delimiting areas. Good maps for a particular area however, may be unavailable. Tactical maps used by the U.S. Air Force cover the entire globe and can be purchased by the general public.

Global positioning system (GPS) receivers have dramatically fallen in price over the last few years, and a good-quality receiver can be purchased for $200 or even less in the United States. Given adequate reception conditions a GPS can furnish accurate latitude, longitude, and altitude data for a site.

Suitable notebooks for recording field data are those used by engineers or surveyors. If a notebook with neutral-pH paper is available, then it is highly preferable to ensure the permanence of the data. A sample data sheet is provided in Appendix 2.

Writing materials should include No. 2 or HB pencils or leads, as well as pens or markers with indelible ink (see Chapter 11).

Baiting

Objectives

Baiting uses food substances to attract foraging ants to points where they may be collected or observed. Tuna or sardine baits are the most common (Fig. 9.2), but foods that are richer in carbohydrates—such as fruit jelly, cookie crumbs, honey, peanut butter, or sugar solu-

tions—are also used alone or in combination with proteinaceous baits. Live or dead insects and seeds are employed for special applications.

This technique is commonly used to estimate the composition and richness of the active ground-foraging ant fauna, to examine ant activity and behavior patterns in studies of community structure, and to estimate the contributions of particular ant species to ecosystem processes such as seed redistribution or scavenging. The abundance of ant foragers at baits may help to measure ecological and behavioral dominance and provides a general measure of ant foraging efficiency (Greenslade and Greenslade 1971). Baits can be set out in different microhabitats at different times and can provide information on habitat use, biotic interactions, and activity patterns on very fine scales (Bestelmeyer 1997).

Many factors influence the species composition and abundance of ants at baits. The species most likely to visit baits are trophic generalists. Ant species with marked preferences for particular items (such as leaf-cutting ants or specialist predators) may not visit artificial baits, but dietary generalists represent a significant proportion of ant faunas worldwide and can be used to examine patterns in ant communities. More specialized groups can be targeted for study by using the appropriate bait; grass seeds, for example, can be used to attract desert harvester ants (Davidson 1977a, 1977b).

Baits are used on the surface of the soil or litter as well as in vegetation and underground. Because the activity of different ant species varies with microclimate, daily and seasonally, baiting performed at different times of the day (and night) or year in the same area will attract foragers of different species or of the same species in varying abundance. Because ants may occupy nest sites for long periods, repeated collections in the same location may attract individuals from the same colonies, and temporal variation in colony activity may be exam-

Figure 9.2. A tuna bait monopolized by *Solenopsis xyloni* in a desert grassland in New Mexico, USA. Photo by Brandon Bestelmeyer.

ined. Furthermore, ant species that differ in foraging behavior and behavioral dominance tend to discover and occupy baits at different times after the baits are placed; submissive and rapidly moving species find baits early but are often later displaced by dominant but slower-moving species (e.g., Fellers 1987). Thus repeated observational samples of a bait over time can reveal behavioral dynamics.

Materials

A bait substance, bait platforms (made of paper, cardboard, plastic, or leaf), vials or twirl bags, ethanol, forceps, an aspirator, and a timer (optional).

Methods

Baits that are pastes or solids are usually preferred because they tend to be more difficult for ants to remove than liquids or particles; thus ants will be present to be collected or observed for longer periods. Tuna or other fish baits, honey mixed with lard, and peanut butter are frequently used bait substances, and they can be deployed in pieces of about 1–2 cm^3. Tuna should be well mixed and should not contain excessive amounts of oil. The bait may be set directly on the ground or on a piece of paper to make the attracted ant species more readily visible. Graph paper is useful for distinguishing some ant species in the field because the squares provide a reference with which to compare worker sizes. The bait placed on a paper platform tends to attract more dominant ants, while the oil around and under the paper will attract smaller and/or less aggressive species.

Alternatively, an impermeable bait platform (e.g., plastic) will restrict ant activity to the bait above the platform, and it will be less likely to blow away in windy conditions. Baits can be placed on the ground, in shrubs and trees, and underground in small containers that are perforated to allow the ants to enter (a string can be attached to the container to allow the buried trap to be recovered; Quiroz-Robledo and Valenzuela-González 1995). The amount of time the bait is observed will vary depending upon the objectives of the study. Care should be taken to avoid disturbance of the baits and foraging trails during observations. If the vegetation must be disturbed, allow at least a day before baiting so that foraging trails may be reestablished.

To collect ants attending baits in a single-sample or "snapshot" fashion, investigators may collect the bait and platform (and also the leaf litter around the bait, when necessary) into a coded plastic twirl bag. When cleaner samples are desired, baits may be quickly placed into a plastic tub and the ants removed with forceps or an aspirator from the bottom of the tub. Fluon or petroleum jelly around the walls of the tub will prevent the escape of fast-moving species. Our observations suggest that 60–90 minutes is usually sufficient time for dominant ant species in an area to discover and recruit to the baits. A small amount of ethanol may be injected into the bags to allow the ants to be separated from the bait and debris under a microscope afterwards; ants may then be transferred to ethanol-filled vials. Acetone can be used to remove tuna oils from the ant specimens. Brandão and Silvestre (unpubl. data) found that circa 1800 bait samples were needed to record 90% of the fauna visiting baits in a Brazilian cerrado site.

For behavioral studies, it is imperative to leave the bait undisturbed for the duration of observations. Baits can be observed repeatedly (e.g., every 20 minutes for 2 hours) to study behavioral dynamics over time (see Fellers 1987). Voucher specimens of unrecognized ant species may be collected from around the bait with forceps. It is important to collect these ants, because some ant species may leave the bait before the end of the observation period. It is best to target individuals that are separated from foraging groups so that other ants will not be alerted by chemical signals released by the victim. It is convenient to place several previously labeled vials on the ground next to baits in advance, so that specimens can be collected in sequential samples. It is important to collect representatives of the different castes of polymorphic species, especially majors of such species as *Pheidole* and *Solenopsis,* which facilitate species identifications. Majors are often more timid than minors, and they may take more time than minors to recruit to baits.

While ants are present at baits, behavioral interactions between individuals of different species and the numbers of individuals of the various species attending the baits can be recorded. Small baits, such as cookie crumbs, can be used to follow ants back to their nests. As noted earlier, baiting at different times of the day or year in a locality can reveal the influence of abiotic features on ant activity and interactions between species. In general, ant activity levels tend to vary from the cooler hours of the day, to the warmer hours (especially in full sun), and at nightfall. Bestelmeyer (1997, unpubl. data) found that soil surface temperatures of 37–40°C represent an important transition between the activity of different species at baits in both North American and South American arid-zone ant communities.

In studies of the impacts of ants on ecosystem processes, dead insects can be used as baits to measure the consumption rates of scavenging or predaceous ants (see Jeanne 1979; Fellers and Fellers 1982; Seastedt and Crossley 1984; Retana et al. 1991; Olson 1992), and seeds can be used to examine the impact of granivorous ants on seed removal and redistribution (see Crist and Wiens 1994). In both cases, the size and density of the items presented may influence the species of ants that will remove them.

Data Output

The data produced by this technique may include richness, composition, relative abundance of ant foragers at individual baits, frequency of occurrence of species at sets of baits, frequency and nature of behavioral interactions, timing and duration of forager activity, and rate and distance of removal of food items.

Evaluation

Baiting is the most common of the techniques used to study ant communities, no doubt because it is very simple and inexpensive and can be deployed rapidly and extensively. Baiting is ideal for work involving behavioral questions. Baits may, however, seriously bias descriptions of community composition (Greenslade and Greenslade 1971). Baits are dietarily selective and may systematically exclude some components of an ant fauna from samples. Because baits tend to be monopolized by dominant, mass-recruiting species (e.g., *Solenopsis;* Fig. 9.2), subordinate and single-foraging ants may be underrepresented at baits relative to their abundances. This problem may be alleviated in part by using several baits at a sample point (Culver 1974). Because ant activity at baits varies with time after bait placement, daily, and seasonally, multiple observations at baits will ensure a more complete representation of the species and the factors that affect ant foraging.

There is evidence that the preference of some ants for proteinaceous or carbohydrate baits may vary seasonally (Stein et al. 1990), although it is unclear how this may affect the data. Brandão and Silvestre (unpubl. data) compared the species composition of ants that visited sardine or tuna and honey-water baits, which mimic protein/fat and sugar sources, respectively. Their results indicate that the nutrient composition of the baits does not significantly affect the composition of species at the baits. Canned sardine or tuna bait may be transported easily along with collecting gear and stored for indefinite periods.

Pitfall Trapping

Objectives

Pitfall trapping involves the placement of open containers in the ground (Fig. 9.3a). Surface-active animals fall unwittingly into these traps and are either killed and preserved in a liquid or "dry-trapped" and allowed to survive after a census.

a

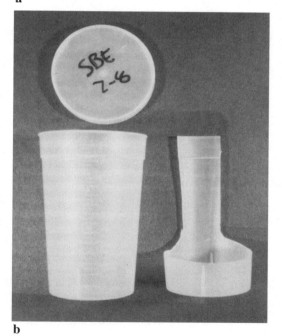

b

Figure 9.3. (a) A pitfall trap placed in desert soil. (b) A polypropylene sample container used as a pitfall trap and a pitfall-trap scoop (right) made from the same container, used to catch and remove debris that falls into the trap while it is being set. Photo by Brandon Bestelmeyer.

This method is used to estimate the abundance and species composition of ground surface–active ants in an area. As in baiting, the abundances of ants in pitfall traps provide a measure of species importance in a community by integrating both forager attributes and colony dispersion patterns (Greenslade 1973). Pitfall trapping may be used to census ants foraging on soil or leaf litter. It may be very difficult to use pitfall traps on rocky surfaces. Pitfall trapping may be performed for short (days) or long (continuously) durations.

Materials

Pitfall traps (containers), pitfall trap scoop, hand trowel or shovel, killing and preserving agent, detergent, and a tea strainer or muslin cloth with additional containers (optional).

Methods

For ants, it is best to use a killing agent in the pitfall trap; otherwise, captured ants will dismember one another and the specimens will be damaged. Several killing agents can be used. Propylene glycol (available in the United States as "environmentally friendly" automobile antifreeze) is an ideal choice because it is slow to evaporate (even when soil surface temperatures exceed 60°C) and is reputed to be nontoxic to vertebrates. The more common antifreeze, ethylene glycol, may also be used, but it is toxic to vertebrates. Ethanol solution may serve as a killing agent, but a few drops of glycerol should be added to retard evaporation. The addition of ethanol to propylene or ethylene glycol may kill ants more quickly and reduce the chance of escape. A drop of unscented detergent added to the killing agent breaks surface tension and may prevent ants from escaping the trap. Ideally the killing agent should not attract or repel ants (or at least not attract or repel species differently); otherwise estimates and comparisons of forager densities may be biased. Ethanol/glycerol (Greenslade and Greenslade 1971) and propyl-

ene glycol (Abensperg-Traun and Steven 1995) are believed neither to attract nor to repel ants; other substances await examination in this regard.

Traps may be plastic or glass containers, such as jars or drinking cups. Polypropylene sample containers (see Appendix 1) make ideal traps because they are durable and flexible and have tightly fitting lids, enabling them to be used to transport and temporarily store the specimens after trapping. Metal containers should be avoided because rust produces a rough surface in the trap that ants can use to escape. In general traps should have clean and smooth interiors (Luff 1975). The diameter of the mouth of the trap has been shown to affect the efficacy of pitfall traps for ants (Abensperg-Traun and Steven 1995). Traps with a very small diameter (18 mm) may bias against larger ant species and collect fewer of the species present in an area than larger traps. A 42-mm-diameter trap was found to perform as well as traps 86 and 135 mm in diameter. Smaller traps (40–70 mm in diameter) are easier to use and best for studies concerned solely with ants. Larger traps may be called for if other taxa are to be trapped.

Traps should be placed so as to minimize the disturbance of the surface around the trap because surface texture conditions may affect ant capture rates. A hand trowel that is only slightly bigger than the trap should be used to dig the hole. Surface features should be returned to normal by hand (e.g., coarse sand, stones, or leaf litter should be replaced).

When possible, traps should be allowed to settle for about a week (with the lids on or the trap inverted) before they are opened, in order to avoid the "digging-in effect" (Greenslade 1973). One manifestation of this effect is an abnormally high capture rate of ants when traps are placed in the ground and opened immediately thereafter. Causes include the penetration of nest galleries in the course of placing the trap and the exploration of novel habitat features by the ants. A settling period ameliorates this effect as the ants become accustomed to the disturbance. The return of natural surface characteristics (e.g., a soil crust) with settling is also desirable.

Traps should be placed in the ground with the lip of the trap flush with the soil or leaf litter surface or a few millimeters below the surface. If the lip is even slightly above the surface, small and/or wary ant species may be undersampled. Soil or leaf litter should completely cover the lip. When setting the trap, a tight-fitting scoop made from another trap container (Fig. 9.3b) may be used to catch soil and litter that falls into the trap so that they can be removed. This will result in cleaner pitfall samples and reduce the time needed to sort. Alternatively, one cup may be placed into the ground and act as a sleeve for a second cup that may be placed and removed easily. Over the trap settling period and especially after precipitation or wind, the surface may fall below the lip. If the soil around the trap is packed well this effect will be minimized. Immediately before trapping (and during trapping, for long-term applications), the condition of the trap lip should be checked.

The killing agent is placed after the trap is set, and it should fill about 25% of the cup's volume. In situations in which soil or litter is likely to fall or blow into the trap, more liquid may be necessary. If rain is likely to flood the trap, a cover may be suspended over it. In ecological studies, covers should not extend beyond the trap circumference in order to avoid changes to the microclimate. Traps placed in depressions or drainages may also flood.

The duration that the traps are left open will depend on the objectives of the study and on logistics. Traps left open longer will collect more ants and more of the species occupying an area. Traps left open for very long periods may deplete populations of foragers or alter foraging paths around the traps (Greenslade 1973). In general, 2 or 3 days seems to be sufficient to capture the ants foraging around the trap and provide a measure of forager abundance. Temperature and humidity have profound effects on ant activity, and in cooler, drier weather a longer duration of trapping may be required. When traps are collected, they may be capped and removed for short-term storage when polypropylene sample containers are used. In other cases, the contents may be poured through a tea strainer to remove excess liquid. The tea strainer is then inverted and the contents washed into a container using 90% ethanol solution. In the case of large pitfall traps, which may collect large quantities of animals, the contents may be poured through a piece of muslin cloth, the cloth tied into a tight ball, and the ball stored in ethanol. In these cases, washing the strained specimens with water before storing may be desirable in order to remove the propylene glycol and debris that may stick to the ants.

Data Output

The data produced by pitfall traps include richness and composition, relative abundance of ant foragers in traps and sets of traps, and frequency of occurrence of species in sets of traps.

Evaluation

The greatest advantage of pitfall traps is that they take little time to place and operate by themselves. Most epigaeic ants are well represented in pitfall traps, especially in open habitats. Andersen (1991b) found that the results obtained from pitfall traps were comparable to those from the relatively unbiased but time-intensive quadrat method (discussed in the next section).

If one wishes to use ant capture frequencies or abundances in pitfall traps as a measure of ant populations, the estimates may be biased by differences in locomotion among ant species (Greenslade 1973; Andersen 1983). Fast-moving species (e.g., *Forelius*, *Iridomyrmex*) will be overrepresented relative to slower-moving species (e.g., *Crematogaster*) in forager abundance studies. Ant species may differ in their ability to climb on trap walls or in their wariness of traps, and this may bias estimates of activity. More importantly, the physical structure of the ground surface may affect ant capture rates (Greenslade 1964; Adis 1979). Heavy litter or numerous stones will reduce ant captures and can confound between-habitat comparisons of forager populations. Ant species may also differ in their deliberate avoidance of pitfall traps (Marsh 1984). Certain castes of ant species needed for identification (e.g., *Pheidole* majors) are often not recorded in pitfall traps, and additional nest collections are required. Because pitfall traps collect only surface-active ants, they are believed not to provide an adequate sample of most leaf litter ants (Olson 1991; Majer 1996).

Quadrat Sampling

Objectives

In this technique, ants are sampled directly by the investigator from within a quadrat-delimited sample area. Like pitfall trapping, this method is used to estimate the abundance and species composition of surface-active ants in an area (Andersen 1991b). Quadrats are best used in open-ground situations at different times of the day or year.

Materials

A prefabricated quadrat, vials, ethanol, forceps, an aspirator, a timer, and data sheets.

Methods

This method is similar to techniques used to measure ground-layer vegetation (see Bonham 1989). A fixed, transportable quadrat made of wood or plastic (e.g., polyvinyl chloride [PVC]) pipe is used to delineate the observation area. The quadrat may be raised slightly off the ground by pegs or nails so as not to interfere

with ant movements. Species of ants seen inside the quadrat or entering the quadrat over a fixed time interval (e.g., 2 minutes; Andersen 1991b) are counted, collected, or both. As with baiting, ant activity in quadrats varies throughout the day, and quadrats can be sampled several times, for example, in the morning, at midday, and at night.

Although a quadrat of 0.5×0.5 m is generally small enough that it can be viewed effectively by a single researcher, if the level of ant activity is high a smaller quadrat may be desirable. Data may be collected in two ways: (1) all the ants in the quadrat may be collected, or (2) the number of ants may be tallied by species onto data sheets. If all the ants are collected, the researcher must be adept with forceps or use an aspirator to remove ants to a vial. When the level of ant activity is very great, it may be impossible to collect all the ants in a quadrat. Furthermore, subsequent samples in the same quadrat may be affected by the loss of ant foragers. If the ants are to be tallied, the researcher must be able to distinguish on sight the ants present in the study site. This will require a considerable amount of preparation for researchers who are unfamiliar with ants. A few representatives of unknown ants may be collected to a vial for identification in the laboratory and given a provisional code in field notes. It is important not to disturb other ants during collecting because natural foraging behaviors may be altered; for example, ants may swarm into the quadrat. During periods of high activity, it may be useful to record ant counts using an abundance scale (e.g., 1, 2–5, 6–20, >20 ants; Andersen 1991b) rather than absolute numbers.

Data Output

The data produced by quadrat sampling include richness and composition, relative abundance, frequency of occurrence in sets of quadrats, and time and duration of activity.

Evaluation

Quadrats provide information that is similar to that of pitfall traps and represent the densities of epigaeic ant foragers more accurately than pitfall traps. The densities of the foragers recorded in quadrats are not influenced by the differential tendency of ants to be trapped (Andersen 1991b). Furthermore, quadrats can be used to examine hourly and daily activity patterns, whereas pitfall trapping sums activity over time. Of course this level of detail comes at the cost of a considerable investment of time in the field and in preparing to identify the ants by sight. Quadrat techniques may be difficult to implement at night when small, yellowish-colored ants are difficult to see. Observations can also be difficult when levels of activity are very high.

Litter Techniques

Objectives

In the two techniques described in this section, a quantity of moist leaf litter (usually all the litter and humus present under a 1×1-m quadrat) is collected and placed in an extraction apparatus. The apparatus compels mobile ants, through disturbance to the litter or through changes in microclimate, to migrate from the litter into a collecting receptacle.

These techniques are designed to measure the abundance and composition of ants inhabiting a volume of leaf litter. Whole colonies of ants nesting in the litter as well as ants foraging in the litter from colonies outside the litter sample are collected. These methods are especially appropriate for use in forest and woodland habitats, where many ant species inhabit the litter layer. However, this method is not particularly successful during very dry periods. When the leaf litter is dry, ants move their nests deep into the soil or up into the vegetation.

Winkler Extraction

In this technique the collected litter is first sifted to remove large leaves and twigs from the sample. The sifted litter is then placed in the Winkler sack to process. During this time, ants from within the litter sample migrate out of the litter as a behavioral response to disturbance of their habitat and eventually fall into a container (see Besuchet et al. 1987; Fisher 1998).

Materials

Winkler extraction requires a litter sifter, a Winkler sack, a quadrat, a ground cloth, large plastic sample bags, plastic cups, twirl bags, vials, ethanol, and in some cases a machete.

Methods

First, the litter sample is collected within a quadrat placed on the ground. The wooden or plastic quadrat should have movable joints (use bolts with wing nuts) and be able to open at one corner so that the frame can be placed around shrubs or trees. The litter should be scooped from the edge of the quadrat toward the center and be removed by hand into the opening of the *sifter* (Fig. 9.4a). Gloves should be used to prevent stings and bites. The litter should be removed from the top of the litter pile to the bottom and put quickly into the sifter. Twigs and clods should be broken open; decayed logs can be minced with a machete to expose and disturb ant nests within them (Fisher 1998). The sifter should be held immediately adjacent to the litter sample to minimize the loss of ants from the sample. Water-soaked litter should not be collected.

The sifter consists of an open-ended sack with a metal ring and attached handle at the top end, a mesh screen and handle located at about one-third the length of the sack from the top, and a bottom end that may be tied shut. The sifter should be long enough so that part of the sack is supported by the ground while being held. Prior to filling the sifter, its bottom end should be tied with two knots, a single and a shoestring knot, so that the sack does not open during the sifting process. The upper part of the

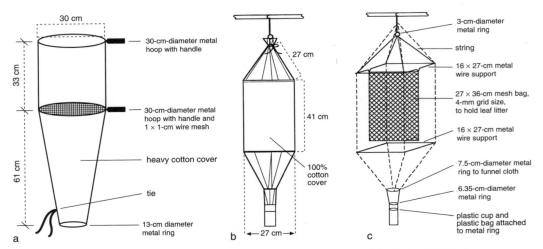

Figure 9.4. (a) Construction of the litter sifter. (b) External dimensions of the "mini-Winkler" sack. (c) Construction of the "mini-Winkler" sack (Fisher 1999a).

sifter is filled with litter that should not quite reach the upper margin. Holding both handles, the worker shakes the sifter to separate detritus and smaller invertebrates into the bottom of the sack while retaining coarse material above the mesh.

The sifter should be shaken thoroughly both laterally and vertically (Fig. 9.5a). The litter in the upper section should be turned over several times in the process. When the litter is very dry it should be shaken briefly, because most of the animals will fall through the mesh quickly and extended shaking will only add more debris to the sample. When the litter is moist, it should be shaken longer so that ants that are stuck to wet leaves may fall through. The sifting process may need to be repeated a number of times for a 1-m² sample. After the sample has been sifted, the top of the sifter sack should be twisted (twice) shut to ensure that animals do not escape through the top.

The *sample bag* should be large enough to hold a single litter sample, and the sample code should be written on it. The bag should be porous and made of synthetic material (e.g.,

nylon) to prevent rot. The contents of the sifter sack are poured into the sample bag through an opening in the bottom of the sifter. Close the sample bag using a tight knot. If the bags are to be stored for days in dry conditions, a little water can be used to moisten the sample. The sample bags are now removed to be processed in the *Winkler sack*.

The Winkler sack (Figs. 9.4b,c and 9.5c) consists of a metal box frame that supports a covering made of canvas or cotton. Litter from each sample bag is separated into one or more 4-mm *mesh inlet sacks,* which are suspended inside the Winkler sack. Ants in the litter migrate out of the inlet sacks and are collected in a receptacle tied to the bottom. The inlet sacks should have stitches in their centers that allow the sacks to maintain a flattened shape, which accelerates the migration of ants from the litter. The receptacle may be a twirl bag or a cup partially filled with ethanol solution. The first step in using a Winkler sack is to find a protected site where it can be mounted. A sack can be suspended from a nail in a wall, a beam in a shed, a pole under a tarp in the field, or a tree branch

a

b

c

Figure 9.5. Leaf litter extraction using the Winkler extractor. (a) Sifting leaf litter. (b) Transferring sifted litter into a mesh inlet sack that will be placed inside the Winkler sack. (c) Winkler sacks hanging from support beams, with researcher collecting excess debris from sacks. Photo by Donat Agosti.

at sites where rain is unlikely. It is important to find a location where the sack will not be tossed about by the wind or bumped by passersby, since any vibration or shock causes additional debris to fall into the receptacle. In preparation for loading inlet sacks, attach a dry receptacle to catch falling debris. Label each Winkler sack according to the sample it is to receive.

The next step is to distribute the contents of the sample bag into one or more inlet sacks (Fig. 9.5b). Prior to filling the inlet sacks, place a large white plastic cloth on the ground, prepare the inlet sacks, and have a vial or two on hand in which to place escaping ants. Open the inlet sacks, pour some material onto the cloth, and immediately fill each inlet sack by hand. Hold the inlet sacks over the leaf litter so that escaping animals fall back to the litter. As each sack is filled, occasionally and gently shake the sack to settle the material. Air spaces in the litter may hinder migration from the sack. Because ants crawl to the top of the litter column before falling out, it is most effective to fill each inlet sack as completely as possible such that only the last sack is partially filled, if need be. Ensure that the inlet sacks are kept flat by the stitching.

After the inlet sacks have been filled, hang them inside the Winkler sacks (Fig. 9.5c). This should be done as quickly as possible. The "mini-Winkler" (Fisher 1999a) shown in Figs. 9.4b,c holds one inlet sack; the standard Winkler sack (see Appendix 1) will accommodate up to four sacks. The inlet sacks should not touch the walls of the Winkler sack. Pour the material that remains on the ground cloth into a cup and place in an inlet sack. Next, pour the material that has fallen into the collecting receptacle into an inlet sack. Add the ethanol solution to the cup and reaffix it to the Winkler sack. Finally, tie the top of the Winkler sack with a single and a shoe-string knot to prevent animals from escaping.

The Winkler sack should be allowed to run for at least 24 and preferably 48 hours. Leaf litter from Brazilian Atlantic rainforest that was allowed to process for 1 day collected about 90% of the species and 70% of the individuals that could be extracted from the sample, and in 2 days about 95% of the species and 85% of the individuals were collected (Delabie et al. 2000). The length of time that each Winkler sack can be allowed to process will depend upon the duration of one's stay in the field, the number of samples to be processed, and the number of Winkler sacks available (see Chapter 10). The rate of extraction of ants from litter samples can be increased by removing the litter to a polyethylene bag and shaking it once during every 24 hours of processing. When the litter is shaken gently and returned to the inlet sack, ants that have settled down in the center of the litter are again agitated, begin to move, and eventually fall out. After 4 days, Delabie and do Nascimento found that samples that were agitated once per day yielded 15% more species and 70% more individuals than unagitated samples. On conclusion of the processing period, remove the collecting receptacle and rinse the contents with ethanol into a labeled vial.

The Berlese Funnel

In the Berlese funnel technique, a quantity of leaf litter is placed directly into one or more of the funnels to process. The funnels are then placed under a lamp or in the sun. As the upper portions of the litter column dry, mobile invertebrates are driven down the litter column to the bottom of the funnel and fall into a collecting receptacle (Southwood 1978). Berlese funnels can be purchased (see Appendix 1), fabricated, or modified from other kinds of funnels, as described later in this section.

Materials

Berlese funnels, a quadrat, large plastic sample bags, plastic cups, a ground cloth, supports for the funnels, vials, ethanol, and a light source (optional).

Methods

Litter is collected as described previously for Winkler extraction.

This section describes a simple, portable, and inexpensive version of the Berlese funnel process. The Berlese funnel may be fabricated from 0.7-mm-thick acetate sheeting. The funnel is constructed such that it may be opened and carried flat or rolled up for transport. The general outline of the funnel pattern is shown in

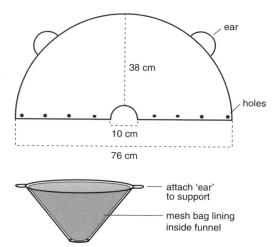

Figure 9.6. The pattern used to create a Berlese funnel (above) and the appearance of the assembled funnel (below).

Fig. 9.6. For setup, the funnel is formed by bringing the two straight edges of the sheet together to overlap slightly, so that the four to five pairs of punched holes are aligned. Paper fasteners are then placed through the holes and spread.

A bag to support the litter in the funnel is fabricated from plastic netting (available at fabric stores) with a mesh size of about 3 mm. The bag is made by attaching the edges of a circular piece of mesh to the upper edge of the funnel with paper clips. The two ears at the outer edge of the funnel are bent and may be used to affix the funnel to a support, such as two chairs or the rungs of a wooden, straight-sided ladder that is supported horizontally by sawhorses or benches (the ladder may hold five or six funnels). Alternatively, a Berlese funnel may be constructed from oil or kitchen funnels, perhaps with some minor modifications (e.g., cutting the tip off the oil funnel).

Before filling the mesh bag in the funnel, place a white plastic cloth or dry cup under the tip of the funnel to catch material that falls through the mesh. As with Winkler extraction,

break up sticks, clods, and decayed logs before placing them into the funnel. After filling, make sure the opening at the funnel tip is not clogged. Next, add material that fell through the mesh to the top of the sample. The litter should be circa 10–12 cm deep in the middle of the funnel. Place a collecting receptacle filled with ethanol under the funnel tip.

If an electric light with a reflector shade (or improvised aluminum foil reflector) is available, suspend the light about 2 cm above the surface of the litter. To reduce fire hazard, do not allow the bulb to touch the litter. A screen may be added to the top of the funnel to keep out flying insects that may be attracted to the light at night. If no electricity is available, place the funnel in bright sun and away from wind and other disturbances. The sample should be allowed to process until the litter is dry or for 2–4 days, depending on the condition of the litter, the temperature, and the humidity. The alcohol should be checked daily and replenished as needed. Do not disturb the funnel to minimize dirt in the sample. After the processing is complete, place the contents of the receptacle into a labeled vial or jar for storage. Add some fresh ethanol to the vial if needed.

Data Output

Both litter techniques produce the following data: richness, composition, relative abundance, and frequency of occurrence among litter samples.

Evaluation

Litter sampling techniques have been relatively little used for ants, and therefore the ants that inhabit litter microhabitats remain largely unknown (Olson 1991; Agosti et al. 1994). Both the Winkler and Berlese methods sample the abundant and diverse leaf litter ant fauna, which is severely undersampled using other methods owing to the ants' cryptic habits and small foraging ranges (Greenslade and Greenslade 1971;

Majer 1996). In ant surveys, Winkler sacks may contribute a relatively large proportion of unique species compared with pitfall traps (Olson 1991). Many individuals and species are often represented in a single litter sample, and Winklers or Berleses are the most efficient way of obtaining extensive samples in litter microhabitats. Nonetheless, larger, active, epigaeic ant species tend to be undersampled because they escape litter samples, and extensive litter sampling may require considerable effort and cost (especially for Winkler extraction).

Delabie and do Nascimento (unpubl. data) have shown that several species of litter ants in Brazilian Atlantic rainforest are undersampled in Berlese funnels relative to Winkler sacks. They attribute these differences to the sifting procedure used with Winkler sacks, which compels the ants to migrate relatively quickly, and to the possibility that some sensitive ants may desiccate and die before leaving the litter in Berlese funnels.

Colony Sampling

Objectives

In colony sampling, ant colonies in a defined area are identified and counted to estimate colony density and monitor changes in populations. Frequently colonies are also mapped so that demographic processes and spatial relationships among colonies (Herbers 1994) and between colonies and environmental features (Crist and Wiens 1996) may be studied. This technique can be used to sample species of ants that nest in an area and are detectable to the investigator in several kinds of habitats.

Materials

A white sorting tray (in litter microhabitats), vials, ethanol, forceps, an aspirator, and materials for mapping (optional).

Methods

In open habitats such as deserts (Schumacher and Whitford 1976; Whitford 1978; Bernstein and Gobbel 1979) some ant colonies may be identified to species by the characteristic structure of the aboveground portion of the nest or by observing worker ants at the nest. Such identification will of course require a knowledge of the local ant fauna prior to using the technique. Nests are identified by searching the sample area, usually along belt (rectangular) transects (e.g., 1 × 50 m [Wisdom and Whitford 1981] or 20 × 1800 m [Johnson 1992]) or by searching a marked study plot using overlapping belt transects (e.g., Chew 1995) and marking every nest encountered within the transects. The width and length of the transects and the size of the plots will depend on the colony densities and the conspicuousness of the nests. Often these studies are limited to common, active, and large-bodied ant species because other species have secretive habits, inconspicuous nest entrances, or both.

In forests with dense leaf litter, most nests are inconspicuous, and the sample area must be searched by excavating the litter and soil. Ants are found nesting in the soil; between rotting leaves; in all sizes of rotting twigs, branches, and logs; in nuts such as acorns; and under rocks. Quantities of litter and soil that are removed from the ground are searched in a white pan to help find the ants (see also the section on intensive sampling). Partial or entire nests may be collected into vials, and the location where the nest was found may be flagged for mapping. This technique is destructive, so temporal patterns may be examined by sampling in nearby plots (e.g., Herbers 1989). A nondestructive alternative is to place baits systematically in plots and follow the trails of ant foragers to locate a subset of the nests in the plot (Herbers 1985; see also the section on intensive sampling). This method is easier to execute but, in some cases, may only sample a subset of the nests because baits are monopolized by nests of dominant species or colonies and because some species are less attracted to the bait than others. Obviously the spatial extent of sampling is limited by the methodology used. For example, Herbers (1989, 1994) studied forest ant communities in 25-m² plots by searching leaf litter and soil for nests. Crist and Wiens (1996) mapped the large and conspicuous colonies of the western harvester ant, *Pogonomyrmex occidentalis,* in shortgrass steppe in areas of up to 146 ha using low-level aerial photography and geographic information system software within which photographs were digitized.

For many studies addressing ecological questions, researchers may consider each nest as an independent sample unit. However, researchers should not confuse nest entrances with nests; a single nest may have several crater- or conelike entrances at variable distances from one another (e.g., *Lasius,* some *Pheidole*). In some population studies, it is important to recognize that the number of nests may overrepresent the number of genetically distinct colonies. This is because some species of ants are polydomous, with individuals of a single colony occupying two or more distinct nests. This challenge may be addressed by transplanting individuals among nests that are suspected to represent a single colony. The introduced workers may be distinguished by dusting them with fluorescent powder (Snyder and Herbers 1991). Agonistic interactions (e.g., threat displays, chasing, biting, swarming) between the transplant and the occupants of the nest indicate that the nests represent distinct colonies. The relationship among aggressiveness, nest relatedness, and polydomy can be quite variable and complex and may require more detailed study (see Banschbach and Herbers 1996).

Data Output

The data produced by colony sampling include richness, composition, colony density, and colony location.

Evaluation

Colony sampling can provide a population-based perspective on ant communities by focusing on the genetically distinct individuals rather than on the foraging components of the colonies. As noted previously, forager-based sampling methods may misrepresent patterns in population structure. Ant colonies provide unique information about the ecological setting because the location of the colony reflects the responses of the foragers as well as the response of the queen, especially through her selection of a nest site (see Johnson 1992). Colony sampling would best serve a single-species study and is essential to develop conservation strategies for endangered ant species. For community studies, researchers should recognize the potential bias against ants that have inconspicuous nests.

Intensive Sampling

Objectives

The primary goal of intensive sampling is to gather data on the total number of ants in an area by searching for and collecting all the ants within fixed plots. This approach allows precise estimation of the number of ant colonies and ant species per unit area, as well as an estimate of the total species richness of the site.

As with some kinds of colony sampling, entire ant colonies may be collected to obtain data on the number of workers, queens, male and female reproductives (alates), and pupae per colony. These "nest series" provide information on the life history characteristics of the ant colonies (e.g., colony size, reproductive status) and are of great use to both taxonomists and ecologists. This method can be used in all habitat types; it is particularly appropriate for structurally complex habitats with abundant leaf litter.

Materials

A plot frame or flagging, a white sorting tray and sifting tray, a bait substance, a shovel and sample bags (for soil samples), vials or twirl bags, ethanol, forceps, and an aspirator.

Methods

Samples are usually taken in 1-m^2 plots set in a linear transect. As in quadrat sampling, the plot can be measured using a meter stick and marked at the corners with small flags, or the plot outline may be delimited with a premeasured 1-m^2 quadrat made of PVC tubing.

Ants are sampled by carefully inspecting all the litter in the plot (see the section on colony sampling). After the litter has been inspected, it is placed in a sifting tray, which consists of a wire mesh screen embedded into the center of a shallow tray and then placed over a deeper tray that measures at least 30 × 30 cm. A metal colander can be used in place of the wire screen. After accumulating some litter in the tray, the worker shakes the tray so that ants and other small animals fall through the screen into the tray below. This additional sifting method ensures that no ants were overlooked while inspecting the litter. Whenever an ant nest is found, the entire nest—including workers, queen(s), alates, and pupae—is collected and preserved in a vial of ethanol. To save time in the field, any ant colonies found in twigs, litter, or logs can be placed into resealable plastic bags and labeled by plot and nest. The ants are later collected from the bags in the laboratory and preserved in one or more vials of ethanol. Ants present in the plot but not associated with a nest are also collected and placed together in a vial labeled "strays." Nest series and strays must be preserved in separate vials so that ants that nest in the plot can be distinguished from ants that may nest outside the plot.

After all litter has been removed, the plot can be baited. Light-colored cookie crumbs are an ideal bait because the crumbs are very visible against the dark leaf litter. As in colony sampling, the cookies, or other bait, are crumbled over the plot to help locate any ant nests in the soil. After 15–60 minutes, the plot is inspected and any ants carrying cookie crumbs are followed to their nest. If a nest is located inside the plot, the nest should be collected, which would involve digging up the nest if it is in the soil. If the nest is located outside the plot, the ants are collected as strays. Some cryptic genera, such as *Basiceros* or *Trachymyrmex,* will feign death if alerted and are nearly invisible until they begin to move again (Romero and Jaffe 1989). If the study plot is undisturbed for 10–15 min, the ants resume activity and may be collected. After the sampling is complete, it is best to return the sorted litter to the plot so that ants and other organisms can recolonize the area.

Several unique ant species may dwell entirely below the soil, and their abundance may vary with soil depth (Harada and Bandeira 1994). Soil-dwelling ants can be censused by retrieving blocks of soil (e.g., 20 × 20 × 30 cm) to sample bags and sectioning the blocks along the vertical axis into subsamples (e.g., 5-cm increments) in order to examine variation in ant composition and abundance with soil depth (Harada and Bandeira 1994). The subsamples may then be searched by hand in a white tray or separated from the soil using flotation techniques (Chapter 11).

Data Output

The data produced by intensive sampling include richness, composition, the abundance of colonies and foragers, and the frequency of occurrence of colonies and foragers in sets of plots.

Evaluation

Unlike the other methods, intensive sampling is able to provide a complete representation of the ant fauna in a sample plot. Colonies within and foragers from outside the plot can be recorded and distinguished from one another. Because the method relies upon visual inspection to record nests, there is a great potential for investigator bias. This method also requires more time than other methods. Although it is very useful for small-scale studies, the extensiveness of sampling may have to be compromised. For faunal surveys, Romero and Jaffe (1989) found that intensive sampling was unlikely to record many species that were not recorded by simpler methods. Some fast-moving foragers may escape the plot while nests are being searched. The destructive nature of the sampling (as with some kinds of colony sampling) may interfere with temporal comparisons.

Direct Sampling

Objectives

Direct or hand sampling involves searching for and collecting ants in different microhabitats within an area. Unlike intensive sampling, in which the objective is to provide a precise estimate of colony or forager density in a relatively small area, direct sampling may be spatially extensive, and the primary goal is to record the number of species inhabiting an area. A minimum of material is needed in this technique, although some experience with ants is required.

Materials

Vials, ethanol, forceps and an aspirator, a white cloth or tray, and a timer.

Methods

Several microhabitats in which different kinds of ants nest and forage should be searched, with searches carried out on bare ground, in leaf litter, on twigs and nuts, under and on shrubs and trees, in epiphytes, at the base and in the roots of grass clumps, under stones, and in decaying logs (especially under bark; see also the sections on intensive sampling and additional techniques for arboreal and herbaceous strata). Twigs, nuts, and logs should be broken open (over a cloth or tray) during the search. A favorite method of Hölldobler and Wilson

(1990) in forests is to clear loose leaves from small plots to expose soil and humus and watch the plot for 30 minutes in order to find small or cryptic ants. When ants are discovered, forceps, an aspirator, or both may be used to collect the ants into vials. As in intensive sampling, nest series should be collected into separate, labeled vials; stray foragers may be collected into a single vial.

As with other techniques, samples taken at different times of the day will include different ant species. To better standardize collecting effort for comparative purposes, the area searched may be delimited and the time taken for the search recorded using a stopwatch. When a colony is discovered and collected (which may take several minutes), the timer should be paused. Searches may be stratified by variables such as habitat or microhabitat type and investigator identity (Longino and Colwell 1997). Hölldobler and Wilson (1990:630) note that an experienced collector can obtain a "virtually complete list of the fauna of a 1-ha site within 1 to 3 days."

Data Output

The data produced by direct sampling include richness, composition, and the frequency of occurrence of species in plots.

Evaluation

Many of the species inhabiting an area can be recorded in relatively little time through direct sampling (Romero and Jaffe 1989). Direct sampling may be spatially extensive, and several microhabitats in an area may be sampled simultaneously. Direct sampling is especially useful for short-term faunal inventory. Abundance, however, is difficult or impossible to record with this technique (although frequency may be used as a surrogate). Considerable expertise is required for this technique to be efficient. Variability in the competence and technique of researchers, as well as differences in habitat structure between areas (Greenslade and Greenslade 1977), reduces the comparability of samples, and direct sampling alone is inappropriate for long-term monitoring. Direct sampling is frequently used as a supplement to other techniques, such as pitfall trapping (e.g., Andersen and Reichel 1994), and may significantly augment the number of species recorded.

Additional Techniques for Arboreal and Herbaceous Strata

Although arboreal and herbaceous microhabitats are used extensively by ants and harbor considerable ant diversity (Wilson 1987), we will not cover in detail the techniques used to sample these strata. For arboreal habitats, insecticidal fogging (Erwin 1983; Majer 1990) or malaise traps (Longino and Colwell 1997) may be used to sample ants high in the forest canopy. To sample ants from the lower parts of trees and herbaceous strata the following techniques may be employed:

1. Sweeping vegetation with insect nets to collect ants (Lynch 1981).
2. Beating vegetation with a stick to dislodge ants onto sheets or trays (Andersen and Yen 1992; Majer and Delabie 1994; Perfecto and Snelling 1995).
3. The use of sticky traps to capture ants on tree trunks or limbs (Majer 1990).
4. Collecting ants by hand, especially by breaking open twigs and branches and searching epiphytes (see the section on direct sampling). Tree-falls after storms offer a good opportunity to collect arboreal species.

Southwood (1978) and Clements (1982) provide some additional discussion of these techniques, and Basset et al. (1997) provide a review of several techniques.

Environmental Covariates

Information on the environment at the localities where ants are collected contributes greatly to the value of any specimen and is necessary for ecological studies. Both regional and local information (see Appendix 2), as well as GPS coordinates, should be recorded whenever possible.

Researchers should consider recording environmental variables that are known to covary with or affect ant distribution and activity (see Appendix 2). These include the following:

1. A habitat classification by vegetation type or dominant plant species, including slope, aspect, and elevation.
2. Information on the type of ant nests.
3. Soil-surface and air temperatures, relative humidity, insolation levels, and wind speed and direction.
4. The percentage ground cover of bare ground, litter, vegetation, rocks, logs, and other potential ant nest sites (measured with a point frame; see Bonham 1989).
5. The depth of the leaf litter, measured with a wire marked off in 0.5-cm units or some other measuring device.
6. Soil type and texture.
7. Vertical vegetation profiles (or foliage height profiles), measured as the number of touches of vegetation on a thin rod at different height intervals above the ground.
8. The amount of overhead canopy cover, estimated using a densitometer or by eye.

Measurements such as canopy cover or vegetation profiles may be made at several points around each ant sampling point. The characteristics of the sample should also be recorded, including the litter volume and density sampled, quadrat sizes, size of the pitfall trap, and bait type. This information will provide a more mechanistic understanding of comparisons that

Table 9.1 Relative Efficacies of Field Techniques Used to Study Ants[a]

Technique	Total Species Richness	Epigaeic Ants	Litter Ants	Forager Abundance	Behavior	Population	Ease	Comparability
Baiting	0	0	0	−	+	−	+	+
Pitfall trapping	0	0	−	0	−	−	+	+
Quadrat sampling	0	+	−	+	+	−	−	0
Litter techniques	0	−	+	−	−	0	0	+
Colony sampling	0	0	0	−	−	+	−	0
Intensive sampling	+	0	+	0	−	0	−	−
Direct sampling	+	+	+	−	−	−	0	−

[a]Entries are based on a subjective 3-point scale: +, good; 0, moderate; −, poor.

Table 9.2 Percentage of Species Recorded Uniquely by Different Techniques in Single Communities[a]

Technique	Percentage Unique	n (area, m^2)	Duration (hours)[b]	Reference
Quadrats	15	12 (0.25)	1–4[c]	Andersen (1991b)
Pitfall traps	30	20	48	
Direct sampling	18.6	1020	nr	Cammell et al. (1996)
Baits	nr	10	na[c]	
Pitfall + direct + baits + Berlese/Winkler	40-82	na	na	Majer (1996)
Pitfall alone	nr	610	168	Majer (1996)
Winkler extraction	46	60 (1.5)	na	Olson (1991)
Pitfall traps	13	40	48	
Beat sheets	38	20		Perfecto and Snelling (1995)
Baits	24	50	na	
Intensive sampling	38	3 (25)	nr	Quiroz-Robledo and Valenzuela-Gonzáles (1995)
Bait traps (in soil)	30	18	72	
Pitfall traps	25	18	72	
Bait traps (on ground)	20	18	72	
Pitfall/bait traps	16	20	48	Romero and Jaffe (1989)
Direct sampling	14	nr	4[c]	
Intensive sampling	3	3 (7.5)	4	

[a]Abbreviations: n, sample size per site or season; na, a measurement is not applicable for a given technique (e.g., area for pitfall sampling); nr, the percentage of unique species was not reported for a particular method.

[b]The duration of sampling events per site or season is given where appropriate.

[c]Sampling duration was divided among different times of day.

are made between the ants of different habitats and ecosystems. Which variables are measured and the degree of detail in the measurements will depend on the importance of those features in the habitat that is sampled. In Appendix 2 we offer a generalized format for a data collection sheet that could be used at each sampling point. This format would be useful when a variety of techniques are used in ant inventories.

Conclusion

Table 9.1 summarizes the utility and disadvantages of the techniques presented in this chapter for several different research foci and practical considerations. Each technique is useful for a particular research priority, and a combination of techniques is usually best. For example, 56% of 75 published studies on ant communities that we reviewed employed two or more methods. Several studies have compared the relative efficacies of different techniques in maximizing the number of unique species records at a site (Table 9.2). Unfortunately these comparisons suffer because (1) the results are confounded when different techniques are spatially separated owing to small-scale changes in ant richness and composition (Andersen and Reichel 1994), and (2) it is difficult to match either the intensity or the extent of sampling in applications of the different techniques, even when the sample sizes are similar (but they are often not similar; see Table 9.2). Conclusive comparisons of the relative efficacies of different methods for recording different kinds of ants await further study.

The use of species-accumulation curves (Chapters 10 and 13) can aid in standardizing comparisons. Nonetheless, it is clear from Table 9.2 and other studies that the use of several techniques adds considerably to the number of species recorded at a site. Baiting, pitfall trapping in open microhabitats, the use of Winkler sacks or Berlese funnels for litter-dwelling ants, and direct sampling are an ideal set of techniques for biodiversity monitoring programs and together form the basis for the ALL Protocol (Chapter 14). This combination of methods will ensure both the comparability of samples and as complete a representation of the ant fauna as can be expected. The success of any sampling program will necessarily depend on the sampling protocol used alongside each method, as well as a careful interpretation of the data that accounts for the limitations of the methodology.

ACKNOWLEDGMENTS

Many of the Ants of the Leaf Litter conference participants shared information that is included in this chapter. In particular, we thank Alan Andersen, Brian Fisher, Jonathan Majer, and Heraldo Vasconcelos. Joan Herbers provided valuable suggestions. Stephanie Bestelmeyer, John Longino, and Scott Miller commented on the manuscript. BTB was supported by a National Science Foundation grant (DEB-95-27111) to John Wiens during the preparation of this chapter.

Chapter 10

Sampling Effort and Choice of Methods

Jacques H. C. Delabie, Brian L. Fisher,
Jonathan D. Majer, and Ian W. Wright

A number of studies have qualitatively assessed the efficacy of diverse methods for sampling rainforest ground-dwelling ants, including pitfall traps (e.g., Adis 1979), Winkler extraction (e.g., Olson 1991), baits (e.g., Fowler 1995), hand collection from quadrats (Room 1975), and "trap-nesting" using artificial nesting sites (Young 1986). However, very few studies have addressed the question of how many samples need to be taken in order to obtain a reasonably complete census of an ant community (e.g., Fisher 1999a). This deficiency of the literature results in part from the fact that the goal of most past ant sampling studies has been to provide a general inventory of the ant fauna of a region (e.g., Wilson 1959; Cover et al. 1990; Veerhagh 1990) rather than a rigorous measure of the underlying

biodiversity in terms of species numbers and relative abundances.

The qualitative approach may well be adequate when the aim is merely to provide a list of the species present. However, if ants are to be used as bioindicators of some aspect of the environment, or if a rigorous census of species is desired, then richness and abundance measures must be described explicitly per unit area or per unit of sampling effort. In such cases, it is necessary to know whether the samples are or are not capturing a reasonably high proportion of the ant species present; if they are not, then it is necessary to estimate what proportion of the total ant fauna is being sampled.

A single sampling method is unlikely to capture all the ants present in the litter or in other parts of the ground stratum. To overcome this problem, researchers often use a combination of complementary sampling procedures. Combinations that have been used to sample rainforest ground-dwelling ants include Winkler extraction of both litter and soil (Belshaw and Bolton 1994a); Winkler extraction plus pitfall traps (Olson 1991); Winkler extraction plus pitfall traps plus hand collecting (Andersen and Majer 1991; Fisher 1996a, 1998, 1999b); pitfall traps plus hand collecting (Jackson 1984); pitfall traps plus baits (Fowler 1995); and Berlese funnels plus baits (Levings 1983). In this chapter we use data from comprehensive studies of the ant communities in Brazilian cocoa plantations to address (1) the optimal combination of sampling methods for maximizing the species count of rainforest ground-dwelling ants; and (2) the relationship between the size of individual litter samples and the number of species obtained by Berlese funnel or Winkler extraction methods.

Methods

Field work was carried out on the grounds of the Centre for Cocoa Research (CEPLAC), Ilheus, Bahia (14°45′S, 39°13′W), Brazil. The CEPLAC site formerly consisted of primary Atlantic rainforest (Mata Atlantica), although most of it is now planted with cocoa. Cocoa plantations provide a habitat that retains many features of the original rainforest, and the ground-dwelling ant fauna retains a high degree of similarity to the fauna of the original habitat (Belshaw and Bolton 1993; Young 1986; Delabie et al. 2000).

Sampling Methods Experiment

A 1-ha plantation of regularly planted 20-year-old cocoa trees, shaded with *Erythrina fusca,* was divided into three rows of six cells, each measuring 23.5 m × 23.5 m. For each sampling method, three sample points were randomly selected in each cell, resulting in a total of 54

samples for each method for the entire plot. Sampling was originally designed to census the ant fauna from the soil, the litter, the ground surface, the tree trunks, and the canopy; the preliminary results have been reported in Delabie et al. (1994). Here we consider only those sampling procedures that are relevant to the soil and litter stratum. The 17 sampling methods used were as follows:

1. Small soil samples. Cubes of soil measuring 15 cm on a side were dug up, broken open, sieved, and then inspected on a white surface so that ants could be manually removed.

2. Large soil samples. As in (1), except that the sides of the cubes were 30 cm across.

3. Berlese funnel samples. Samples of litter measuring 1 m² were collected and placed in a funnel for 24 hours.

4. Winkler extraction samples. As in (3), except that the litter was sieved and then placed in Winkler sacks for 24 hours.

5. Pitfall traps (24-hour). A 75-mm-internal-diameter pitfall trap, containing water and a drop of detergent, was placed out for 24 hours.

6. Pitfall traps (7-day). As in (5), except that the traps contained a mixture of ethanol and glycerol and were placed out for 7 days.

7. Sardine bait (4-hour). Small pieces of sardine were placed on a square of tissue paper, and the ants that were attracted were collected after 4 hours.

8. Sardine bait (24-hour). As in (7), except that the baits were inspected after 24 hours.

9. Meat bait (24-hour). Small pieces of meat were placed on a square of tissue paper, and the ants that were attracted were collected after 24 hours.

10. Cassava flour bait (4-hour). A small pile of coarse cassava flour was placed on a

square of tissue paper, and the ants that were attracted were collected after 4 hours.

11. Cassava flour bait (24-hour). As in (10), except that the baits were inspected after 24 hours.

12. Sugar bait (4-hour). A small amount of dilute sugar solution was placed on a square of tissue paper, and the ants that were attracted were collected after 4 hours.

13. Sugar bait (24-hour). As in (12), except that the baits were inspected after 24 hours.

14. Orange peel bait (4-hour). A small piece of orange peel was placed on a square of tissue paper, and the ants that were attracted were collected after 4 hours.

15. Orange peel bait (24-hour). As in (14), except that the baits were inspected after 24 hours.

16. Dead wood inspection. A rotting branch near the sampling point was cut open and the ants within were collected.

17. Dried cocoa pod inspection. A rotting cocoa pod near the sampling point was cut open and the ants within were collected.

A number of additional sampling methods were used to collect ants from other strata in the plantation. These included chemical knock-down and beating of trees, meat and sardine baiting in trees, inspection of dried cocoa pods on trees, inspection of fallen epiphytic bromeliads, and direct observation over fixed time intervals. The ant collections resulting from these methods are outside the scope of this chapter and will be reported in a subsequent publication.

Extended Sampling Experiment

To investigate the effect of number of samples on the number of species obtained, an extended set of soil and litter samples were collected in a nearby plantation of 60-year-old cocoa. This area differs from the 20-year-old plantation described previously in two important ways:

(1) the cocoa is grown in a less "manicured" way, is shaded by around 15 species of native overstory rainforest trees, and thus constitutes a more structurally diverse habitat than the 20-year-old plantation; and (2) no pesticides have been applied to it for 30 years. In this plantation, 500 Winkler extractions were collected from 1-m^2 samples of litter taken from randomized points in a 0.87-ha area.

Sample Size Experiment, 60-Year-Old Brazilian Cocoa Plantation

To investigate the influence of litter sample size on the number of species obtained, a set of samples was taken from the same 60-year-old plantation described previously, but using different litter sample sizes. Samples of 0.01 m^2, 0.04 m^2, 0.25 m^2, and 1 m^2, each replicated 20 times, were collected and extracted by Berlese funnels over 24 hours. An identical sampling regimen using Winkler extraction methods was also undertaken, except that additional samples of 2 m^2 were also taken.

Data Analysis

All ants from each experiment were initially sorted to the level of morphospecies, then identified to genus and, where possible, to species. For each sampling method, we constructed a matrix of individual ant species by individual samples and determined the frequency of each species in the samples. Standard diversity analyses were then carried out on the resulting data sets. See Chapter 13 and Magurran (1988) for additional background on the statistical analyses described in this section.

The incidence-based coverage estimator (ICE) (Lee and Chao 1994; Chazdon et al. 1998) and the first-order jackknife estimator (Heltshe and Forrester 1983) are nonparametric approaches for estimating species richness in the local community from which the samples were taken, that is, for estimating how many species (including those not collected) are present in the sampled community. Calculation of the ICE is based on the number of species found in ten or fewer sampling units, whereas calculation of the jackknife is based on the observed frequency of unique species. To estimate what proportion of the total species richness was collected by each of the methods employed in the current study, plots of cumulative species-per-sample curves were generated in which species accumulation was plotted as a function of the number of samples taken. Three values were plotted for each succeeding sample: the observed number of species, the ICE of the total number of species present, and the first-order jackknife estimate of the total number of species present. For species-accumulation curves, sample order was randomized 100 times and the means of the ICE and jackknife estimates were computed for each succeeding sample station using the program EstimateS (Colwell 1997; see also Colwell and Coddington 1994; Chazdon et al. 1998).

The asymptote of a species-accumulation curve (i.e., the value that the curve approaches as a limit) is interpreted as the total number of species present at the sample site, including those that were not collected. In the current study, the asymptote of the observed species-accumulation curve was calculated with EstimateS (Colwell 1997) using the two-parameter Michaelis-Menten (M-M) equation (Colwell and Coddington 1994) and the maximum likelihood method of Raaijmakers (1987), which is based on the Eadie-Hofstee transformation of the M-M equation. The observed number of ant species and the proportion of the M-M asymptote represented by this number were evaluated for different sample sizes for each of the sampling methods.

The numbers of species collected by different combinations of the various sampling methods have important implications for future sampling studies. To elucidate the optimal combination of

sampling methods tested in the sampling methods experiment, first a table was prepared listing the total number of species sampled by each method. The method that sampled the most species was identified, and combinations including one and two additional sampling procedures were assessed in order to elucidate the optimal combinations of two and three sampling methods for maximizing the species count.

To evaluate the influence of the size of the litter sample from which ants were collected by Berlese funnel or Winkler extraction, the mean number of individual ants and the number of species per extraction were calculated; the total number of species from 20 samples of a particular area was also calculated.

Results and Discussion

Interpretation of species richness estimates should take into account a number of factors. A species-accumulation curve is specific to the area of the survey, the season or year, and the collecting techniques employed. The use of additional collecting methods, or a survey in a different area or season at the same site, would most certainly collect additional species. The actual number of species in an area at a given time is of course finite, but, in most cases, exhaustive sampling is not physically or logistically possible. If an observed or estimated species-accumulation curve indicates a decrease in the rate of species accumulation across the number of samples collected, then, for the particular methods employed, that number of samples is arguably adequate for estimating the species richness in the area or transect surveyed. Conversely, if the curve continues to rise rapidly for the number of samples collected, then more intensive sampling may be necessary to obtain an adequate measure of the diversity at that site.

The number of samples sufficient for achieving a high level of species completeness is thus practically defined as the point at which the accumulation curve shows an adequate decrease in species accrual. The principal problems with this "sufficient-sampling" definition are the lack of a single asymptote for diverse taxa and the difficulty in quantifying "an adequate decrease in species accrual." One practical solution to the latter problem is to sample until a certain percentage—say 80%—of the estimated species are obtained, based on ICE and jack-knife estimates of the total number of species that occur in the plot or transect. An alternative approach is to continue to sample until additional sampling effort is predicted to achieve a negligible increase in the number of species sampled. In this approach, the increase in species obtained with additional sampling must be weighed against both the cost of sorting and identifying additional specimens and the relative importance of identifying the full complement of the species at the site. Under either approach, species-accumulation curves can be calculated by randomizing sample accumulation order and using asymptotic or nonasymptotic functions (Colwell and Coddington 1994), and the predicted species richness values can be extrapolated using the ICE, jackknife, and M-M estimation techniques.

Sampling Methods Experiment

The 17 sampling methods yielded a total of 134 ant species. In the total of 918 samples (17 methods × 54 samples per method), most of these ground-dwelling ant species are extremely rare, with 43 species collected only once, 18 species collected twice, and 34 species collected in up to 10 samples; only 39 species occurred in 11 or more samples.

The number of species sampled by the 17 different methods is shown in Table 10.1. Winkler extraction sampled the greatest number of species (63), followed by Berlese funnels (48), inspection of dead wood (45), small soil samples (42), and pitfall traps. Pitfall traps run for 7

Table 10.1 Actual Number of Ant Species Sampled by the 17 Methods Described in the Text

Sampling Method	Number of Species	Rank[a]
Winkler extraction samples	63	1
Berlese funnel samples	48	2
Dead wood inspection	45	3
Small soil samples	42	4
Pitfall traps (7-day)	40	5
Pitfall traps (24-hour)	27	6
Large soil samples	26	7
Sardine bait (24-hour)	20	8.5
Orange peel bait (24-hour)	20	8.5
Sardine bait (4-hour)	19	10.5
Orange peel bait (4-hour)	19	10.5
Sugar bait (4-hour)	18	12
Dried cocoa pod inspection	17	13
Cassava flour bait (4-hour)	16	14
Meat bait (24-hour)	15	15
Cassava flour bait (24-hour)	14	16
Sugar bait (24-hour)	11	17

[a]Methods are ranked from 1 (most species sampled) through 17 (fewest species sampled).

days yielded more ant species than those run for 24 hours (40 versus 27 species). Large soil samples yielded only 26 species and dead cocoa pods only 17 species, while the species counts from the baiting methods never exceeded 20. Baiting for longer periods of time did not necessarily yield more species; indeed, with the cassava and the sugar baits, a lower number of species was obtained after the longest baiting period.

The species-accumulation curve plots are shown for the individual methods in Figs. 10.1a–q, and for all methods combined in Fig. 10.1r. These graphs also show jackknife and ICE estimators of species richness based on successively larger numbers of samples. The agreement between the predicted M-M asymptotic values and the actual maximum value encountered in the 54 samples is shown in Table 10.2. This table

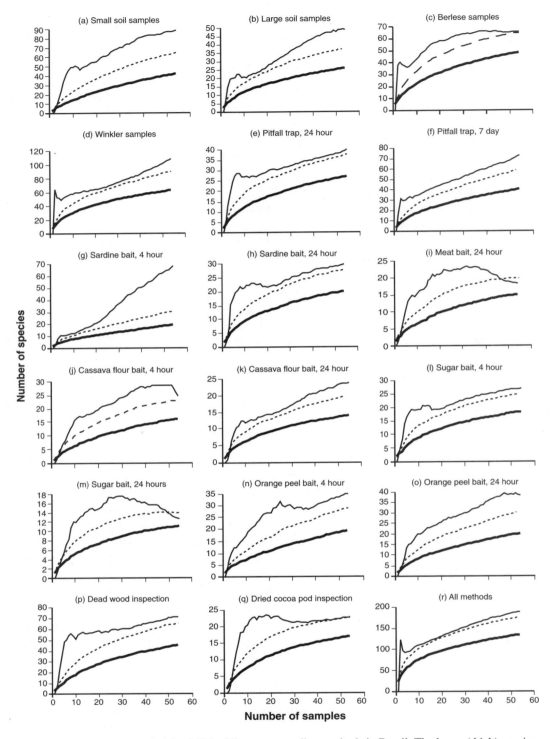

Figure 10.1. Assessment of each of 17 leaf litter ant sampling methods in Brazil. The lower (thick) species-accumulation curve plots the observed number of species as a function of the number of stations sampled. The upper curves display the nonparametric first-order jackknife estimator (dashed) and ICE (solid). The estimated total species richness is based on successively larger numbers of samples from the data set (Heltshe and Forrester 1983; Lee and Chao 1994). Curves are plotted from the means of 100 randomizations of sample accumulation order.

Table 10.2 Observed Number of Ant Species Evaluated at Different Sample Sizes for Each of the 17 Sampling Methods[a]

| Methods | Observed Species Richness after: | | | | | Estimated Species Richness[b] | | |
	10 Samples	20 Samples	30 Samples	40 Samples	All (54) Samples	ICE	Jack-knife	M-M
Small soil samples	14.0 (20.9)	22.4 (33.5)	25.3 (37.9)	29.6 (44.3)	42 (62.8)	88.5	64.6	66.9
Large soil samples	11.9 (39.1)	16.4 (53.9)	20.0 (65.8)	22.7 (74.5)	26 (85.5)	49.0	37.8	30.4
Berlese funnel samples	21.8 (39.4)	31.2 (56.4)	37.8 (68.4)	42.4 (76.7)	48 (86.8)	65.7	64.7	55.3
Winkler extraction samples	29.3 (42.2)	41.1 (59.1)	49.1 (70.6)	55.0 (79.1)	63 (90.6)	108.1	90.5	69.5
Pitfall traps (24-hour)	12.2 (35.9)	17.4 (51.1)	20.8 (61.1)	23.9 (70.2)	27 (79.4)	40.0	37.8	34.0
Pitfall traps (7-day)	17.3 (36.5)	24.5 (51.7)	30.0 (63.2)	34.4 (72.4)	40 (84.3)	72.8	59.6	47.5
Sardine bait (4-hour)	7.3 (34.6)	10.5 (49.4)	13.2 (62.2)	15.8 (74.6)	19 (89.8)	68.5	30.8	21.2
Sardine bait (24-hour)	9.1 (37.3)	13.0 (53.4)	15.8 (64.7)	17.8 (73.2)	20 (82.1)	29.8	27.9	24.4
Cassava flour bait (24-hour)	6.1 (36.4)	8.9 (52.8)	10.8 (64.1)	12.2 (72.7)	14 (83.2)	23.8	19.9	16.8
Cassava flour bait (4-hour)	6.2 (27.5)	9.3 (41.8)	11.8 (52.8)	13.9 (62.2)	16 (71.6)	24.6	22.9	22.4
Meat bait (24-hour)	6.5 (35.0)	9.7 (52.4)	11.8 (64.2)	13.4 (72.6)	15 (81.3)	18.3	19.9	18.4
Sugar bait (4-hour)	8.7 (41.0)	12.0 (56.7)	14.3 (68.0)	16.0 (75.9)	18 (85.4)	27.0	24.9	21.1
Sugar bait (24-hour)	5.2 (39.3)	7.3 (54.9)	8.9 (66.9)	10.0 (75.5)	11 (83.0)	12.7	13.9	13.3
Orange peel bait (4-hour)	7.1 (31.0)	10.8 (47.1)	13.8 (60.4)	16.1 (70.0)	19 (82.9)	34.9	28.8	22.9
Orange peel bait (24-hour)	8.1 (31.8)	12.0 (47.2)	14.8 (58.1)	17.2 (67.5)	20 (78.5)	38.2	30.8	25.5
Dead wood inspection	17.6 (27.9)	27.6 (43.6)	34.1 (54.0)	39.2 (62.0)	45 (71.3)	71.0	65.6	63.1
Dried cocoa pod inspection	7.1 (32.7)	11.2 (51.2)	13.6 (62.4)	15.2 (69.8)	17 (77.9)	22.8	22.9	21.8
All methods	75.0 (55.0)	95.8 (70.3)	110.1 (80.7)	121.5 (89.1)	134 (98.3)	188.9	176.2	136.4

[a]Number of species represents the mean of 100 randomizations of sample pooling order.

[b]ICE, incidence-based coverage estimator; jackknife, first-order jackknife estimator; M-M, Michaelis-Menten asymptote (the percentage of the M-M asymptote is given in parentheses in the first five columns).

also shows the ICE and jackknife asymptote values, as well as the predicted proportions (as percentages) of the M-M asymptotic value that would be obtained if 10, 20, 30, 40, and all 54 samples were taken.

Depending on the sampling method, the percentage of the M-M asymptote value that was obtained varied from 20.9 to 42.2, 33.5 to 59.1, 44.3 to 70.6, 37.9 to 79.1, and 62.8 to 90.6 when 10, 20, 30, 40, and 54 samples were collected, respectively. In all cases, Winkler extraction obtained the highest percentage of the asymptote value and small soil samples the lowest.

There is reasonable agreement between the percentages of the asymptotes obtained for the various sampling methods, which were on average 34.6, 50.4, 61.8, 69.8, and 81.0 for 10, 20, 30, 40, and 54 samples, respectively.

For all methods combined, 70% of the asymptote value was collected by 20 samples, whereas for each individual method, a much lower percentage (33.5–59.1%) was collected by 20 samples, with some methods obviously performing much better than others. Thus, in the context of the total survey, increasing the number of samples from each individual method did not necessarily add new species to the total census, although for particular methods, species that were rarely collected by one method were common in another.

Considering that the study area was sampled by 17 different methods and that a relatively large number of samples was taken, the total species count probably represents a nearly complete census. This conclusion is additionally supported by an asymptote value of 136.4 species (Fig. 10.1r) for the curve for all methods combined, which implies that the combination of 54 samples from all 17 methods obtained 98.3 percent of the asymptote.

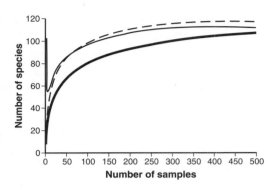

Figure 10.2. Assessment of Winkler sampling methods in Brazil. Lines follow the same convention as in Fig. 10.1.

Extended Sampling Experiment

The species-accumulation curves for the 500 Winkler extraction samples are shown in Fig. 10.2. As above, this graph also shows ICE, jackknife, and M-M estimators of species richness based on successively larger numbers of samples. The agreement between the predicted M-M asymptotic value and the actual maximum value encountered in the 500 samples is shown in Table 10.3. This table also shows the ICE and jackknife asymptotes, as well as the predicted percentages of the M-M asymptotic value that would be obtained if 10, 20, 30, 40, 100, 200, 300, 400, and 500 samples were taken.

If we were to compare the M-M asymptote value of 136.4 species for this extended sampling experiment (carried out in the 60-year-old cocoa plantation) with the asymptote value of 69.5 species obtained for only the Winkler extraction samples from the sampling methods experiment (carried out in the 20-year-old cocoa plantation), we might be led to conclude that ant species richness is higher in the 60-year-old cocoa plantation. In fact, this disparity is probably an artifact of the greater number of Winkler extraction samples that were taken in the extended sampling experiment, in which an average of 67.7 species was collected with 54 pooled samples. This value is quite close to the observed total of 63 species for 54 Winkler sack samples in the sampling methods experiment. When the species-accumulation curve for the 54 Winkler sacks in the sampling methods experiment is extrapolated out to 500 samples using the logarithmic equation of Soberón and Llorente (1993), 109.9 species are predicted. This is also quite close to the 107 species observed in the extended sampling experiment. But perhaps the most telling argument against a significant difference in species richness between the 20- and 60-year-old plantations is the similarity in richness estimates of 134.6 species for the former (based on all sampling methods combined) and 136.4 for the latter (based on the full complement of 500 Winkler extraction samples). It is particularly noteworthy that the first 100 samples yielded over three-quarters of the species that were ultimately sampled by this procedure, and that the last 200 samples yielded very few additional species.

Sample Size Experiment

For both the Berlese funnel and Winkler sack samples, number of individual ants, total number of ant species, and mean number of ant species per sample generally increased with increasing litter sample size. However, although

Table 10.3 Observed Number of Ant Species Evaluated at Different Sample Sizes for the Extended Winkler Sampling Experiment[a]

	Observed Species Richness after Following Number of Samples:								Estimated Species Richness[b]		
10	20	30	40	100	200	300	400	All (500)	ICE	Jackknife	M-M
36.5	50.0	56.3	62.1	79.5	92.7	100.1	104.5	107	112.0	117.0	103.6
(35.2)	(48.2)	(54.3)	(60.0)	(76.7)	(89.5)	(96.6)	(100.8)	(103.3)			

[a]Number of species represents the mean of 100 randomizations of sample pooling order.

[b]ICE, incidence-based coverage estimator; jackknife, first-order jackknife estimator; M-M = Michaelis-Menten asymptote (the percentage of the M-M asymptote is given in parentheses in the first nine columns).

the number of individuals increased quite dramatically with increasing sample size, the increase in mean number of species per sample was far less pronounced. Increasing the sample size from 0.25 m² to 1 m², and certainly from 1 m² to 2 m², is associated with a very limited return for a costly additional investment in effort.

Based on the results of the sample size experiment, when sufficient sampling devices are available it is probably generally more efficient to take a greater number of smaller samples than to take a lesser number of larger samples. Since a single Winkler sack can generally hold the sieved litter from a 1-m² sample, this is probably the most appropriate sample size for most situations.

Complementarity of Sampling Methods

Table 10.4 lists the combinations of two and three sampling methods that produced the largest numbers of ant species in the sampling methods experiment. In each case, only combinations that produced the four highest species counts are shown. Winkler extraction is an element in all combinations, along with inspection of dead wood, small soil samples, pitfall traps (7-day), and Berlese funnels. The various two-

method combinations captured species totals ranging from 59 to 65% of the M-M asymptote calculated for all methods combined (Table 10.2). Species totals for three-method combinations ranged from 73 to 77% of the M-M

Table 10.4 Combinations of Two and Three Sampling Methods That Obtained the Maximum Number of Ant Species in the Sampling Methods Experiment

Combination of Sampling Methods	N[a]
Winkler sack samples + small soil samples + inspection of dead wood	105
Winkler sack samples + inspection of dead wood + pitfall traps (7-day)	104
Winkler sack samples + pitfall traps (7-day) + small soil samples	103
Winkler sack samples + small soil samples + Berlese funnel samples	99
Winkler sack samples + inspection of dead wood	88
Winkler sack samples + small soil samples	87
Winkler sack samples + pitfall traps (7-day)	84
Winkler sack samples + Berlese funnel samples	80
Total number of litter species sampled by all methods	134
Total number of species from all strata	167

[a]N, number of species.

asymptote calculated for all methods combined. A similar exercise (results not shown) was performed with combinations of four sampling methods. Generally speaking, the combinations that maximized species counts are simply permutations of the methods listed in Table 10.4. The combination of methods that produced the maximum species count (117 species) was Winkler sack samples + small soil samples + inspection of dead wood + pitfall traps (7-day). This value of 117 species is 86% of the asymptote value calculated for all methods combined (Table 10.2).

Conclusion

The choices of what collecting methods to use and how many samples to collect are dependent on the intended species completeness of the proposed inventory, that is, on what proportion of the ant fauna the inventory intends to survey. The results presented here demonstrate that Winkler extraction is the most efficient method for surveying leaf litter ants and therefore that this method should be included in all ground-dwelling ant inventory protocols. If a second method is also to be used, we recommend pitfall traps. The total number of samples to collect should be determined both by the alpha diversity of the inventory site and by the level of species completeness that is necessary to achieve the project's goals, but, based on the results reported here, in most situations we recommend taking 20 1-m^2 Winkler samples for areas comparable in size to the 1-ha 20-year-old cocoa plantation.

Chapter 11

Specimen Processing

Building and Curating an Ant Collection

John E. Lattke

The proper preparation of ant collections is just as important as the collection of specimens in the field. Unlike some groups of organisms, such as birds and mammals, which can be identified in the field, ants must be carefully preserved and prepared for identification in the laboratory. The use of good preservation and preparation techniques is critical to collection quality and serves to facilitate the identification of species. Poor collecting and curation practices can substantially diminish the value of research collections.

This chapter outlines the process of assembling an ant collection and maximizing its value. The emphasis is practical, and the chapter focuses on a few issues critical to the maintenance of high collection quality. This discussion is intended to apply to large and small collections alike. General aspects of keeping an insect collection are discussed in a number of texts (e.g., Martin 1977; Borror et al. 1989; Upton 1991), and the reader should consult these as well. We hope our contribution will provide food for thought for practicing ant systematists and ecologists alike. For newcomers to myrmecology, we hope to provide some help in ensuring the value of their contributions to museum collections, and to the study of ant systematics and diversity in general.

Sorting Ant Specimens

The first step in ant specimen preparation is to sort the material collected from the field. After field work, ant specimens are usually contained in bags, vials, and jars and are mixed with soil,

155

bait, other organisms, and miscellaneous organic matter. Separating ant specimens from such a mix can be tedious work, but methods do exist to permit a faster search by eliminating the dirt and mineral matter from the sample. It is important to remove the ant specimens from this other material as soon as possible in order to prevent damage to specimens by abrasive particles and to avoid the formation of coatings of clay and mineral salts on specimens.

During the sorting process it is extremely important that the collection data, particularly the field number, always be kept with the specimens, and that the samples not be pooled before all have been identified. It is best to prepare and place labels in the vials and petri dishes first when working with samples.

Salt Water Extraction

The salt water method of extracting ant specimens from other materials is highly recommended for its simplicity and low cost. The process is simple. Slowly heat water in a beaker on a hot plate, adding generous amounts of salt until the solution becomes saturated and no more salt will dissolve. The water should be hot but not scalding, and never boiling. Empty the sample with ants (e.g., a vial) into a graduated cylinder of no more than 4 cm diameter and drain off the alcohol. Add the saline solution to the sample, cover the top, and slowly turn the cylinder upside down a few times. Dirt and other inorganic material will sink while most organic material, including the ant specimens, will float to the top. Tapping the cylinder will help to dislodge larger items that may be suspended by air bubbles or have adhered to the sides of the cylinder.

Allow 15 seconds for settling, then quickly decant floating matter onto a fine mesh of plastic or a metal strainer and rinse it with alcohol. Repeat the process two or three times, rinsing well with alcohol each time. Take the material from the mesh and place it in a petri dish with

alcohol. Then sort the ant specimens from the other organic material. Sometimes plant parts, such as roots or bits of decayed wood, will still abound, but at least the dirt and sand will be gone.

The material remaining at the bottom of the cylinder should also be checked by rinsing it well in a strainer, as heavier ants and other insects sometimes may sink to the bottom. If alcohol is at a premium, the initial rinsing of specimens may be done with warm water, but alcohol rinsing should be performed before storage of the separated specimens to avoid dilution of the alcohol in the storage vial and the danger of deterioration.

Manual Sorting of Ants from Debris

Either after salt water extraction or directly after the collection of field samples, some manual sorting of ant specimens will be required. Manual sorting is usually performed with the aid of a dissecting or stereoscopic microscope. Initially the field samples (including ant specimens) should be poured into a petri dish and spread out, forming a layer that does not totally obscure the dish bottom. Alcohol can be added to the sample to dilute the mixture. After spreading out the sample, one may choose to wait about 15 minutes for the silt and sediments to settle if the solution is too cloudy. The dish is then inspected by systematically viewing each part under a microscope. A petri dish with a grid greatly facilitates the task; alternatively a piece of paper with a grid marked on it can be taped to the bottom of the dish.

Ant specimens should be manually picked out with forceps and transferred to individual vials of alcohol. Ants should preferably be handled with soft forceps or watchmaker's forceps, taking care not to squeeze them excessively in order to avoid damage. Sorting can also be done with a fine brush.

Separating Ants from Other Arthropods

Since a number of insects and other arthropods mimic ants and may cause confusion during the sorting process, we recommend collecting and storing all arthropods that resemble ants for later verification. Several groups of wingless Hymenoptera (wasps and other groups) may be confused with ants. Identifying these specimens may pose a problem later on in the identification process, as keys to ant subfamilies and genera will not work. *Hymenoptera of the World: An Identification Guide to Families* (Goulet and Huber 1993) is an excellent help in determining whether a specimen is or is not an ant. Other insects or invertebrates found in the samples should also be removed from samples and stored separately for future study by other researchers.

Identifying Ant Specimens to Morphospecies

Since biodiversity data are often analyzed by relying on the presence or absence of species, accurate sorting and identification of ants at the species level is important. Although there have been substantial advances in taxonomic work on ants over the past decade, identification of tropical ant specimens to species is still challenging owing to the lack of identification aids for many genera. Limited availability of relevant literature, lack of recent revisions and expertise in some groups, and the backlog of work for most ant specialists make the recognition of morphospecies by nonspecialists a necessary part of biodiversity studies (see, e.g., Beattie and Oliver 1994).

Identification of ant specimens begins with dividing specimens from each sample into morphospecies. Individuals are grouped into morphospecies according to distinct morphological characteristics without reference to taxonomic classifications.

The challenges involved with species and morphospecies identification vary with the scope of each project. One advantage for many non–ant specialists working with environmental monitoring, conservation evaluation, and ecological research is that they will be dealing with little geographic variability in morphology because of the sampling of relatively small areas. Taxonomists generally must look at material covering the whole geographic ranges of species and genera, and must deal with greater variation. Although Oliver and Beattie (1996a) found little difference between ant morphospecies determined by nontaxonomists and biological species determined by a specialist, their sampling methods tend to underestimate this difference.

Sorting to morphospecies will involve the use of a stereoscopic microscope. A good-quality scope as well as good illumination will help make long hours of staring through the lens more bearable and decrease the chances of error due to eye fatigue.

Initially, ant specimens from each sample should be put into a single vial. Then, depending upon the number of ant specimens in the sample, the sorted material can be divided into separate vials for future work, or it may all be sorted to morphospecies in one session. Initial sorting is usually conducted in petri dishes. Putting material into petri dishes before putting it away into vials permits easier control by the principal researcher.

Although it is possible to identify ant species in alcohol-filled petri dishes, it will generally be best, and it is strongly recommended, to mount a series of three specimens of each of the species in question. Distinguishing characteristics of ant species can best be viewed on dry, mounted specimens, for they are often masked by the alcohol. At least three to ten ants from each morphospecies should be mounted to document a geographic record, depending on how many samples are being collected. If some common species are present and dominant, it is not necessary to mount more then three specimens per sample or survey. It is strongly advised that all the remaining, nonmounted specimens be kept in one vial, in case taxonomic problems emerge later. It is not necessary to keep each sample in a separate vial; all the samples from one survey can be combined into one vial.

Specimen mounting and morphospecies sorting will go hand in hand, and it may be necessary to go through several cycles of separation, mounting, and comparing to make sure that all the material has been accurately sorted. Ideally at the end of the sorting one should have mounted specimens from each species present in the sample, even though the vast majority of specimens will remain in alcohol.

The time and resources invested in such a project will depend upon the scope of the project and its objectives (Chapters 10, 13, and 14). If the study involves the separation and counting of every ant and the mounting and labeling of representatives of each species, a principal researcher and a full-time assistant will need at least a month of work for 50 mini-Winkler samples with abundant and diverse ant material (B. Fisher, pers. comm.). If only species numbers are necessary, two trained part-time students can sort and mount 50 samples within a week (J. Delabie, pers. comm.).

Mounting Ant Specimens

The proper preparation of mounted ant specimens is key to identification. Poorly mounted specimens can rarely be identified because diagnostic characters are frequently obscured by other body parts or by glue. The following protocol describes the preparation of a standard mount that facilitates examination of the specimen and enhances its value. The technique requires practice to accomplish successfully, but the resulting specimens, easy to compare and examine, make the time and effort worthwhile. Direct pinning of ants (putting the insect pin through the ant body) is not usually done,

except for the largest of ants, because of the propensity of the ant cuticle to fracture when penetrated by the pin. Mounted specimens can be stored indefinitely if kept away from moisture, temperature extremes, light, and insect pests.

For small ants, three specimens from the same nest series, or three specimens from the same morphospecies, may be mounted together on the same pin. These three specimens should preferably be of different castes (e.g., worker, soldier, queen). Space must be left on the pin below the specimens for the locality data labels. It is advisable to include a note on the label indicating the origin of the specimen (i.e., nest series or traps).

Learning this technique can be difficult, and it is recommended that trainees first practice with large specimens and then learn to mount progressively smaller ants.

Ant specimens are prepared by mounting them on a paper triangle, generally referred to as a point (Fig. 11.1). Ants are glued to the tip of a small triangle of stiff cardboard or bristol board of neutral pH. Dimensions of the triangle should be no more than 10 mm long and circa 2 mm wide at the base. Points can be easily made using a specialized point puncher (similar to a paper hole puncher) available from entomological supply sources. A water-soluble glue should be used; it should be stored in a sealable petri dish while working to avoid excessive drying.

The selected specimens are taken out of the alcohol and manipulated with fine forceps so the legs are directed ventrally and away from the body. With the specimen's head facing toward the left the ventral areas of the meso- and metacoxa are left relatively exposed so the tip of the triangle can easily touch them. Before gluing they should be put on absorbent paper to dry. Then the series of ants, duly associated with their collection data, may be organized into columns or rows on an index card for expedient, assembly-line processing.

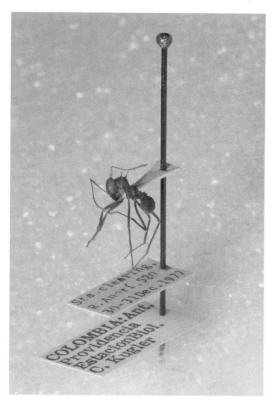

Figure 11.1. An ant mounted on a point and pinned with labels. Photo by Ted Schultz.

care not to bend the triangle or damage the specimen.

An alternative method is to mount ants directly onto a prepinned triangle. Using this method, up to ten ants can be spread with their heads to the right (as seen from the researcher's point of view), and if possible with the legs spread out. A cardboard triangle is first positioned on a pin (Fig. 11.1). Then the pin is stuck into a piece of cork or Styrofoam and placed under the microscope so that the tip of the triangle is clearly in view. A fine drop of glue is then put on the tip of the triangle, just big enough to glue the ant's alitrunk (thorax) on the tip. The ant is picked up with two pairs of forceps, one holding the right front leg and the other holding either the left or right hind leg. The ant is then set on the triangle, so that it is at the very edge of the tip and the petiole is freely visible. Depending on the consistency of the glue, the specimen may be held for few seconds to prevent it from falling off the pin. The legs should be arranged so that the researcher has a full view of the alitrunk outline and the ventral part of the petiole, but care must be taken not to pull off the legs in the process.

The finished product is an ant transversely mounted on the apex of the triangle, with its head pointing to the right when the triangle is pointing away from the researcher. Its head, waist, and gaster project freely and the ventral surfaces are visible; its legs should be directed downward so as not to obscure the rest of the body (Fig. 11.1). If the glue is dry it may be possible to manipulate some parts of the ant's body into a flat plane, although if the specimen itself is totally dry this may break it. Broken parts are preferably glued onto the same triangle or onto a separate triangle on the same pin. Trying to reglue parts together will usually result in the specimen being obscured by a mass of glue.

Dirt on specimens may hinder observation. It can be removed by briefly dipping the specimens into dilute acetic acid or potassium

A tiny amount of glue is placed on another index card, the triangle is grasped by the base with forceps, and a small amount of glue is allowed to adhere to a single side of the apex of the triangle. The smaller the ant, the smaller the amount of glue needed. Alternatively, one may use two types of glue, one thinned with water for use with small specimens and the second full strength for larger specimens. The tip of the triangle is then delicately maneuvered through the upward-pointing legs so as to touch only the meso- and metacoxa and glue the specimen. It will be necessary to mount small ants using a stereoscope. Once the glue is dry the triangle with the ant is picked up with forceps, and an entomological pin is run through its base, taking

hydroxide solutions. This will remove most mineral salts and oily residues, but ants must be thoroughly rinsed in water immediately after exposure to cleaning solutions. Organic solvents (e.g., xylene, toluene, hexane) remove oily residues but are dangerous to inhale or contact. These compounds should be used only in well-ventilated areas or under a laboratory hood while wearing gloves. Strong soap solutions can be useful for oily residues, but once again thorough rinsing afterward is necessary.

Dissecting Ant Specimens

Dissections of ants are sometimes helpful in order to obtain a good understanding of the body parts. When making a dissection, first soak the ant specimen in 10% potassium hydroxide solution at 80°C (for one to several minutes, depending on size and degree of sclerotization, until the ant falls to the bottom of the vial), then rinse several times with water and transfer the ant to 70% ethanol, which prevents it from floating during dissection. After dissection, put the dissected parts into 100% ethanol and, if the parts are very delicate, rinse the ant with xylene, which hardens the cuticle. Finally, dry the parts on a piece of paper towel and mount the parts on a cardboard triangle so that the important structures point toward the viewer. For minute preparations, special tools may be constructed using minutiae (very fine entomological pins used for small insects) mounted on toothpicks or wooden matches; the tips of the minutiae can be slightly bent to form hooks or other useful shapes.

For labial and maxillary palp counts, it is sometimes necessary to dissect the mouth parts out of the head. This is best done with the specimen in ethanol. The mouth parts should be dissected out from the ventral side of the head, with one pair of forceps holding the head and the other gliding with the two arms of the forceps along the side of the buccal cavity (mouth opening) into the head capsule, then holding firm to the mouth part, which will not easily be damaged by pressure as it is torn out. The wet mouth part is then put on a piece of paper towel and allowed to sit until all the ethanol is evaporated. During that time, it is best to manipulate the mouth parts so that all the palps are sticking into the air. This is also the best time to count the segments. The mouth part is then mounted on the same cardboard triangle as the head. The palpal count can then be noted on a colored (preferably green) label in the form "PF = 5,4" (palpal formula = 5 maxillary and 4 labial palps).

Labeling Ant Specimens

Labeling is perhaps the most important part of specimen preparation. Without the pertinent field data presented on labels, biological specimens are worthless. Its label is basically an abstract of the most vital locality data for each ant sample.

Materials

Labels should be written on fine, 100% rag or neutral-pH card stock, similar in gauge to that used for index cards, that will not let ink bleed. Many apparently fine card stocks may be cheaply purchased at stationary and office supply stores, but those not of a neutral pH will start to deteriorate on point-mounted specimens within 40–50 years and become brittle, easily breaking and falling apart upon manipulation of the specimens. Avoidance of such unsuitable stocks is especially critical for collections in humid tropical areas.

India ink is the time-proven standard for writing labels, but excellent labels can also be printed with a laser or dot matrix printer capable of making letters 4 or 5 points in size. Laser-printed labels are apparently safe for point-mounted specimens but should not be used for specimens preserved in alcohol, although dot matrix printer labels are adequate for this purpose. (Nevertheless, each new brand of print cartridge should be tested beforehand.) If labeling is to be done manually, the handwriting must be legible! Accurate locality data can be rendered useless if the information cannot be read.

Label Size

For point-mounted specimens the labels should be approximately 7 mm wide and 15 mm long. The size reflects a compromise between the amount of information to be included and the ability to store and manipulate the specimens, as excessively large labels can easily damage other specimens.

Position and Order of Labels

Ideally a pinned specimen should have no more than two or three labels. For mounted ants, labels are pinned underneath the ant specimens (Fig. 11.1). Labels must all be consistently oriented parallel to the longitudinal axis of the point or pinned specimen in such a way that one does not have to change the direction of the storage drawer with each specimen label to be read. The standard for point-mounted ants is that the label is read with the point directed to the left; for specimens directly on a pin the label is read with the specimen's head directed to the left. When perforating the label with the pin, it should be jabbed close to the right-hand margin for point-mounted specimens and close to the middle for directly pinned specimens. In both cases care should be taken not to obliterate important data with the pin itself.

For specimens preserved in alcohol, labels can be inserted directly into the vials with the ants. Multiple copies of labels in alcohol vials are helpful when additional specimens are to be removed for mounting.

Label Information

The principal label (uppermost on the pin) should have no more than five lines and should

VZLA, Sucre
Las Melenas 9.7 km NW
800 m, 10° 41′N, 62° 37′W
10-V-1993, J. Lattke, leg 4457

Primary Forest
Leaf litter, Winkler

Atta laevigata
det. 20-V1-1993, E. O. Wilson

Figure 11.2. Sample labels (not to scale) for mounted and alcohol specimens.

include the following standard information in this order (Fig. 11.2):

First line: country, state, department or
    province (abbreviated)
Second line: specific locality
Third line: altitude, latitude and longitude
Fourth line: date, collector, collection
    number

Brevity dictates that the country, state, and locality be abbreviated; standard abbreviations that can be easily found in gazetteers should be used. An abbreviation should not be so truncated that someone else will have difficulty in interpreting the name. The locality is the descriptive name of the collection site. A site can also be pinpointed by its direction and distance from a more prominent reference point, such as a large town. In abbreviating the date, the month should always be expressed as a Roman numeral and the year cited in full. The top label in Fig. 11.2 provides an example.

A second label, pinned underneath the principal locality label, may contain ecological information, such as habitat type and microhabitat description (e.g., vegetation, rotting log), and the collection method (Fig. 11.2). Keep in mind that additional information can be accessed from the specimen database or notes through the collection number. Species and morphospecies identifications, the name of the person who identified the species, and the date identified should be presented on the last (second or third) label, since identifications may change but locality and collection data do not (Fig. 11.2).

Associated Data Records

The best and safest storage of information on specimens is in publications based on the specimens and in which they are referred to using unique field data or sample numbers. All collection data should also be entered into a specimen database.

Identifying Ants to Subfamily and Genus

Before ants can be separated into morphospecies, two important steps must be followed. First, knowledge of the important features of an ant's external anatomy must be confirmed before ants can be separated and identified. See Chapters 5 (especially Fig. 5.1) and 12 and also Hölldobler and Wilson (1990) and Bolton (1994) for basic descriptions of ant morphology.

Second, ant specimens, already sorted to morphospecies and mounted, are generally first identified to subfamily and then to genus before they are sorted into morphospecies within each genus (i.e., "morphospecies 1" will become "genus X morphospecies 1"). Identification up to the genus level and the most common morphospecies quickly becomes an easy task. Excellent keys to ant subfamilies and genera of all parts of the world are available and are fairly easy to use. Resources for determining ant genera are listed in Chapter 12.

Separating Morphospecies Using Characters

General recommendations are given in this section to assist nonspecialists in search of differences for species separation as well as for managing specimens and facilitating comparisons. They are all merely suggestions and should not be interpreted as a cookbook recipe, since the criteria for determining a species in one genus may not be valid in another. Criteria for species determination can differ from genus to genus, and the ant taxonomy literature should be consulted (Chapter 12) for clues to the criteria most commonly used, even if the references are not recent or from the same geographical region. Many morphological characteristics, especially in the case of older literature, may not necessarily be correct, but at least the results will be consistent with what is known, and they may eventually be corrected when revisions are carried out.

It is also strongly recommended that each worker keep his or her own notebook where characters used to separate each of the morphospecies are recorded. Ideally the notes should also include sketches of traits that are difficult to describe, such as hair patterns or shapes.

Morphological Differences between Castes

Morphological differences between ant castes (workers, soldiers, queens, females and males), as well as polymorphism within castes, create an additional challenge for ant identification. Collecting entire nest series—containing all sizes of workers, soldiers (if present), females, and males—is the best way to learn about the variation within an ant species and to relate the castes to one another. Especially in local surveys, this allows one to match all the different castes fairly quickly, and it also adds a measure of satisfaction to the collecting process, as an understanding of the biology of the species begins to develop. For many species, one can obtain samples of other castes, such as males and soldiers, by keeping a fraction of the workers and some brood from a nest in an artificial colony in the laboratory.

Specimens collected in pitfall traps or Winkler samples are often difficult to associate. In some cases it is relatively easy to match workers with soldiers and queens, but sometimes the castes look totally different. In general, the following guidelines apply:

Queens generally will resemble the workers, especially the majors, to an extent that pairing them from a non-nest sample is usually not problematic. Typically one finds that a queen has larger compound eyes, ocelli, a larger mesosoma with more segments and sutures as well as wings or wing stumps, and usually a larger gaster than the workers. Differences in sculpturing are usually not very great.

Males are wasplike and dissimilar from their female counterparts, so they are usually difficult to match with their conspecifics when taken disassociated from their nestmates. Normally males have much larger eyes, a short antennal scape, a small head relative to the mesosoma, and an elongate gaster (often with the genitalia protuding from the apex). It is often impossible to identify the males even to genus level. Currently there are no good keys to ant males.

In the case of *worker* polymorphism it is usually the major (larger) workers that furnish the most reliable characteristics for species separation, because the minor castes of some species from the same genus may present negligible differences among themselves. Workers from incipient (newly formed) nests on average are smaller than those from mature nests and have lighter coloration. Table 11.1 lists the genera in which polymorphism is present in the workers.

Table 11.1 Ant Genera with at Least One Species in Which the Worker Caste Is Divided into Physical Subcastes[a]

Ponerinae: *Megaponera*
Myrmeciinae: *Myrmecia*
Dorylinae: *Dorylus, Eciton*
Ecitoninae: *Cheliomyrmex, Labidus, Nomamyrmex*
Pseudomyrmecinae: *Tetraponera*
Myrmicinae: *Acanthomyrmex, Acromyrmex, Adlerzia, Anisopheidole, Atta, Cephalotes, Crematogaster, Daceton, Machomyrma, Messor, Monomorium, Oligomyrmex, Orectognathus, Pheidole, Pheidologeton, Pogonomyrmex, Solenopsis, Strumigenys, Zacryptocerus*
Aneuretinae: *Aneuretus*
Dolichoderinae: *Azteca, Iridomyrmex, Liometopum, Tapinoma*
Formicinae: *Camponotus, Cataglyphis, Euprenolepis, Formica, Gesomyrmex, Melophorus, Myrmecocystus, Myrmecorhynchus, Notostigma, Oecophylla, Proformica, Pseudhomomyrmex, Pseudolasius*

[a]From Hölldobler and Wilson (1990:318).

Choosing Characters

A quick overall look at an ant's body will usually permit preliminary separation of specimens using obvious traits, such as size, color, presence or absence of denticles, structure of the petiole and postpetiole, and odd mandibular shapes. This step permits the division of large samples of specimens into smaller, more manageable, lots. One can expect to find some variation in almost any trait, and trying to assess the limits of infraspecific variation is the crucial task for fine sorting. Color can be quite unreliable, so it should always be used in combination with other characteristics and not by itself. The final sorting calls for attention to finer anatomical details that will mean looking at more restricted areas of each specimen.

Rarely will one find a morphospecies that can be distinguished on the basis of just one outstanding character; a combination of three or more is usually needed. To help keep track of the morphospecies one may write down diagnostic characters or illustrate them on an index card. It is generally a good idea to keep a running list of the characteristics used to separate morphospecies so that specimens from different samples can be compared and grouped.

Study each body part from different angles in an effort to detect useful characters, such as shape, projection, excavation, pilosity, sculpturing, sutures, and sulci. To have a clearer image of any general body shape, it helps to use background lighting, so that only a silhouette is seen. Dirty specimens may present totally distorted silhouettes and sculpturing patterns—a common source of error. Cleaning may uproot pilosity, and care should be taken with the use of hairs as characters in such a case. Pilosity may differ between species, but more delicate hairs may be easily abraded, reducing their diagnostic value in some cases. When studying pilosity it helps to distinguish between the very short, fine hairs that form a base pilosity and the longer emergent hairs; their angle of inclination (using the cuticular surface as a reference) may be characteristic.

Some sculpturing may be difficult to assess because of a shining surface that reflects too much light. A strip of Mylar or chalk paper placed very close to the specimen between the light source and the ant will reduce glare and permit distinction of details of cuticular sculpturing. When working with lengths and widths, a comparison between the dimensions of one body part and another is helpful; this is why indexes are often used in keys and descriptions. *A Glossary of Surface Sculpturing* (Harris 1979) is a good introduction to the various types of sculptures and their terminology.

Specific Characters

Figure 5.1 provides a general diagram of the body parts of an ant.

HEAD. The shape of the head itself may be distinct: Is it wider anterad than posterad? Are the compound eyes part of its silhouette or not? Look for the presence of distinctive sculpturing that affects the cephalic shape, such as spines or crests. Mandibles are frequently useful for specimen determination. The amount and disposition of dentition should be noted, but dental abrasion does occur in older specimens. The clypeus frequently presents useful characters, and studying differences such as the shape of its anterior margin and sculpturing is recommended. How far back does it extend between the antennal lobes? Antennal scapes can differ in their relative length to the head; see how far back they extend beyond the posterior cephalic border. This distance can be quite obvious in some specimens, or one may gauge the distance in apical widths of the scape itself in close calls. The frontal carinae may have a distinct shape, especially the external margins; how far back do they prolong themselves? The eyes may differ in shape, size, and position on the head. Their presence or absence may vary within a species, such as in the army ants (ecitonines), where the minors are eyeless but the majors may retain a single-faceted eye. The ventral side of the head may reveal differences in the hypostomal teeth. The occipital corners frequently have distinct dentition, lobes, or other sculpturing of use.

MESOSOMA OR ALITRUNK. Each of the four parts (pro-, meso-, and metanotum and propodeum) that make up the mesosoma may have differences in sculpturing from the others even though impressed lines do not clearly separate them. In a dorsal view take note of the presence or absence as well as the development of the promesonotal suture and metanotal sulcus. The shape of the mesosoma in lateral view is a useful indicator of differences; note convexities and angles between different parts. The pronotum may present angles or denticles along its ventral margin that may not at first be noticed. The anteroventral edge of the mesopleura is frequently bordered by a carina that may present differences in height and width. How are the propodeal spiracle openings oriented? What is their diameter or position on the body? A frequently overlooked spot is the declivitous propodeal face when it is surrounded by teeth. The legs may differ by the amount of pilosity or in the ratio of their length to width, and they may sometimes have characteristic spines on the tibia or tarsi.

PETIOLE AND POSTPETIOLE. The petiole will vary in the presence or absence of a peduncle and in its general shape. The length-width ratio of the petiolar node in a dorsal view may be helpful. The subpetiolar process—a lobe or denticle that may or may not be present on the anteroventral petiolar border, and that can be shaped in various ways—is frequently hidden from view by the legs.

GASTER. Differences in shape, especially in lateral view, may be quite distinctive, as may sculpturing. The ventral area close to the union with the postpetiole, as well as the rest of the first gastric sternum, may have useful characters that are difficult to observe owing to the closeness of the postpetiole (or more usually because of sloppy mounting). Attention should be paid to differences in sculpture between the basal and apical parts of the same tergite.

Identifying Ants to Species

Identification of ant specimens to species, beyond morphospecies, is a science in itself, one that may take up to 10 years of study and practice to master. Fortunately, many studies of ant diversity can be carried out without species identification (Chapter 7). If one's study is limited to a geographically restricted area, identification to morphospecies, for which the genus name is secured, should suffice. For analyses over larger areas, species identification is important in order to make comparisons.

Without accurate species identifications, one would have literally to compare all the morphospecies from multiple research groups.

However, identifying ants to the species level will greatly enhance any ecological study by linking the species richness and diversity data to biological data on the species of interest. Species names are like gateways to the enormous amount of knowledge accumulated in publications over the years. In many cases, a species name allows one to associate the specimen and diversity data with information on the biology, ecology, or distribution of the species. It has recently been argued that without ecological and biological information about the species under study, interpretation of species diversity data is incomplete and may even be misleading (Lawton et al. 1998; Goldstein 1999). For example, several studies have found that ant species richness often increases with increasing levels of habitat disturbance but that the composition of the ant species changes (Chapter 7). Without knowledge of the biology of the species sampled, these changes in species composition cannot be interpreted.

A species name also allows one to search digital and printed databases, such as Formis (Porter 1999), with over 20,000 bibliographic records, or the social insects Web site (Chapter 1). This information enables further analysis and interpretation of the species data.

Before diving into species identification, one should be certain to have familiarized him- or herself with the basic morphology of ants (Chapter 5; Hölldobler and Wilson 1990; Bolton 1995b). Those who may still feel insecure should begin by choosing some of the larger ants from various groups and preparing drawings to compare various parts.

Species identifications are facilitated by first preparing a list of species known to occur in the area of interest. This can be done using the catalogue of the ants of the world (Bolton 1995b), which allows one to search for the type localities by country, through lists of local faunas, or by visiting local or regional collections and going through their records. This quickly becomes an enormous task, but the more data that are compiled and shared, the easier it will be for the next generation of inventories to proceed. Local lists can be found on (and should be submitted to) the social insects Web site (http://research.amnh.org/entomology/social_insects/).

Assistance from Ant Taxonomists

Relationships developed with a major ant collection or with an ant taxonomist will undoubtedly make identification of ant specimens a much easier task. It is inevitable that a researcher will eventually accumulate specimens that should be looked at by an ant taxonomist in order to clarify dubious identifications. Before proceeding to send a large number of ants to a specialist, one should first contact that person and ascertain if he or she has the time to examine them. Chapter 12 lists institutions with practicing ant taxonomists. Given that taxonomists are usually inundated with material, one will increase the chances of cooperation if the sample size is reasonable, the deadline for the determinations is not too tight, and duplicates are supplied, enabling the specialist to deposit them in his or her own institutional collection (Table 11.2). It is always a plus when the specimens happen to belong to the taxonomist's particular group(s) of interest.

It is best to first develop a synoptic collection of ants from the study before sending any specimens to a specialist for identification. The specimens should be sorted to morphospecies, and then only a few specimens of each morphospecies should be sent. If one has little prior understanding of ants, it might be wise first to send specimens of the ten most common

Table 11.2 Rules for Submitting Specimens for Identification[a]

1. Do not assume that specialists desire more specimens for their own sake; usually they are interested only in establishing new records or in additional species. The specialists are spending valuable time to make identifications. At consultant wages, this could be expensive. Therefore a request for identification is equivalent to asking for a substantial donation of time, either from the individuals involved or from their employing institution.
2. Never ship specimens for identification without making prior, detailed arrangements with the specialist, including:
   a. Return shipping costs.
   b. Time frame within which specimens are to be returned.
   c. Number and identity of specimens to be retained by the specialist.
   d. Where type specimens will be deposited (if applicable).
3. All specimens submitted for identification must be:
   a. Properly prepared and preserved.
   b. Provided with exact locality data.
   c. Sorted to genus if possible.
4. Never send bulk collections or unsorted collections with a request that the material be picked over to find things of interest.
5. Remember that a refusal to make identifications of large masses of material does not necessarily indicate a lack of interest, but sometimes merely a lack of time or facilities.
6. Under certain circumstances, a specialist may request a fee for providing an identification. This practice has become nearly universal for court cases, commercial activities such as pest control operations, or environmental impact studies. Make sure that both you and the specialist agree on a fee, if any, before the identifications are made.

[a]From Arnett and Samuelson (1986).

morphospecies from the study. This approach will not overwhelm the specialist, and it will also allow time to adjust the method of preparation to meet the requirements of the specialist. This procedure will result in early identifications for the most common ants. It also shows that one is willing to spend some time sorting through the material and is willing to invest one's own scarce time in identification as well. Most taxonomists will not touch material lacking collection data and will likewise keep well away from sloppily mounted specimens.

Instead of sending specimens to an ant specialist, one may consider visiting a major ant collection to try one's own hand at identification. Proper preparation for such a visit should include preparation of and familiarity with a reference collection of dry-mounted specimens in one's collection. Make a list of priorities based on the most common ants and chose the 20–50 most common morphospecies in the research program. Take at least three pins of each morphospecies. Be sure to bring along a few extra pins of each species (properly labeled) that can be left for the specialist's ant collection.

Identifications are most easily carried out with the ant specialist guiding one through the key characteristics of each species. If the specialist does not have much time, one can also compare one's specimens to those in the specialist's collection. Once a verified identification has been made, another label should be added to the pin, with the species name, the name of the person who identified the specimen, and the date it was identified. If one compares one's specimens with the type specimen (the specimen used for originally naming the species), a colored label noting "Compared with type, [your name, date]" should be added. Specimens compared to a type specimen will be the most important future reference specimens in one's growing collection.

In looking for reference publications, it might be best to prioritize by first choosing keys to general or regional faunas, then revisions of keys, then faunal lists, and finally the original species descriptions (Chapter 12). Growing knowledge of the literature will familiarize one with the specialists in a particular group, with whom one might begin a fruitful relationship (Chapters 5 and 12, and the social insects Web site).

Building up a Morphospecies Reference Collection

Specimens of the identified morphospecies from each sample should be mounted and labeled in order to build up a reference collection, as discussed previously. This will permit comparison of the species collected from various sites and samples. Simply relying on an index card or fiche with some characters or illustrations as the only reference to a particular morphospecies will not do, since the difficulties that arise as new species are identified are best solved by keeping mounted samples of each morphospecies for comparison. The arrangement of a reference collection should best suit the needs of the particular project. In some cases it may be preferable to maintain reference collections separated by study site or project, thus reducing the amount of morphological variability owing to sampling from widely separated populations and facilitating comparisons.

Regardless of how collections are separated or joined, each one should have the ants arranged at least according to subfamily, genus, and species. The simplest and least complicated arrangement is alphabetical order starting with each subfamily, then putting all of its genera into alphabetical order, and then putting the species within each genus into alphabetical order. When only morphospecies are known, they are placed after the determined species of that genus and deposited in numerical order. This strategy permits easy retrieval and depositing of specimens by nonspecialists.

Eventually some groups of ants may grow in size to the point that trying to match so many morphospecies will become quite difficult. At this point one can attempt to divide each genus into groups of species that are similar in one or more diagnostic characters.

A fiche system can be grouped into nested sets of fiches: a set that guides to the subgroups of one genus and a set that guides to the species belonging to each subgroup. Alternatively a computerized database can be implemented, functioning as an electronic key.

Building up a Proper Ant Collection

The development of proper taxonomic collections is beyond the scope of this book—and probably also beyond the interest and capacity of individual researchers and projects. The challenges are huge, especially in guaranteeing long-term funding and stable conditions for maintenance and curation beyond the individual researcher's working life.

However, the fundamental link between research collections and the results of inventory and biodiversity assessment studies lies in the deposition of voucher specimens in major ant collections, and in the integration of research collections at the end of a project or a scientist's career.

Though it is tempting to maintain an ever-growing ant collection, a few points should be considered before proceeding with such a strategy. The fact that one's investment, in both time and money, grows along with the collection has already been mentioned. With time, and a greater number of successful projects, more and more species in the collection will be described. In well-studied areas, a collection will therefore come to include few new species, whereas in such largely unstudied areas as Madagascar, up to 95% of all the species might be new to science. Following the laws of most countries, at least the primary type specimens must be deposited in the country of origin.

Holotypes and, to a lesser degree, paratypes (specimens used to describe the species) play a key role in taxonomy, as they are the final authoritative reference for any species name. Thus they should be well preserved, accessible to scientists at any time, and, if possible, available for loan. It is a wonderful feeling to discover a new species, but housing the type specimens for it is a great responsibility.

It is therefore strongly recommended that the researcher establish a strong relationship with one or more of the major ant collections in his or her region (Chapter 12) before undertaking an ant diversity study. There are benefits to both partners: researchers gain access to new specimens, often in large numbers, which are often very rare in major collections and thus of high value to the collections. Researchers who deposit voucher specimens actively contribute to the growth of these institutions. Major collections usually have staff capable of identifying specimens for outside researchers, and they are often able to exchange specimens for other species of the region, or closely related species.

Specimen Storage

Since only a sample of ants will be mounted, an additional group of specimens in alcohol will accompany the collection of mounted specimens. Specimens in alcohol should preferably be kept in an area separate from the dry collection.

Field numbers should readily be visible in alcohol vials and on mounted specimens so that the retrieval of a particular sample is relatively easy. Specimens may be ordered according to the field number for each sample, since each is distinct and permits the grouping of ants from the same area. The alcohol collection can also complement the dry collection when more specimens must be consulted for identification or

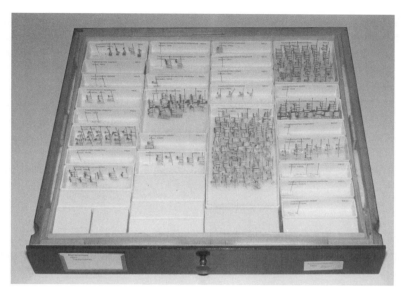

Figure 11.3. Mounted ants arranged in unit trays in a collection drawer. Photo by Ted Schultz.

new analysis, or if specimens are to be deposited in other collections.

The alcohol used for storing ants is ethanol at a concentration of 90% (the concentration should never drop below 70%; keep in mind that adding specimens will lower the concentration). Cheaper alcohol such as isopropyl alcohol will adversely affect the specimens. DNA analysis normally works best with specimens preserved in an ethanol concentration that is as high as possible. Thus preserving specimens in originally 95% pure ethanol is best. However, too high a concentration often stiffens specimens, making it more difficult to mount them properly later.

Vials for reliable long-term storage may be hard to come by in some countries. Vials designed for other purposes (e.g., centrifuge tubes) can be used, although the quality of the glass and lids is generally lower. Stoppers may deteriorate with time and need replacing. Cotton should never be used with specimens as the fibers entangle them, making observation difficult.

An alcohol collection should be checked at least once a year. If vials are almost empty, do not just add new ethanol. Rinse the vials several times with ethanol and then add new ethanol. The risk of desiccation of individual vials can be reduced by storing the vials in a larger jar that is also filled with ethanol. In case of complete desiccation of specimens, do not move them unnecessarily. Open the vial and put it into a wet chamber. This is normally made of a tightly closed box that has sterilized quartz sand on the bottom, which is kept humid with distilled water to which fungicide has been added. The specimens are kept there overnight; afterwards they are relaxed and perfect for handling. Two other techniques can also be used: either soak the specimens in hot water for few minutes or soak them in a solution of ammonia for a few hours.

Dry specimens should be stored in drawers that are as airtight as possible. If drawers are to be made locally, be very precise about the measurements in order to preserve uniformity in the storage system and allow drawers to fit interchangeably into all storage cabinets. Unit trays are very strongly recommended, as they allow the movement of ants around in trays without moving individual pins (Fig. 11.3). Unit trays are commonly made out of cardboard with a foam bottom. They come in various sizes and are generally arranged together so that they add up to the size of the collection trays. Drawers and unit trays can be easily custom made or purchased from an entomological supply company (see Appendix 3).

The bottom of unit trays was originally made out of pressed peat, but today polyethylene or other foam is used. It is important that the foam not retain pin holes but be self-healing after pins are removed. This detail can save quite some time over the long term.

Pins should be of stainless steel, especially when they are to be used in humid tropical areas. Size 3 pins are preferred by many workers, as they are sturdy enough not to bend when replaced, although this is to some extent a question of taste and availability.

Finally, dry collections are best kept in rooms with climate control to protect against excessive heat, cold, and humidity. Excessive heat in combination with humidity makes specimens prime targets for fungi, insects, and mites; common pests include museum beetles (*Anthrena* spp.), silverfish, and house lice. Damage can be prevented by adding some insecticide (e.g., camphor) to each box and especially by checking the collection at regular intervals, perhaps two or three times a year. Infested boxes should be taken away to be treated, making the process less of a health risk for workers in the laboratory, if the collection is housed in the same rooms as the laboratory.

Specimen Shipping

To ship specimens to specialists or collections, use sturdy cardboard boxes padded with Styrofoam chips to a thickness of at least 12 cm around the enclosed insect box. Use wide sticky tape to seal off all possible entrances to the box and cut off access by unwanted visitors, such as other insects, during the trip. If possible ship by an express courier (e.g., Federal Express, UPS, DHL), and do not forget the proper labels, including the return address. All appropriate customs forms should accompany international shipments to avoid delays at customs.

Vials containing specimens should be tightly bound together and sealed in a plastic bag to prevent spillage of ethanol if a vial should break or leak. Dry-mounted ant specimens should be firmly pinned in unit trays that are then secured within a cardboard box. Unit trays should be braced so that they do not move around in the box. Pin the ants as far down into the tray bottoms as possible with the help of a pair of strong forceps. If labels or triangles with specimens mounted on them are loose, either use thicker pins (size 3 pins usually prevent this problem) or (and for larger specimens this should be done as well) brace the specimen with a pin on each side of the cardboard triangle just in front of the specimen, so that it cannot rotate and thus fall off or destroy neighboring specimens.

Databases

The advent of personal computers, laser printers, and the Internet has made building a collection and managing its related data much simpler than in decades past. Once a new record has been entered into a database, it is easy to print labels, to search for specimens according to a particular locality or research question, to add images or drawings, to print distribution maps, or even to perform the daily bookkeeping of the collection. Maintaining an updated database also facilitates data analysis and allows one to check on questionable identifications or data easily.

Databases are available in many different formats. Depending on the size of the project, a simple spreadsheet program such as Excel may be sufficient. Such a program allows for the input, sorting, and export of data and the printing of labels. More advanced database programs normally require extra effort to learn and customize. BIOTA and BioLink are two that have been specifically developed for the purpose of specimen and species data handling and might well serve the needs of most of the readers of this volume.

With a little patience and practice, the art of preparing and identifying and specimens can be mastered by anyone. A well-labeled reference collection serves as a solid baseline for ant studies and allows for comparisons to other sites and studies. By following the steps outlined in this chapter, the researcher can sort, mount, and identify ants relatively easily and rapidly.

Chapter 12

Major Regional and Type Collections of Ants (Formicidae) of the World and Sources for the Identification of Ant Species

C. Roberto F. Brandão

This chapter lists the main depositories of formicid type specimens, provides a database of regional collections of ants, and comments on the main resources for identifying ant genera and species. One of the reasons for studying and cataloguing diversity is that it is the major prerequisite for understanding how biological systems work (May 1990). Catalogues are unquestionably indispensable to the support of systematics and other biological research.

The synoptic classification of ants by Bolton (1995a) is based on his new catalogue of the world's fauna (Bolton 1995b). Deleting taxa known to be currently in press but not published by the limit date of 31 December 1993, Bolton (1995a) accepts as valid 16 subfamilies, 59 tribes, and 296 genera, and he lists 9538 described extant species. Bolton (1995b) also lists

the suprageneric, genus- and species-rank taxa known only from fossils.

The world catalogue brings together all names published thus far for the extant and fossil ant fauna, listing every publication on ant taxonomy and related fields of study from 1758 to the present. The estimated size of the world ant fauna, however—including those still to be described—elevates this number to a total of 20,000 species, or possibly more (Hölldobler and Wilson 1990).

Thus at least half of the world's ant fauna remains to be described, and it will mostly be found in the tropical part of the world, as one of the few relatively indisputable generalizations in community ecology is that a latitudinal gradient in biological species richness and diversity exists from the temperate regions to the tropics

(Kusnezov 1957; Jeanne 1979; Gadagkar et al. 1990; Chapter 8).

The recent classifications of the Formicidae into subfamilies (Baroni Urbani et al. 1992; Grimaldi et al. 1997) provide a fresh framework for the much-anticipated reappraisal of several taxa in different subfamilies. In the past, species and genera were added to the subfamilies with little concern for the overall effects on other genera. At the tribe rank, classification is chaotic, having recently been challenged in the Dolichoderinae and Formicinae. In the Dolichoderinae, even the recognition of the traditional tribes has not proven useful in classifying these ants (Shattuck 1994). The tribal organization of the Myrmicinae is in complete disarray and awaits a complete overhaul. This situation can be extended to other taxonomic levels, as several taxonomic changes in the past resulted in the formation of paraphyletic genera, the formation of several unsatisfactory monotypic genera, and the instability of generic and subgeneric concepts. Thus the generic definition within subfamilies is frequently obscure (Bolton 1995b), and the same name may represent entirely different concepts to different authors.

The number of generic revisions, monographs, and faunistic studies has increased steadily during the 1980s. Yet, as Hölldobler and Wilson (1990:21) state, "like a mosaic lacking just enough pieces so that the pattern remains obscure, the classification of the world fauna still lacks satisfying coherence and practical utility."

Regional surveys of local ant faunas often include keys for the identification of species within the country or territory in question. Bolton (1995b:2) does not recommend the use of faunal surveys antedating 1950, as they are "clogged with unavailable names and infraspecific taxa and are of limited use by modern standards."

Ward et al. (1996) published a comprehensive bibliography of ant systematics, including references that treat the taxonomy, evolution, and comparative biology of ants and also those dealing with morphology, genetics, physiology, biochemistry, social behavior, identification, phylogeny, biogeography, ecology, and faunistics, from Linnaeus (1758) to 1995. They list 8185 entries, of which 8109 are literature records and the remainder cross references. Where there is a discrepancy between the publication dates in Bolton (1995b) and Ward et al. (1996), the latter date may be considered definitive.

Published Resources

Bolton's (1994) identification guide includes a glossary of external morphological terms (including those specific for use in ant taxonomy), full diagnoses of workers of the 16 recognized extant subfamilies, an illustrated guide and updated keys for the identification of ant subfamilies and genera, and relevant literature for most genera. This guide presents scanning electron micrographs of both a full-face view and a body profile for the worker caste of nearly every described genus, being the first visual atlas of the Formicidae. The work will make it possible to separate and name ant genera without too much trouble for years to come, provided the reader is aware of several problems in the generic classification outlined by the author. In particular, the ground-dwelling ant fauna may reveal new taxa, and the use of existing keys, designed for described taxa, may not help to disclose the new ones. In addition, an existing genus may not have been well defined, or it may be necessary to restudy its limits.

Having assembled this information, Bolton (1994) discusses the need for a detailed taxonomic study of fossil forms embedded in amber and rock. He also comments on problems of classification, even at the level of subfamily. A good proportion of genera are now represented by monophyletic units, although there are still some unrecognized synonymies, genera that may be para- or polyphyletic, or genera that are poorly characterized or that are still defined by different concepts in different zoogeographical regions.

Very user-friendly keys to ant subfamilies and genera are presented in Hölldobler and Wilson (1990). These keys are accompanied by drawings of each genus. A comprehensive list of identification guides for each genus is presented in Table 5.1.

For the New World, MacKay and Vinson (1989) produced a guide to the species identification of the ants, commenting on the power and validity of species identification keys for most genera occurring in the Americas. This and Creighton's (1950) review of the ants of North America, along with information found in Kempf's (1972) catalogue and the Brandão addendum of 1991, may aid in locating the sources to be used for the study of the New World Formicidae. (See Chapter 5 for more resources.)

Where the ground-dwelling ant fauna is concerned, some special difficulties may arise. The study of this seldom-collected segment of the fauna may reveal new taxa or indicate the need for better definitions of others.

Electronic Resources

A new and important source of information regarding ants is the growing number of Internet World Wide Web pages. In particular, the social insects Web site, http://research. amnh.org/entomology/social_insects/, included, as of late January 2000, the following information: phylogeny; on-line catalogues (Neotropical Ponerinae and Australian ants); Torre Bueno's and Harris's glossaries of terms relating to surface sculpture; visual guides to ant morphology and anatomy; inventories (Central Park [New York City], Saint Louis region, Michigan state, major habitats in southern Bahia [Brazil], *Acacia* ants, Costa Rica, Arabian Peninsula, and Lakekamu [New Guinea]); collection invento-

ries (American Museum of Natural History [AMNH], School of Environmental Biology, Curtin University of Technology [CURT], and Fundación e Instituto Miguel Lillo, Universidad Nacional de Tucumán [IMLA] types); keys (pictorial guide to Japanese ants, keys to Atlantic forest and Madagascar genera, and keys to species of *Glamyromyrmex, Leptothorax (Myrafant), Neostruma, Nomamyrmex, Pheidole, Simopelta, Smithstruma, Strumigenys,* and *Trichoscapa* of Costa Rica); Ward's technique for mounting ants for museum collections; images of CURT and Museu de Zoologia da Universidade de São Paulo (MZSP) ant collections (including the MZSP types); and slide shows on the ant colony cycle and food storage, as well as movies offering a 360° view of an ant. The site includes important links to related Web pages and links to popular, educational, and pest control sites.

Collection Resources

The levels of curation of entomological collections in temperate zones and of those in tropical countries are generally unequal; furthermore, the levels of curation across collections in tropical countries are grossly uneven as well. Thus any generalizations must be viewed critically. For the sake of efficiency, collaboration and aid must account for these differences, in order to avoid inefficiently utilizing limited resources. A shared mission among collections is necessary—one that supersedes myopic nationalistic visions and that complements each institution's weaknesses and shortcomings. Such cooperation is absolutely essential to the mission of surveying the planet's biodiversity before it is gone forever.

Fortunately, natural history museums and collections have a long-standing tradition of cooperation, resulting from the shared realization that no collection can possibly be "complete" in every field of knowledge, especially given the revolution in the concept of species originating in the early 1940s. The Modern Synthesis made clear that collections are not mere depositories of stamplike typological entities, but information tools that document the variation of species throughout their distribution. Moreover, specimens carry a history in themselves. How, where, by whom, and in which circumstances the specimens were gathered are recorded in catalogues, notebooks, field diaries, and other sources of information that are vital components of natural history collections.

Table 12.1 lists the major public collections of ants by country. Sources for this information include Arnett and Arnett (1990), Arnett et al. (1986, 1993), Gaedike (1995), Heppner and Lamas (1982), Horn and Kahle (1935a, 1935b), Horn et al. (1990), Hudson and Nichols (1975), Rohlfien (1979), Sachtleben (1961), Entomological Society of Canada (1978), Watt (1979), and Williams (1978). In the following discussion, the four-letter acronyms are based upon those of Arnett et al. (1993), but some additions were necessary. The format follows those authors' rules for the construction of new acronyms and avoids the use of acronyms already employed for other institutions.

Stored in newly renovated rooms, the ant collection of the Museum of Comparative Zoology (MCZC) of Harvard University houses perhaps the richest assemblage of ant types in the world. Here are stored the famous collections of William L. Brown Jr., William M. Wheeler, and Edward O. Wilson, as well as parts of the William Mann and Marion R. Smith collections. The extensive Berlese collections recently made by Robert Anderson and Stewart Peck have been deposited in the MCZC. It also contains the Bruno Finzi collection and recent material collected by Gary Alpert, Stefan Cover, and Philip Ward. Parts of the Wheeler collection were divided between the MCZC and the American Museum of Natural History (AMNH) in New York City, and parts of the Mann and

Smith collections were divided between the MCZC and the Smithsonian Institution (USNM) in Washington, D.C.

The Natural History Museum in London (BMNH) holds important Asian and African collections (although some New Guinea [Maffin Bay] material collected by Kenneth Ross is in the California Academy of Sciences [CASC]) and most types of the species described by Horace Donisthorpe and more recently by Barry Bolton. Frederick Smith's collections were divided between the BMNH (most of the Neotropical material) and the Hope Entomological Collections of the University Museum, Oxford (OXUM) (Asian collections), which also holds Charles Thomas Bingham's and W. Cecil Crawley's Asian ants. The Museo Civico di Storia Naturale "Giacomo Doria" (MCSN) in Genoa has, among other smaller collections, that of Carlo Emery, which has been organized according to the *Genera Insectorum* by Carlo Menozzi. The Museum d'Histoire Naturelle (MHNG) in Geneva has the Auguste Forel collection, preserving the organization Forel preferred. It also houses collections of leaf litter ants from Southeast Asia, Sri Lanka, India, Australia, New Zealand, West Africa, and Chile. The Naturhistorisches Museum (NHMB) in Basel has parts of the Forel and Felix Santschi collections, along with more recent material, studied in part by Cesare Baroni Urbani.

The Heinrich Kutter and Daniel Cherix collections are in the Musée Zoologique (MZLS), which also houses important Afrotropical and Middle Eastern collections. At the Museum National d'Histoire Naturelle (MNHN) in Paris is deposited the collection that has been known as the "old collection" (ancienne collection), with specimens studied by Pierre André Latreille and reorganized in part by Carlo Menozzi, as well as material collected more recently, including the Maltese ant fauna, studied by the curator, Dr. Janine Casewitz-Weulersse.

Table 12.1 Major Formicidae Type and Regional Collections of the World[a]

Argentina

IMLA, Fundación e Instituto Miguel Lillo, Universidad Nacional de Tucumán, Miguel Lillo 251, 4000 Tucumán
   (Dr. Fabiana Del Cuezzo); e-mail instli@satlink.com

MACN, Museo Argentino de Ciencias Naturales, Avenida Angel Gallardo 470 (c.c. 220, suc. 5), 1405 Buenos Aires
   (Dr. Arturo Roig Alsina)

MLPA, Museo de la Plata, Universidad Nacional de la Plata, Paseo del Bosque, 1900 La Plata

Australia

AMSA, Australian Museum, P.O. Box A285, Sydney South, NSW 2000

ANIC, Australian National Insect Collection, CSIRO, P.O. Box 1700, Canberra City, ACT 2601 (Dr. Steven Shattuk)

AWAP, Agriculture Western Australia, 3 Baron-Hay Court, South Perth, Western Australia 6151 (fax 61-9-3683223;
   e-mail johnvs@apb.agric.wa.gov.au)

CURT, School of Environmental Biology, Curtin University of Technology, GPO Box U 1987, Perth, Western Australia
   6001 (Dr. Jonathan Majer; e-mail imajerj@info.curtin.edu.au)

MAMU, Macleay Museum, University of Sydney, Sydney, NSW 2006

MVMA, Museum of Victoria, 71 Victoria Crescent, Abbotsford, Victoria 3067

QMBA, Queensland Museum, P.O. Box 3300, South Brisbane, Queensland 4101

SAMA, South Australian Museum, North Terrace, Adelaide, South Australia 5000

TERC, Tropical Ecosystems Research Centre, Division of Wildlife and Ecology, CSIRO, PMB 44, Winnellie, NT 0821
   (Dr. Alan Andersen; e-mail Alan.Andersen@terc.csiro.au)

WAMP, Spider and Insect Collection, Western Australian Museum, Francis Street, Perth, Western Australia 6000

Austria

NHMW, Naturhistorisches Museum Wien, Postfach 417, Burgring 7, 1040 Vienna

Belgium

FSAG, Collections Zoologiques, Faculté des Sciences Agronomiques, 5030 Gembleaux

ISNB, Institut Royal des Sciences Naturelles de Belgique, 29 rue Vautier, B1040 Brussels

MRAC, Musée Royal de l'Afrique Centrale, Leuvensesleenweg 13, B3040 Tervuren

Brazil

CPDC, Centro de Pesquisas do Cacau, Comissão Executiva do Plano de Lavoura Cacaueira (CEPLAC), C.P. 7, Itabuna,
   BA 45600 (Dr. Jacques Delabie; e-mail delabie@nuxnet.com)

DZIB, Departamento de Zoologia, UNICAMP, C.P. 1170, Campinas, SP 13100 (Drs. W. W. Benson and P. S. Oliveira;
   e-mail pso@unicamp.br)

DZUP, Museu de Entomologia Pe. Jesus Santiago Moure, Universidade Federal do Paraná, C.P. 19020, Curitiba, PR
   81531-970

FIOC, Fundação Instituto Oswaldo Cruz, Avenida Brasil 4365, C.P. 926, Rio de Janeiro, RJ 21045-900

IBSP, Instituto Biológico, Secretaria de Agricultura, Avenida Conselheiro Rodrigues Alves 1252, São Paulo, SP 04604
   (Prof. Eliana Bergamasco)

IBUS, Instituto de Biologia, Universidade Federal Rural do Rio de Janeiro, Antiga Rodovia, Rio-São Paulo Km 47,
   Seropédica, RJ 23460 (Dr. Antonio Mayhé-Nunes; e-mail amayhe@ufrrj.br)

INPA, Instituto Nacional de Pesquisas da Amazonia, Estrada do Aleixo 1756, C.P. 478, Manaus, AM 69011
   (Dr. Celio Magalhães; e-mail celiomag@inpa.gov.br)

MPEG, Museu Paraense "Emílio Goeldi," C.P. 399, Belém, PA 66000 (Dr. Ana Y. Harada;
   e-mail ayharada@museu-goeldi.br)

MZSP, Museu de Zoologia da Universidade de São Paulo, Avenida Nazaré 481, São Paulo, SP 04263-000
   (Dr. C. Roberto F. Brandão; phone 55-11-2743455; fax 55-11-2743690; e-mail crfbrand@usp.br)

QBUM, Museu Nacional, Quinta da Boa Vista, São Cristovão, Rio de Janeiro, RJ 20942

UFVB, Museu de Entomologia, Universidade Federal de Viçosa, Viçosa, MG 36570 (Drs. Evaldo Vilela [e-mail
   evilela@mail.ufv.br] and Teresinha Della Lucia)

Table 12.1 continued

Canada

CNCI, Canadian National Collection of Insects and Arachnids, c/o Crop Protection Program (ECORC), Agriculture and Agri-Food Canada, Ottawa, Ontario KIA OC6 (Dr. John Huber, Hymenoptera Unit curator)

CPMQ, Department of Biology, Université Laval, Quebec PQ GIK 7P4

LEMQ, Lyman Entomological Museum and Research Laboratory, McDonald College, McGill University, St. Anne de Bellevue, Quebec H9X 3M1

Chile

MNNC, Museo Nacional de Historia Natural, Casilla 787, Santiago

UCCC, Departamento de Zoologia, Facultad de Ciencias Naturales y Oceanograficas, Universidad de Concepción, Casilla 2407, Apartado postal 10, Concepción, Chile (Dr. Andres Angulo Ormeso; fax, voice 56-41-240280; e-mail aangulo@halcon.dpi.udec.ch)

China

CFRB, Forest Research Institute, Chinese Academy of Forestry, Wan Shou Shan, Beijing 100091 (Drs. Wang Changlu and Dr. Wu Jian)

CNHP, Beijing Natural History Museum, 126 Tien Chaio Street, Beijing (Dr. Hong You-chong) (fossils)

GNUC, Department of Biology, Guangxi Normal University, Guilin 541004, Guangxi (Dr. Zhou Shan-yi)

ISAS, Kunming Institute of Zoology, Academia Sinica (Chinese Academy of Sciences), Kunming 650107, Yunnan

IZAS, Institute of Zoology, Academia Sinica (Chinese Academy of Sciences), 19 Zhongguancun Lu, Haidian 100080, Beijing (Dr. Yuan Decheng)

KFBG, Kadoorie Farm and Botanic Garden, Lam Kam Road, Tai Po, New Territories, Hong Kong Special Administrative Region (Dr. John R. Fellowes)

SNUC, Institute of Zoology, Shaanxi Normal University, Xi An, P.O. Box 191, Shaanxi 710062, (Dr. Zheng Zhemin)

Colombia

CELM, Colección Entomológica "Luís María Murillo," Instituto Colombiano Agropecuario, Tibaitabá. Apartado aéreo 151123 (El Dorado), Bogota (Dr. Ingeborg Zenner-Polania; phone 91-2672710)

IHVL, Instituto Humboldt, Villa de Leyva, Apartado aéreo 8693, Santa Fé de Bogotá, D.C. (Fernando Fernández C.)

UNCB, Museo de Historia Natural, Instituto de Ciencias Naturales, Universidad Nacional de Colombia, Apartado 7495, Santa Fé de Bogotá, D.C.

UNCM, Museo de Entomologia "Francisco Luís Gallego," Facultad de Ciencias, Apartado aéreo 3840, Medellín

Costa Rica

INBC, Instituto Nacional de Biodiversidad, Apartado 22-3100, Santo Domingo de Heredia, 3100 Heredia

MUCR, Museo de Insectos, Universidad de Costa Rica, Ciudad Universitaria, San José

Cuba

IZAC, Instituto de Zoologia, Academia de Ciencias de Cuba, Capitolio Nacional La Habana 2, Ciudad de la Habana 10200

MHNC, Museo Nacional de Historia Natural, Capitolio Nacional La Habana 2, Ciudad de la Habana 10200

Denmark

ZMUC, Universitetets Zoologiske Museum, Universitetsparken, DK-2100 Copenhagen

Ecuador

QCAZ, Museo Zoológico de la Pontifícia Universidad Católica del Ecuador, 12 de Octubre y Roca, Apartado 2184, Quito (Dr. G. Onore and Patricio Ponce; e-mail varsovia@puce.edu.ec)

Finland

UZMH, Zoologiska Muset, Universitets Helsinki, P. Rautatiek 13, SF-00100 Helsinki

Continued on next page

Table 12.1 continued

France
ENSA, École Nationale Supérieure Agronomique, Toulouse
LBIT, Laboratoire de Biologie des Insectes, 118 route de Narbonne, F-31077 Toulouse
MNHN, Musée National d'Histoire Naturelle, 45 rue Buffon, 75005 Paris (Dr. J. Casewitz-Weulersse)

Germany
DEIC, Deutsches Entomologisches Institut, Fachhochschule Eberswalde, Schicklerstrasse 5, 16255 Eberswalde
EMAU, Ernst-Moritz-Arndt-Universität Greifswald, Zoologisches Institut und Museum, Johann-Sebastian-Bach-Strasse 11/12, 17489 Greifswald
SMNG, Staatliches Museum für Naturkunde Görlitz, Am Museum 1, 02826 Görlitz (Dr. Bernhard Seifert)
SMNK, Staatliches Museum für Naturkunde Karlsruhe (formerly Landessammlungen für Naturkunde Karlsruhe [LNK]), Erbprinzenstrasse 13, 76133 Karlsruhe (Manfred Verhaagh; e-mail verhaagh_smnk@compuserve.com)
SMNS, Staatliches Museum für Naturkunde Stuttgart, Rosenstein 1, 70191 Stuttgart
SMTD, Staatliches Museum für Tierkunde Dresden, Augustusstrasse 2, 01067 Dresden
ZFMK, Zoologisches Forschungsinstitut und Museum Alexander Koenig, Adenaueralle 160, 53113 Bonn (Dr. Karl-Heinz Lampe) (holds an important collection of myrmecophilous beetle taxa; contact Dr. Michael Schmitt)
ZMHB, Museum für Naturkunde der Humboldt-Universität Berlin, Invalidenstrasse 43, 10115 Berlin
ZMUH, Zoologisches Institut und Museum der Universität Hamburg, Martin-Luther-King-Platz 3, 20146 Hamburg
ZSMC, Zoologische Staatssammlung, Münchhausenstrasse 21, 81274 Munich (Dr. Erich Diller)

Ghana
UGLA, University of Ghana, Legon, Accra

Greece
ZMAA, Zoological Museum of the University of Athens, Athens

Hungary
HNHM, Hungarian Natural History Museum, Baross utca 13, H-1088 Budapest
SUEL, Bakowyi Természettudományi Muzeum/Bakony Museum, Zirc, Rakoczi ter 1, H-8420

India
NZSI, National Zoological Collection, Zoological Survey of India, 34, Chittaranjan Avenue, Calcutta 700 012
UASB, Departament of Entomology, University of Agricultural Sciences, Bangalore, Karnakata 560-065 (Drs. Rhagavendra Gadagkar [e-mail ragh@ces.iisc.ernet.in] and Musthak-Ali)

Indonesia
MBBJ, Museum Zoologicum Bogoriense, P.O. Box 110, Jalan. Juanda 3, Bogor, Java

Israel
TAUI, Zoological Museum, Tel Aviv University, Tel Aviv 69978

Italy
IEGG, Istituto di Entomologia "Guido Grandi" dell'Universitá degli Studi di Bologna, via Filippo Re 6, 40126 Bologna (phone 051-354161)
MCSN, Museo Civico di Storia Naturale "Giacomo Doria," via Brigata Liguria 9, 16121 Genoa (Dr. Valter Raineri; phone 010-564567)
MHNT, Museo Civico di Storia Naturale, Piazza A. Hortis 4, 34123 Trieste (phone 040-301821).
MRSN, Spinola Collection, Museo Regionale di Scienze Naturali, via Giolitti 36, 10123 Turin (phone 011-432061/73; fax 011-4323331)
MSNM, Museo Civico di Storia Naturale, Corso Venezia 55, 20121 Milan (phone 02-62085405)
MSNV, Museo Civico di Storia Naturale, Lungadige Porta Vittoria 9, 37129 Verona (phone 061-213356)

Table 12.1 continued

Japan

DBUT, Department of Biology, College of Arts and Sciences, University of Tokyo, Komaba, Meguro-ku, Tokyo 153-8902 (Dr. Mamoru Terayama; e-mail terayama@pop.fa2.so-net.or.jp)

EUMJ, Entomological Laboratory, Ehime University, Matsuyama

ITLJ, Laboratory of Insect Systematics, National Institute of Agro-environmental Sciences, 3-1-1 Kannondai, Tsukuba, Ibaraki Prefecture 305 (Terayama types)

KUEC, Entomological Laboratory and Institute of Tropical Agriculture, Faculty of Agriculture, Kyushu University, Hako Zaki, Higashi-ku, Fukuoka 812 (Yamatsu's material and Ogata types) (Dr. Kazuo Ogata; e-mail kogata@agr.kyushu-u.ac.jp)

KUIC, Department of Biology, Faculty of Agriculture, Kagoshima University, Uera Ta-cho, Kagoshima 890 (Dr. Seiki Yamane)

LECJ, Laboratory of Ecosystem Conservation, Obihiro Univerisity of Agriculture and Veterinary Medicine, Inada-cho, Obihiro 080-8555 (Dr. Keiichi Onoyama)

MNHA, Museum of Nature and Human Activities, Yayoigaoka, Sanda, Hyogo 669-13 (Dr. Yoshiaki Hashimoto)

NSMT, National Science Museum (Natural History), Hya Kunin-cho 3-23-1, Shinjuku-ku, Tokyo 141

UOPJ, Entomological Laboratory, University of Osaka Prefecture, Museum of Natural History, Sakai, Osaka 593

Kenya

NMKE, National Museum of Kenya, P.O. Box 40658, Nairobi

Malaysia

SMSM, Sarawak Museum of Natural History, 93566 Kuching, Sarawak

Mexico

INEC, Instituto de Ecologia, A. C., Km 2.5 antigua carretera a Coatepec, A. P. 63, Xalapa, 91000, Veracruz (Dr. Patricia Roja)

MCMC, Museo de Historia Natural de la Ciudad de Mexico, Apartado postal 18845, Delegación Miguel Hidalgo, Mexico 11800, D.F.

UNAM, Instituto de Biologia, Universidad Nacional Autónoma de México, Apartado postal 70133, Mexico 04510, D.F. (Dr. Julieta Ramos Elorduy)

The Netherlands

NHME, Natural History Museum, de Bosquetplein 6-7, Post-bus 882, 6200 AW Maastricht

ZMAN, Instituut voor Taxonomische Zoologie, Zoologisch Museum, Universiteit van Amsterdam, Plantage Middenlaan 64, 1018 DH Amsterdam

New Zealand

NZAC, New Zealand Arthropod Collection, DSRI, Landcare Research New Zealand, Private Bag 92170, Auckland 1001 (phone 64-9-8496330; fax 8497093)

PPCC, Plant Protection Centre Collection, Lynfield Agricultural, P.O. Box 41, Auckland 1

Nicaragua

SEAN, Museo Entomológico, SEA, Apartado aéreo 527, León (Dr. Jean-Michel Maes)

Panama

MIUP, Dr. Graham B. Fairchild Museo de Invertebrados, Universidad de Panama, Estafeta Universitaria, Panama (Dr. Diomedes Quintero-Arias)

Peru

MUSM, Museo de Historia Natural "Javier Prado," Universidad Nacional Mayor de San Marcos, Avenida Arenales 1267, Apartado postal 14-0434, Lima 14

Continued on next page

Table 12.1 continued

Poland
ZMPA, Instytut Zoologiczny, Polska Akademia Nauk, Wilcza 64, 00-679 Warsaw

Portugal
MMFM, Museu Municipal do Funchal, Madeira

Russia
IEME, Institute for Evolution, Morphology, and Ecology of Animals (also cited in some publications as Institute of Evolutionary Animal Morphology), Leninsky Prospekt 33, Moscow 117071
ZMAS, Department of Entomology, Zoological Institute of the Russian Academy of Science, Universitetskaya, Naberzhnayal B-164, Saint Petersburg (Dr. Vladimir Tobias; e-mail tvi@zisp.spb.su)
ZMUM, Zoological Museum, Moscow State University, Bolshaja Nikitskaja 6, Moscow 103009 (Dr. Alexander Andropov; e-mail entomol@zoomus.bio.msu.su)

Singapore
NMSC, Department of Zoology, National University of Singapore, Kent Ridge, S-0511 Singapore (includes material formerly at the Raffles Museum, the Singapore National Museum, and the University of Singapore)

South Africa
SAMC, South African Museum, P.O. Box 61, Queen Victoria Street, Cape Town 8000 (Dr. Hamish G. Robertson; phone 21-246330; fax 21-246716; e-mail hroberts@samuseum.ac.za)
TMSA, Transvaal Museum, P.O. Box 413, Pretoria 0001

Spain
DBAG, Departamento de Biologia Animal, Universidad de Granada, Spain
MCCB, Museo de la Ciència, Fundació la Caixa, Teodor Roviralta 55, 08022 Barcelona (Dr. Jorge Wagensberg; e-mail jwagensberg.fundacio@lacaixa.es; fax 34-3-4170381) (Dominican amber fossils)
MNMS, Museo Nacional de Ciencias Naturales, Paseo de la Castellana 84, Madrid (merged with the Instituto Español de Entomologia)

Sweden
NHRS, Naturhistoriska Riksmuseet, 10405 Stockholm (Dr. Per Inge Persson)
UZIU, Upssala Universitets, Zoologiska Museet, P.O. Box 561, 75122 Uppsala

Switzerland
ETHZ, Eidgenössische Technische Hochschule-Zentrum, Universitätsstrasse 2, CH-8006 Zurich
MHNC, Musée d'Histoire Naturelle, La Chaux-de-Fonds
MHNG, Musée d'Histoire Naturelle, route de Malagnon, Case postale 434, CH-1211 Geneva 6 (Dr. Daniel Burckhardt)
MZLS, Musée Zoologique, Place de la Riponne 6, Lausanne (Dr. D. Cherix)
NHMB, Naturhistorisches Museum, Augustinergasse 2, CH-4001 Basel
NMBS, Naturhistorisches Museum, Bernastrasse 15, CH-3005 Bern

Trinidad and Tobago
UWIC, Department of Biological Sciences, University of West Indies, Saint Augustine (Dr. Christopher Starr)

Ukraine
UASK, Institute of Zoology, Ukrainian National Academy of Sciences, B. Khmelnitsky Street 15, 252601 Kiev 30 (Dr. Alexander Radchenko; e-mail rad@usenc.kiev.ua)
ZMKU, Zoological Museum, Academy of Sciences of Ukrania, Vladimirskaya SS, Kiev

United Kingdom
BMNH, The Natural History Museum, Cromwell Road, South Kensington, SW7 5BD London (Mr. Barry Bolton)
CMLU, City Museum, Municipal Building, L51 3AA Leeds (Dr. Cedric Collingwood)
OXUM, Hope Entomological Collections, University Museum, Parks Road, 0XI 3PW Oxford

Table 12.1 continued

United States of America

ABSC, Archbold Biological Station, P.O. Box 20570, Lake Placid, Florida 32852-2057 (Dr. Mark Deyrup; phone 1-813-465-2571; fax 6991927)

AMNH, American Museum of Natural History, Central Park West at 79th Street, New York, New York 10024 (Dr. James Carpenter; phone 1-212-873-1300; (http://research.amnh.org/~agosti/social_insects/sihp1.html) (includes amber fossils)

ANSP, Department of Entomology, Academy of Natural Sciences, 19th and the Parkway, Philadelphia, Pennsylvania 19103 (phone 1-215-299-1189)

BPBM, Bernice P. Bishop Museum, P.O. Box 19000A, Honolulu, Hawaii 96819

BUWT, Department of Biology, Baylor University, Waco, Texas 76703

CASC, California Academy of Sciences, Golden Gate Park, San Francisco, California 94118 (Dr. Wojciech J. Pulawski; phone 1-415-221-5100)

CEMU, Cleveland Museum of Natural History, 1 Wade Oval Drive, University Circle, Cleveland, Ohio 44106 (phone 1-614-231-4600)

CIDA, The Orma J. Smith Museum of Natural History, Albertson College of Idaho, Caldwell, Idaho 83605 (Dr. William H. Clark; phone 1-208-459-5507)

CUIC, Department of Entomology, Cornell University, Ithaca, New York 14850

DEFW, Bell Museum of Natural History, University of Minnesota, Minneapolis, Minnesota 55455

DENH, Department of Entomology, University of New Hampshire, Durham, New Hampshire 03824

DEUN, Division of Entomology, University of Nebraska State Museum, W436 Nebraska Hall, Lincoln, Nebraska 68503

ESUW, Rocky Mountain Systematic Entomology Laboratory, University of Wyoming, P.O. Box 3354, University Station, Laramie, Wyoming 82071 (phone 1-307-766-5338)

FMNH, Field Museum of Natural History, Roosevelt Road and Lake Shore Drive, Chicago, Illinois 60605 (Dr. Alfred F. Newton; phone 1-312-922-9410)

FSCA, Florida State Collection of Arthropods, 1911 34th Street SW, P.O. Box 147100, Gainesville, Florida 32614 (Dr. M. C. Thomas; phone 1-904-372-3505)

INHS, Illinois Natural History Survey Insect Collection, 607 East Peabody Drive, Champaign, Illinois 61820 (Kathleen R. Methven; phone 1-217-244-2149; fax 1-217-333-4949; http://www.inhs.uiuc.edu)

ISUI, Department of Zoology and Entomology Collections, Iowa State University, Ames, Iowa 50010-3222 (Dr. Robert E. Lewis; phone 1-515-294-1815)

KSBS, State Biological Survey of Kansas Invertebrate Collection, 2045 Constant Avenue, Campus West, University of Kansas, Lawrence, Kansas 66044 (phone 1-913-864-4493)

LACM, Los Angeles County Museum of Natural History, 900 Exposition Boulevard, Los Angeles, California 90007 (Drs. Roy Snelling and Brian V. Brown; phone 1-213-744-3363; http://www.lam.mus.ca.us/lacmnh/departments/research/entomology)

MCZC, Museum of Comparative Zoology, Harvard University, 26 Oxford Street, Cambridge, Massachusetts 02138 (Dr. Edward O. Wilson, Dr. Gary Alpert, Mr. Stefan Cover; phone 1-617-495-2464)

MEMU, Mississippi Entomological Museum, Mississippi State University, Drawer EM, Mississippi State, Mississippi 39762 (Dr. Richard L. Brown; phone 1-601-325-2085)

NCSU, Department of Entomology, North Carolina State University, P.O. Box 5215, Raleigh, North Carolina 27607 (phone 1-919-737-2833)

NDSU, North Dakota State Reference Collection, Entomology Department, North Dakota University, Fargo, North Dakota 58102 (phone 1-701-237-7902)

NYSM, New York State Museum, Biological Survey, 3132 Albany Cultural Education Center, Albany, New York 12230 (phone 1-518-473-8496)

OSUC, Ohio State University Collection of Insects and Spiders, 1735 Neil Avenue, Columbus, Ohio 43210 (Dr. Norman Johnson; phone 1-614-292-6839)

PSUC, Frost Entomological Museum, Pennsylvania State University, University Park, Pennsylvania 16802

RUVA, Radford University, Radford, Virginia 24142 (Dr. Charles Kugler)

Continued on next page

Table 12.1 continued

SEMC, Snow Entomological Museum, University of Kansas, Lawrence, Kansas 66044 (Dr. James S. Ashe; phone 1-913-864-3065)

SIUC, Southern Illinois University Entomology Collection, Research Museum of Zoology, Southern Illinois University, Carbondale, Illinois 62901 (Dr. J. E. McPherson; phone 1-618-536-2314)

SRSS, Southern Research Station, Forest Insect Research, U.S. Department of Agriculture, 2500 Shreveport Highway, Pineville, Louisiana 71360

SWRS, Southwestern Research Station of the American Museum of Natural History, Portal, Arizona 85632 (Dr. Wade C. Sgerbrooke; phone 1-602-558-2396)

TAMU, Department of Entomology, Texas A&M University, College Station, Texas 77843 (phone 1-409-845-9712)

TTCC, The Museum, Texas Tech University, Lubbock, Texas 79409 (phone 1-806-742-2828)

UAIC, Department of Entomology Collection, University of Arizona, Tucson, Arizona 85721 (phone 1-602-621-1635)

UCDC, The Bohart Museum of Entomology, University of California, Davis, California 95616 (Dr. Philip S. Ward; phone 1-916-752-0486)

UCMC, University of Colorado Museum, Campus Box 218, Boulder, Colorado 80309-0218, (curator: Dr. M. Deane Bowers; collection manager: Virginia Scott; phone 1-303-492-6270; 1-303-fax 492-4195; scottv@spot.colorado.edu)

UCMS, Department of Ecology and Evolutionary Biology, University of Connecticut, Box U-43, Storrs, Connecticut 06269-3034 (Dr. Carl W. Rettenmeyer; phone 1-203-486-4460)

UCRC, Entomological Teaching and Research Collection, University of California, Riverside, California 92521 (Dr. Saul I. Frommer; phone 1-714-787-4315)

UMMZ, Museum of Zoology, University of Michigan, Ann Arbor, Michigan 48109-1079 (contact: Dr. Paul B. Kannowski)

USDA, United States Department of Agriculture Fire Ant Project Collection. P.O. Box 14565, Gainesville, Florida 32604 (Dr. Daniel P. Wojcik).

USNM, United States National Museum of Natural History, Washington, D.C. 20560 (Dr. Ted Schultz; phone 1-202-357-1311; fax 1-202-786-2894)

VPIC, Virginia Polytechnic Institute and State University, Blacksburg, Virginia 24061-0542 (phone 1-703-231-6773)

WFBM, W. F. Barr Entomological Collection, University of Idaho, Moscow, Idaho 83843 (Dr. Frank W. Merickel; phone 1-208-885-7543)

WSUC, James Entomological Collection, Department of Entomology Collection, Washington State University, Pullman, Washington 99163 (phone 1-509-335-3394)

WVUC, Arthropod Collection, Room G176, Agricultural Sciences Building, West Virginia University, P.O. Box 6108, Morgantown, West Virginia 26505

Uruguay

UYIC, Museo de Entomologia, Departamento de Artropodos, Facultad de Humanidades y Ciencias, Tristán Narvaja 1674, C.C. 10773 Montevideo

Venezuela

IZAV, Instituto de Zoologia Agricola, Facultad de Agronomia, Universidad Central de Venezuela, Apartado postal 4579, Maracay 2010A, Aragua

USBC, Departamento de Biologia de Organismos, Universidad Simón Bolivar, Apartado postal 80659, Caracas, Miranda

Zimbabwe

NMBZ, National Museum, P.O. Box 240, Centenary Park, Bulawayo

---

*ª*Acronyms in alphabetical order by country; see text. *Italicized* acronyms refer to those not found in Arnett et al. (1993) and proposed here using their criteria.

The Gustav Mayr collection is in the Ernst-Moritz-Arndt-Universität (EMAU) in Greifswald and the Naturhistorisches Museum Wien (NHMW) in Vienna; the Erich Wasmann collection is deposited in the Natural History Museum (NHME) in Maastricht and the Massimiliano Spinola collection in the Museo Regionale di Scienze Naturali (MRSN) in Turin. The William Nylander collection is deposited in the Zoologiska Muset, Universitets Helsinki (UZMH) in Helsinki. For Italian collections, readers are referred to the complete list of entomological collections and authors published by Poggi and Conci (1996).

The George Arnold and Andre J. Prins collections are deposited in the South African Museum (SAMC).

In North America the most important collections besides the MCZC are those at the Los Angeles County Museum of Natural History (LACM), with the collections of Arthur C. Cole Jr., Thomas Wrentmore Cook, and William S. Creighton; the North Dakota and Nevada collections of George C. and Jeanette Wheeler; Daniel Janzen's collection of Central American *Acacia*-inhabiting ants; and many thousands of samples collected and studied by Roy R. Snelling; the USNM, holding parts of the Julian F. Watkins II army ants and portions of the Mann and Smith collections; and the AMNH, holding most of the Theodore C. Schneirla Panamanian Ecitoninae documents. Several smaller collections are also important, in particular the Robert Gregg collection, deposited in the Field Museum of Natural History (FMNH) in Chicago, which also houses some 14,000 samples of ants recovered by Berlese extraction of leaf litter. Gregg based his book *Ants of Colorado* on a collection now deposited in the University of Colorado Museum (UCMC) in Boulder. The Orma J. Smith Museum of Natural History, Albertson College of Idaho (CIDA) in Caldwell includes a part of the Watkins army ants, with field notes and a worldwide reference collection of ants,

and it will eventually receive the P. E. Blom and William H. Clark personal collections and libraries. The Illinois Natural History Survey Insect Collection (INHS) has 20 drawers of pinned material and a full cabinet of ants in alcohol, mostly from the American Midwest.

In South America, Argentina and Brazil have the best ant collections. The Thomas Borgmeier, Walter Kempf, Karol Lenko, and Hermann von Jhering collections are all deposited in the MZSP in São Paulo, along with more recent collections by C. R. F. Brandão and collaborators. The Cincinnato R. Gonçalves collection has been divided between the Museu Nacional, Quinta da Boa Vista (QBUM) and the Instituto de Biologia, Universidade Rural Federal do Rio de Janeiro (IBUS), both in Rio de Janeiro. The collection at the Centro de Pesquisas do Cacau (CPDC), although quite new, is expanding rapidly, holding mostly material from the Atlantic forest in eastern Brazil. The Niko Kusnezov collection, mainly from Argentina, is deposited in IMLA in Tucumán, Argentina, whereas the Carlos Bruch and Angel Gallardo collections are in the Museo Argentino de Ciencias Naturales (MACN) in Buenos Aires (duplicates in the Museu de de la Plata, Universidad Nacional de la Plata [MLPA]). Cuba's Instituto de Zoologia, Academia de Ciencias de Cuba (IZAC) collection in Havana has the Pastor Alayo and Juan Gundlach material.

Asian and African material is mostly deposited in European and North American museums, especially in the MCZC and BMNH. Other important sources are the Musée Royal de l'Afrique Centrale (MRAC) in Tervuren, the NHME in Maastricht (the Jozef K. A. Van Boven and Erich Wasmann collections of ants and ant guests from all over the world), and the smaller collections listed subsequently.

Australian ants are deposited in several collections, but most species can be found in the Australian National Insect Collection (ANIC) in Canberra City.

Dr. Gennady M. Dlussky kindly sent me the following information on Russian and former Soviet ant collections. The Zoological Museum, Moscow State University (ZMUM) has the largest ant collection of the former USSR, and in his opinion the best collection of ants of the Palearctic region. It consists of Victor Ivanovich Motschulski's collection; collections of ants from Central Asia assembled by Aleksyei Pavlovich Fedtschenko and described by Mayr; collections of ants from Austria, South America, and Australia, presented by Mayr with some syntypes of his species; Nikolai Victorovich Nasonov's collection; part of Mikhail Dmitrivich Ruzsky's collection; the collection of Konstantin Vladimirovich Arnol'di (previously it was in the Institute for Evolution, Morphology, and Ecology of Animals [IEME] in Moscow, but at the end of his life Arnol'di presented it to the Zoological Museum); and Dlussky's own collection. Paratypes of Emery, Forel, Mayr, Santschi, Wheeler, and many other myrmecologists are also at the ZMUM. At the Institute of Zoology of the Ukrainian National Academy of Sciences (ZIKU) are housed Vladimir Affanassievich Karavaier's collection (mainly from the Ukraine and Oriental region) and Alexander G. Radchenko's collection, as well as many paratypes of Emery, Santschi, and others.

Myrmecologists have never worked in the Department of Entomology of the Zoological Institute of the Russian Academy of Science (ZMAS), and all material has been determined and described by visitors. In 1903 the collection was studied by Forel, who described some species. He also presented to the museum a collection of ants from Madagascar. In 1914 the large collection assembled by Petr Kuzmich Kozlov from Tibet and Southern Gobi was described by Ruzsky. Part of Ruzsky's personal collection was also deposited there (most of Ruzsky's types have been lost). Most material was determined by Arnol'di, Dlussky, and Radchenko.

No one knows the location of the collection of N. N. Kuznetzov-Ugamsky. It may be in the Museum of Nature in Tashkent; however they have no entomologists on staff at present. Some types of his species are in Kiev, Moscow, and St. Petersburg. Types of Alina Kupianskaja are in the Biologo-Pocvennyj Institute in Vladivostok.

Currently several regional ant collections are under construction that will certainly be instrumental in the determination of distributional patterns and in studies of distributional variation within taxa.

Major collections holding type specimens, especially those described in the last century or at the beginning of this one, are listed in Table 12.1. This list, favoring extant fauna, is in alphabetical order and is taken mainly from Arnett et al. (1993), but it has been combined with information from recent papers by several researchers): Agosti, Alayo, Alpert, Baroni Urbani, Benson, Bolton, Brandão, Brown, Cagniant, Casewitz-Weulersse, Cerdá, Collingwood, Delabie, Deyrup, Diniz, Dlussky, Dubois, Espadaler, Fernandez, Fowler, Francoeur, Harada, Ipinza-Regla, Kohout, C. Kugler, Lattke, Longino, MacKay, Moffett, Nunez, Ogata, Pisarski, Radchenko, Shattuck, Snelling, Taylor, Thompson, Tinaut, Trager, Umphrey, Ward, Watkins, and Wilson. The references used have been carefully listed in Ward et al. (1996) under the cited names and are not reproduced here. The list favors extant fauna. This information has been kindly checked and amended by colleagues from several institutions, listed in the Acknowledgments.

Collectors and curators may not necessarily wait for a specialist to request material for study. Instead they often write to systematists and ask them to identify specimens. As with the borrowing of specimens, certain rules should be followed in making such requests for identification; these are outlined in Arnett and Samuelson (1986).

Private Collections

The information given in this section has been taken from several different recent references in the ant taxonomy literature. It is not easy to ascertain whether the cited collections are still in private hands or have by now been deposited in official institutions. Interested readers should contact the collection holders.

Major private collections cited in the literature are those of W. Buren, A. Buschinger, F. Castaño (Colombia), R. Chew, C. A. Collingwood (Skipton, U.K.), G. Delye, J. L. M. Diniz (UNESP, São José do Rio Preto, SP, Brazil), M. DuBois (Clemson University), X. Espadaler (Bellaterra, Spain), F. Fernández C. (Colombia), A. Francoeur (Chicoutani, Canada), R. Gregg, D. Kistner (Chico, California, USA), R. Klein, Masao Kubota (Odawara City, Japan), R. Lavigne, J. Longino, J. Lynch, W. and E. MacKay (University of Texas, El Paso), T. Nuhn, S. O. Shattuck, A. Tinaut (Departamento de Biologia Animal y Ecologia, Universidad de Granada, Spain), J. Trager, G. J. Umphrey (University of Guelph, Ontario, Canada), P. Ward, G. and J. Wheeler, and D. Wojcik. Another fast-growing and interesting private collection in Germany is that of A. Schulz (Leichlingen).

Although quite dispersed and extremely heterogeneous by all conceivable criteria, Table 12.1 demonstrates that the number of ant collections available worldwide is impressively large when compared with other collections of similar taxa. The wealth of biological information residing in these assemblages is by no means negligible. A special effort should therefore be made by the ant research community to improve not only the collections themselves but also—and more importantly—the quality and accessibility of the information associated with the collection specimens.

I am sure that many gaps still remain in this world list of ant collections, and I would be grateful to receive additions and corrections from readers.

ACKNOWLEDGMENTS

I am deeply indebted to several colleagues who answered my requests for information regarding collections: Alan Andersen (TERC), Cesare Baroni Urbani, Brian V. Brown (LACM), the late William L. Brown Jr. (CUIC), Daniel Burckhardt (MHNG), William H. Clark (CIDA), Fabiana Cuezzo (IMLA), Gennady M. Dlussky (ZMUM), John Fellowes (KFBG), Stuart Fullerton (ABSC), Robert Jeanne, Jonathan Majer (CURT), Alfred F. Newton (FMNH), Barry O'Connor (UMMZ), Andrés Angulo Ormeso (UCCC), Alexander Radchenko (UASK), Hamish G. Robertson (SAMC), Virgina Scott (UCMC), Roy R. Snelling (LACM), Mamoru Terayama (DBUT), Gary Umphrey, Kees van Achterberg (RMNH), John van Schagen (AWAP), and Manfred Verhaagh (SMNK). Dr. Mirian D. Marques, Christiane I. Yamamoto, and an anonymous reviewer read and improved the manuscript.

Chapter 13

What to Do with the Data

John T. Longino

At their most basic, the data from biodiversity surveys consist of specimens to which are attached ecological and taxonomic attributes. Ecological attributes might include date of collection, locality, habitat, quadrat number, or collection method. Taxonomic attributes might be species identification, higher taxa to which the species belongs, or perhaps a predefined functional group. This chapter describes how these data can be organized, visualized, and analyzed. The procedures outlined here are not unique to leaf litter ants, and I present only a cursory overview of a subject about which volumes have been written (e.g., Pielou 1975, 1984; Southwood 1978; Ludwig and Reynolds 1988; Magurran 1988; Hayek and Buzas 1996).

Organizing the Data

Imagine an example in which an investigator takes a litter-soil sample in a patch of forest, extracts the ants in a Berlese funnel, and mounts one or a few specimens of each different species in the sample. The investigator can present the results as a species list. Now imagine that the investigator has taken two samples instead of one. Some species will be common to both samples; others will be unique to one or the other. These results can be presented as a matrix with two columns and as many rows as there are species in both samples combined. The presence of a species in a particular sample is indicated by a check mark. Data such as these are presence-absence data or incidence data.

Alternatively the investigator may choose to count the number of individuals of each species in each sample. The cells of the matrix would then contain abundance data rather than presence-absence data. Our investigator may take ten replicate samples in old-growth forest and another ten in managed forest nearby. The species-by-sample matrix now has the columns, which represent different samples, organized in two groups. Species-by-sample matrices such as these are the fundamental data structure for ecological sampling (Pielou 1984). Replicate samples are represented by columns and species are represented by rows (or vice versa). The cell contents may be presence-absence data or they may be abundances. The replicate samples may have no particular order or grouping, or they may be stratified or grouped in various ways.

For very large data sets, a matrix remains the conceptual structure of the data, but not the best way to actually store them. A large matrix, most of the cells of which are empty, is a cumbersome way to store data. Data are better stored as a list, with each row representing a nonzero cell of the matrix. Thus each row contains species name or code number, sample number, any ecological grouping variables (e.g., old growth versus managed forest), and abundance.

Some species will be common in the data set, and others will be rare. Terminology for rare species will become important in some of the analyses discussed later. *Singletons* are species known from a single specimen, and *doubletons* are species known from two. *Uniques* are species that occur in only one sample (regardless of their abundance within the sample), and *duplicates* are species known from two samples (Colwell and Coddington 1994; Coddington et al. 1996; Silva and Coddington 1996; Chazdon et al. 1998).

Caveats to Ant Sampling

Caveat 1: Ants Are Spatially Clumped

Some analysis methods assume that within a spatially and temporally defined community all individuals in the sampling universe have an equal probability of being sampled. In other words, knowing the identity of an individual in a sample should not influence the probability of observing other members of the same species in that sample. This assumption is nearly always violated because spatial aggregation of organisms appears to be the rule rather than the exception (L. R. Taylor et al. 1978). Ants are social and are strongly aggregated when sampling methods capture colonies or portions of colonies. For this reason presence-absence data may be preferred over abundance data in analyses. An extreme example is a pitfall trap that catches 10,000 army ants. In terms of number of individuals, the army ant species might dwarf the abundance of all other species combined, whereas in terms of number of samples in which it occurred the army ant species might be among the rarest.

Caveat 2: Obtaining an Unbiased Sample of Arthropods (Including Ants) Is Nearly Impossible

Ideally the relative abundances of species in a sample should reflect the relative abundances in the community. This is possible if one takes a huge number of small, random volumetric samples from the environment and observes every arthropod in the sample. Arthropods are small and intricately embedded in other biotic and abiotic components of the environment (e.g., soil, wood, leaf litter, foliage, air). Direct searching for arthropods is extremely labor intensive and makes it possible to characterize only very small areas. Rarely is such an exhaustive search practical.

More often, arthropods are sampled by concentrating them from larger areas or volumes. This concentration is influenced by the behavior of individuals, which varies between species and thus introduces almost insurmountable problems of bias (Poole 1974). Baiting attracts ants from surrounding areas and preferentially samples species that are generalized omnivores with highly developed recruiting ability. It undersamples specialized predators and ants that forage beneath the litter. Sifting litter concentrates large volumes of bulk litter, preferentially sampling species that (1) are not quick enough to escape the litter-gathering process, (2) can be dislodged from large litter fragments to which they cling, (3) are not crushed by the sifting process, and (4) readily drop from the suspended litter when it is in the extraction bag. Pitfall traps may undersample sit-and-wait predators and species that can cling to vertical surfaces. Intensive manual searching of litter plots (with its concomitant high cost) comes closest to unbiased community characterization, but even in this case the search must be extremely thorough and painstaking so as not to miss extremely small (circa 1 mm long) and cryptic litter ant species. More often than not, small cryptic ants will be undersampled.

Study Objectives

How important these caveats are will depend on the objectives of the study. Study objectives can be portrayed as a set of questions asked of a data set. Here I discuss some of those questions and how to answer them. I use an example data set (Table 13.1) to illustrate the analysis methods that I discuss. This is a real data set, produced by an arthropod survey project in a lowland rainforest in Costa Rica (Project ALAS; see Longino 1994; Longino and Colwell 1997). Each "sample" is the combined ants from 13 soil-litter cores, taken over a 13-month period from the perimeter of a 10-m-radius circle and

extracted in Berlese funnels. The soil-litter cores were 14.5 cm in diameter and 10 cm deep. Sixteen samples are shown: eight from old-growth forest and eight from second-growth forest. The values in the table are the numbers of adult workers.

What Is the Rate of Species Accumulation in the Sampling Program?

This question alone has no pretensions of describing community characteristics. The question has relevance to what is called "strict inventory" (Longino and Colwell 1997), in which a goal is compiling the largest possible species list for the least effort. Strict inventory is practiced by taxonomists who wish to sample many taxa efficiently for museum study.

The rate of species-accumulation is observed with a species-accumulation curve (Soberón and Llorente 1993). A species-accumulation curve has some measure of effort on the horizontal axis and cumulative number of species on the vertical axis (Fig. 13.1). Examples of effort measures include number of samples, number of individuals observed, time spent collecting, time required to process and identify specimens, and monetary cost of the inventory process. To obtain a species-accumulation curve from a species-by-sample matrix in a spreadsheet, first accumulate abundance across rows and then replace each nonzero value with 1 (this can be done by dividing each value by itself plus 1, then rounding). The column sums will be the species-accumulation curve (Table 13.2). A particular ordering of samples produces a particular species-accumulation curve. The last point on the curve will be the total number of species observed among all the samples. Changing the order of samples may change the shape of the curve but not the endpoint. A smoothed or average species-accumulation

Table 13.1 Example Data Set from Berlese Samples

Ant Species	Sample Number															
	1	2	3	4	5	6	7	8	9	10	11	12	13	14	15	16
Species 1	0	9	0	0	0	0	1	0	0	0	1	5	1	0	0	0
Species 2	0	0	0	0	0	0	0	0	1	0	0	0	1	0	0	0
Species 3	0	0	0	2	3	0	0	3	0	0	0	0	0	0	0	0
Species 4	1	0	0	0	0	0	0	0	0	0	0	7	0	0	0	1
Species 5	0	0	0	0	2	0	0	1	0	0	0	1	0	0	1	1
Species 6	0	0	2	12	0	0	0	0	0	0	0	0	0	2	2	0
Species 7	0	0	0	0	0	0	0	0	38	0	0	0	0	0	0	0
Species 8	0	0	0	0	7	0	0	0	0	0	0	0	0	0	0	0
Species 9	0	0	0	0	0	0	0	0	8	3	3	0	0	1	0	0
Species 10	4	0	0	0	0	27	0	0	0	3	0	1	2	0	0	8
:																
Species 107	0	0	0	0	0	0	0	0	4	0	0	0	0	0	0	0

curve can be produced by repeatedly randomizing sample order, calculating a species-accumulation curve for each randomization, and averaging the resultant curves (Fig. 13.1). (This and many other analyses illustrated here were carried out with the program EstimateS [Colwell 1997].) The curve for a highly undersampled fauna will appear nearly linear, with each new sample adding many new species to the inventory. The curve for a thoroughly sampled fauna will reach a plateau, with few or no species being added with additional sampling.

In addition to observing the current rate of species accumulation, one may wish to predict the results of additional sampling. Projecting a species-accumulation curve allows one to esti-

Table 13.2 Calculating a Species-Accumulation Curve[a]

Ant Species	Sample Number															
	1	2	3	4	5	6	7	8	9	10	11	12	13	14	15	16
Species 1	0	1	1	1	1	1	1	1	1	1	1	1	1	1	1	1
Species 2	0	0	0	0	0	0	0	0	1	1	1	1	1	1	1	1
Species 3	0	0	0	1	1	1	1	1	1	1	1	1	1	1	1	1
Species 4	1	1	1	1	1	1	1	1	1	1	1	1	1	1	1	1
Species 5	0	0	0	0	1	1	1	1	1	1	1	1	1	1	1	1
:																
Species 106	1	1	1	1	1	1	1	1	1	1	1	1	1	1	1	1
Species 107	0	0	0	0	0	0	0	0	1	1	1	1	1	1	1	1
Sum	19	36	47	56	62	68	75	79	92	93	100	101	104	106	106	106

[a]Abundances are summed across rows and then each nonzero value is replaced with 1. The column sums are the observed species-accumulation curve for a particular sample order.

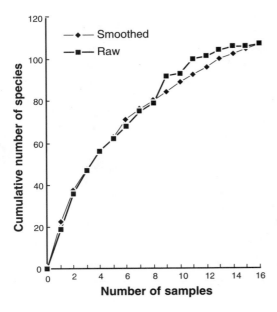

Figure 13.1. Raw versus smoothed species-accumulation curves. A raw curve is produced from a particular order of samples. A smoothed curve is an average of multiple raw curves from random reorderings of the samples. Data are from 16 Berlese samples of litter-soil cores (see text).

bility of encountering additional species never declines to zero, and species-accumulation curves never reach a plateau. The equation for a logarithmic curve is

$$S(t) = \frac{\ln (1 + zat)}{z}$$

where t is the measure of effort, such as time or number of samples; $S(t)$ is the predicted number of species at t, and z and a are curve-fitting parameters. Using the data from a smoothed species-accumulation curve (the $S(t)$ and t values), the parameters a and z can be estimated using a nonlinear curve-fitting procedure in a statistical analysis program. For example, using the smoothed curve from the Berlese data, a fitted logarithmic curve has $a = 28.46$ and $z = 0.023$. The fitted curve is nearly identical to the smoothed curve ($r^2 >$ 0.99).

Longino and Colwell (1997) modified the logarithmic equation to

$$t_s - t_{s-1} = \frac{e^{zs} - e^{z(s-1)}}{za}$$

This equation shows the number of samples (or other measure of effort) needed to add the sth species to the inventory. At 107 species, the number obtained in the 16 Berlese samples, the cost of adding another species is still less than one additional sample, but it is increasing rapidly (Fig. 13.2). Looking at inventory progress in this fashion allows one to develop "stop-rules," invoked when the cost of adding an additional species rises above some threshold.

mate the effort needed to add a particular number of species to the inventory or to increase the species list by a particular percentage. Soberón and Llorente (1993) discuss a variety of mathematical models that can be fit to species-accumulation curves. It is often the case that numerous models will fit an observed curve more or less equally well, yet diverge widely when used to make projections. The choice of a particular model must be based on assumptions about the sampling conditions. For large and poorly known faunas, Soberón and Llorente recommended the logarithmic model, in which the probability of encountering additional species declines as an exponential function of the size of the species list. With this model, the proba-

Other models of species-accumulation assume that the probability of encountering additional species in an inventory eventually reaches zero. These asymptotic models can be used to estimate species richness, and they are discussed in the section on richness estimation.

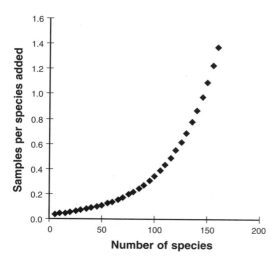

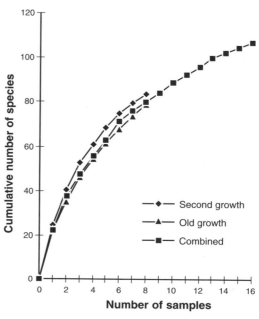

Figure 13.2. Cost in samples of adding an additional species to an inventory. As the number of species already captured in the inventory increases, the cost of adding additional species increases exponentially. This "cost" curve is derived from a logarithmic curve (see text) fitted to the species-accumulation curve from the Berlese samples (Fig. 13.1). The observed species-accumulation curve reaches 107 species after 16 samples. The cost curve predicts that adding a 17th sample to the inventory would add about two species.

Figure 13.3. Within-habitat versus combined species-accumulation curves for the Berlese data (see text). Inventory efficiency (steepness of species-accumulation curves) does not differ greatly between old-growth and second-growth forest habitats, as revealed by eight Berlese samples. In addition, stratifying eight Berlese samples by forest habitat does not improve inventory efficiency.

How Can Species-Accumulation Rates Be Maximized?

When undertaking an inventory, choices have to be made about which sampling methods to use and whether to stratify sampling with respect to habitat variables or over time. How does one evaluate whether stratifying by habitat or using different methods is actually beneficial to an inventory? Comparing species-accumulation curves is one method of evaluating inventory efficiency (Longino and Colwell 1997). If the species-accumulation curve is much steeper in one habitat than another, concentrating inventory effort in the more productive habitat is advised. If two habitats have similar species-

accumulation curves and very low species overlap (high complementarity *sensu* Colwell and Coddington 1994), then a combined species-accumulation curve, drawing samples randomly from both habitats, will be steeper than either within-habitat species-accumulation curve. In such a situation, stratifying samples across the two habitats is advised. If the combined curve is not steeper (it can lie below curves for the richest single samples), then there is less advantage to stratifying.

For example, the Berlese samples can be partitioned into those from old-growth forest and those from second-growth forest. Within-habitat

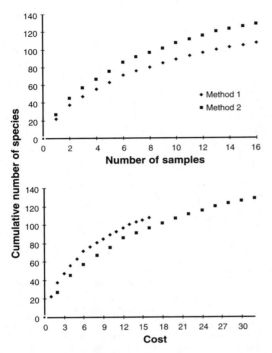

Figure 13.4. Comparing methods that differ in cost. Method 2 may appear more efficient based on number of samples (upper figure), but if method 2 samples cost twice as much as those for method 1, it is actually less efficient (lower figure).

methods, a common currency should be used. Ideally that currency is the direct monetary cost of each sample. Instead of plotting cumulative species against number of samples, plot cumulative species against the cost of obtaining them (Fig. 13.4). Calculate the cost by multiplying the number of samples by the average cost per sample. Proxies of direct monetary cost, such as sample processing time or number of mounted specimens, may also be used (Longino and Colwell 1997).

Is One Group of Samples More Diverse Than Another?

To answer this question, one must first define "diverse." Magurran's review (1988) is a full and highly readable treatment of ecological diversity and its measurement. A graphical depiction of ecological diversity is a rank abundance plot (Fig. 13.5). All the species in a sample are ranked from most abundant to least abundant. Each species has a rank (1 = most abundant species, 2 = second most abundant species, and so on), which is plotted on the horizontal axis, and an abundance, plotted on the vertical axis. Two separate features of this curve are considered components of diversity: (1) the total length of the curve, meaning the number of species in the sample, and (2) the evenness in abundance, meaning the general steepness of the slope going from most to least abundant species. More even distributions (shallower slope) are defined as more diverse.

Numerous measures of diversity somehow reduce this distribution to one number, being variously influenced by species richness, species evenness, or both. In spite of a voluminous literature directed at the development of diversity indexes, many ecologists believe they have failed to add much to our understanding of community ecology. It is difficult to claim that a diversity value is an estimate of a community parameter, one that can be compared to similar

species-accumulation curves are quite similar, showing that one habitat is not especially more productive of ant species new to the inventory than the other (Fig. 13.3). The combined curve is not steeper than the within-habitat curves, and so if only 8 samples are to be taken there is no advantage to stratifying by forest age.

Comparing methods is somewhat problematic. One can apply the method described earlier, comparing combined and separate species-accumulation curves. However, if the samples differ greatly in some measure of cost, then comparing species-accumulation curves based on number of samples may be meaningless. To compare the inventory efficiency (the steepness of the species-accumulation curve) of different

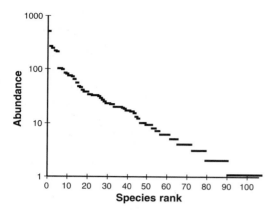

Figure 13.5. Rank abundance plot from Berlese data (see text). The 107 species are ranked from most abundant on the left to least abundant on the right. Abundance is expressed on a log scale.

estimates from other communities or other studies. Thus a diversity value for a single sample has little worth. However, instead of estimating community parameters, diversity indexes can be used to assess differences between groups of samples. This approach follows Taylor's dictum that diversity indexes are only as good as their ability to discriminate the effects of relevant environmental variables (L. R. Taylor 1978). For example, for the Berlese data we may want to know if samples from old-growth forest somehow differ from samples from second-growth forest. We can calculate the value of a diversity index for each sample and compare the sets of values using a t test or Mann-Whitney U test.

Common diversity measures are sample species richness (S), alpha (α), the Shannon index (H'), the Simpson index (D), and the Berger-Parker index (d). These measures vary in how they are influenced by the species abundance distribution. Species richness, a measure that ignores evenness, is strongly influenced by the often long tail of rare species. "Dominance" indexes, such as the Simpson and Berger-Parker, are strongly influenced by the relative abundance of the few most abundant species.

The Shannon index is influenced by both species richness and the dominant species. Alpha is influenced by the species of intermediate abundance and is relatively insensitive to the rarest and most abundant species.

Alpha is calculated by first estimating x from the iterative solution of

$$\frac{S}{N} = \frac{(1 - x)\,(-\ln(1 - x))}{x}$$

where S is the number of species in the sample and N is the number of individuals, and then calculating alpha from

$$\alpha = \frac{N(1 - x)}{x}$$

The Shannon index is calculated as

$$H' = -\sum p_i \ln p_i$$

where p_i is the proportion of individuals in the ith species.

The Simpson index is calculated as

$$D = \sum \left(\frac{n_i\,(n_i - 1)}{N(N - 1)} \right)$$

The higher D the lower the diversity, so the reciprocal of D is often used so that a higher number means higher diversity.

The Berger-Parker index is calculated as

$$d = \frac{N_{\max}}{N}$$

where $N_{\max}$ is the number of individuals in the most abundant species. As in Simpson's index, higher d means lower diversity, so the reciprocal is often used.

Returning to the Berlese data, Table 13.3 shows diversity indexes calculated for each sample. Sample species richness, alpha, and the Shannon index show weak trends toward second growth being more diverse, but the differences are not significant. The two indexes that

Table 13.3 Diversity Indexes from Berlese Samples

	Number of Individuals	Species Richness	Alpha	Shannon Index	Reciprocal Simpson Index	Reciprocal Berger-Parker Index
Old Growth Samples						
1	104	19	6.82	2.02	4.49	2.36
2	226	26	7.59	2.19	5.98	3.37
3	156	22	6.98	2.34	6.97	3.25
4	250	22	5.81	2.02	4.66	2.60
5	132	15	4.35	1.89	4.35	2.49
6	472	32	7.77	2.17	5.47	3.42
7	198	24	7.15	2.02	3.91	2.11
8	122	15	4.50	1.35	2.08	1.45
Mean (standard deviation)	208 (119)	22 (5.7)	6.37 (1.34)	2.00 (0.30)	4.74 (1.47)	2.63 (0.69)
Second Growth Samples						
9	405	37	9.92	2.61	7.68	3.27
10	187	20	5.68	2.10	5.79	3.46
11	274	31	8.99	2.48	7.99	4.49
12	120	23	8.45	2.14	4.25	2.14
13	347	20	4.61	1.80	4.13	2.97
14	108	23	8.95	2.56	8.97	3.86
15	60	18	8.72	2.39	7.70	3.16
16	112	21	7.63	2.24	6.03	3.11
Mean (standard deviation)	202 (126)	24 (6.5)	7.87 (1.82)	2.29 (0.27)	6.57 (1.79)	3.31 (0.68)
t Test	ns[a]	ns	ns	ns	$P < 0.05$	ns

[a]ns, Not significant.

emphasize the effects of dominant species show definite trends toward second growth being more diverse, with the reciprocal Simpson's index being significantly different.

Comparing diversity indexes across habitats or other environmental partitions is the most common use of sample data. For example, Kaspari (1996a) used sample species richness to show how leaf litter ants respond to disturbance. Levings (1983) used species richness and a modified Shannon index to investigate the effects of seasonality, site, and year on leaf litter ants. Roth et al. (1994) used a variety of sample diversity statistics to assess the effect of land management history on ground-foraging ants.

Exercise caution in assuming that significant differences in sample diversity are reflections of the same differences at other sampling spatial scales or in the real community. Imagine a scenario such as that depicted in Fig. 13.6. Habitat A might have higher within-sample diversity than habitat B but lower overall community diversity. Species-accumulation curves should be examined to see if this pattern is occurring. Levings (1983), Roth et al. (1994), and Oliver

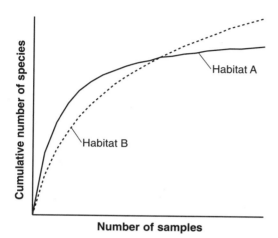

Figure 13.6. Hypothetical communities with contrasting species-accumulation curves. Habitat A has higher within-sample diversity (e.g., species richness) than habitat B, but it reaches an asymptote more quickly. A within-sample analysis alone may conclude that habitat A is more diverse.

and Beattie (1996b) augment their sample diversity analyses with species-accumulation curves for different sites. In each case the curves provide a powerful and intuitive visual confirmation of the results from within-sample diversity statistics.

Do Relative Abundance Distributions Conform to Biological or Statistical Models?

A decades-long tradition has been to compare sample relative abundance data to mathematical distributions. Some of the distributions are based on particular biological models; others are "statistical" models judged purely on goodness of fit. Three of the most common distributions used to fit relative abundance data are the geometric series, the log-series, and the lognormal. Each has a preferred plot—a method of plotting the data that most clearly demonstrates

the goodness of fit of the data to the model (Magurran 1988).

A sample that fits a geometric series produces a rank abundance plot in which each species abundance is a constant proportion of the preceding species abundance. If a rank-log abundance plot is used, the species fall along a straight line. In contrast, a log-series or lognormal distribution is nonlinear in such a plot. Geometric series distributions have lower evenness than log-series or lognormal. A biological mechanism that could produce a geometric series is the niche preemption hypothesis, in which the first species to arrive at a site monopolizes k percent of the available resources, the next species monopolizes k percent of the remaining resources, and so forth. Species-poor communities sometimes exhibit relative abundance distributions that fit a geometric series. The Berlese data do not conform well to a geometric series (Fig. 13.5); the curve is somewhat concave rather than straight.

A sample that fits a log-series is dominated by a few very common species, similar to the geometric series, but also has many rare species. The preferred plot for the log-series is a frequency histogram for which the horizontal axis is species abundance and the vertical axis is number of species. The highest point of the curve will always be the species known from singletons, with a steep monotonic decrease in numbers of species with higher abundances. Alpha, one of the parameters of the log-series distribution, has been touted as one of the best diversity indexes, mainly because of its low sensitivity to sample size (L. R. Taylor 1978). Magurran (1988) gives recipes for calculating the log-series distribution and evaluating goodness of fit of sample data.

Most species are of intermediate abundance in a sample that fits a lognormal distribution (Fig. 13.7). The preferred plot is a frequency histogram like the plot for the log-series, except that the horizontal axis (abundance class) is a

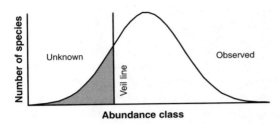

Figure 13.7. Lognormal distribution of species abundance. As sampling intensity increases, the veil line moves to the left, revealing more of the relative abundance distribution. The area under the exposed portion of the curve on the right is the observed species richness. The area under the entire curve is the estimated total species richness of the community. Estimating species richness using this method requires large data sets that clearly reveal a mode (the highest point of the curve) and closely fit a lognormal distribution.

log scale. In such a plot the lognormal, as the name implies, forms a normal distribution. Preston (1948) proposed that if a community is undersampled, only the rightmost part of the curve is revealed. He coined the term *veil line* for a vertical line dividing the lognormal distribution into two portions, the rightmost portion being the more abundant species revealed by sampling and the leftmost portion being the less abundant species remaining to be sampled. He proposed that as sampling increases the veil line moves to the left, revealing more and more of the curve. Thus sample data may be interpreted as a truncated lognormal distribution (Pielou 1975).

In practice, only very large data sets have revealed lognormal distributions. If the mode of a lognormal curve (the highest point in the distribution) is not revealed, it is practically impossible to distinguish a truncated lognormal from a log-series (Magurran 1988). Lambshead and Platt (1985) argue that the shape of the lognormal distribution should be independent of sample size and that there is no evidence that the veil line moves to the left as sample size

increases. Hughes (1986) even suggests that some of the observed lognormal distributions could be caused by species misidentifications and sampling errors.

In constructing frequency histograms of species abundances, the traditional practice is to use $\log_2$ for the horizontal axis, so that each abundance class represents a doubling of the previous one. In constructing observed distributions from real data, abundance classes are defined and the number of species in each abundance class tallied. Ideally abundance is a continuous variable such as biomass or cover, but typically abundance is number of individuals. Problems arise when fitting discontinuous abundance data to a continuous distribution such as the lognormal. Different methods have been proposed for defining abundance classes. Ludwig and Reynolds (1988) describe a common way, which is to define abundance classes as 0–1 individuals, 1–2, 2–4, 4–8, 8–16, and so on. For each species that straddles abundance classes (i.e., with abundance 1, 2, 4, 8, 16, and so on), 0.5 is added to the tally for each of the adjacent abundance classes. A problem with this method is that singletons are split between two abundance classes. The lowest abundance class will contain half the singletons, and the second lowest abundance class will contain half the singletons plus half the doubletons. This method forces the second lowest abundance class to have more species than the lowest in all cases, and it thus gives the false impression that the mode of a lognormal distribution has been revealed (Colwell and Coddington 1994).

Magurran uses an alternative method of defining abundance classes. She defines the lowest abundance class as the sum of all the singletons and doubletons, the next lowest as species with abundance 3 or 4, the next 5–8, then 9–16, 17–32, and so on. This method does not generate a "pseudo-mode" at the second abundance class. The Berlese data set shows that the same data can appear radically different

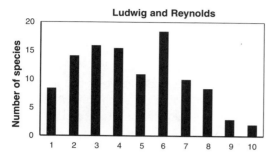

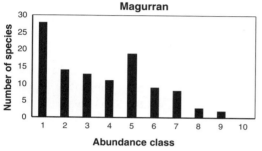

Figure 13.8. Contrasting relative abundance distributions for the Berlese data, differing only in how the abundance classes are defined (see text). Each abundance class is a doubling of the previous one. Differences in how singletons and doubletons are distributed in the lowest abundance classes dramatically alter the shape of the distributions.

when plotted using the two different abundance class definitions (Fig. 13.8).

Coddington (pers. comm.) finds fault with both the Ludwig and Reynolds and Magurran methods. The Ludwig and Reynolds method correctly assigns species to abundance classes by splitting ties into adjacent abundance classes, but it is flawed because it includes the 0.5–1 abundance class. This abundance class is undersampled because the lowest measurable abundance is 1 (the singletons). Magurran's method is flawed because it combines the two lowest abundance classes, and does not account for ties. The Ludwig and Reynolds method underestimates the lowest abundance class; the Magurran method overestimates it. Coddington proposes a modification of the Ludwig and

Reynolds method: use 1–2 as the lowest abundance class instead of 0.5–1. In other words, the lowest abundance class contains half the singletons plus half the doubletons. To see the effect of this method on the Berlese data, ignore the leftmost bar in the upper graph of Fig. 13.8

May (1975) has shown that the lognormal distribution is common in both biological and nonbiological applications (e.g., the distribution of wealth in the United States). The lognormal distribution can be produced by combining the effects of many independent variables, each of which can have any underlying distribution. Thus a lognormal distribution of biological community data may reveal only that many unknown and independent factors are contributing to the observed sample distribution. Alternatively, Sugihara (1980) provided evidence that biological community data fit certain lognormal distributions too well to be explained by multiple independent factors, and he proposed a particular model of community structure that predicted the distributions he observed.

The general problem remains that most data sets are equally well explained by many competing models. Even model distributions as fundamentally different as the log-series and the lognormal have been difficult to distinguish using sample data.

What Is the Species Richness of a Community?

Conservation biologists and environmental planners may be called upon to evaluate or rank different sites for their conservation value and to monitor changes in conservation value over time. Although not the sole criterion in determining conservation value, community species richness is often considered one of the most important (Gaston 1996). Thus obtaining reliable estimates of species richness is an important goal.

Some biological communities, such as islands and ponds, have well-defined boundaries. But most communities are not precisely defined, and so the richness of the community cannot be either. Because sampling is often area based (using, e.g., quadrats, or sampling distributed along transects), as sample size increases the area sampled does too. Ultimately this is a species-area phenomenon, and one expects species richness to be an ever-increasing function of sample area (Rosenzweig 1995). However, it may be appropriate to treat communities as though they were discrete. Colwell and Coddington (1994) take this approach, proposing that biodiversity be partitioned into two parts: the species richness of local communities and the complementarity—the dissimilarity—among these communities. When discrete, bounded communities are assumed, species-accumulation curves rise owing to increasingly accurate sampling, not species-area effects, and species richness is considered a finite community parameter.

A desirable attribute of a richness estimator is that it be independent of sample size (above some minimum sample size). For example, if one treats sample species richness as an estimator of community richness, then a species-accumulation curve shows how the estimator changes with sample size. If the curve is still climbing, then sample species richness is an underestimate of community species richness. If the curve has stabilized (reached an asymptote) above a particular sample size, then sample species richness is deemed an adequate estimate of community species richness for that sample size. Pielou's pooled quadrat method (Pielou 1966, 1969, 1975; generalized by Magurran 1988; applied by Lamas et al. 1991; Colwell and Coddington 1994) is a generalization of this approach to any richness estimator (or any index of diversity). To use the method, calculate the richness estimate based on the first sample, then on the first two samples pooled, then on the first three samples pooled, and so

forth. Plot the estimate as a function of number of pooled samples. Just as a raw species-accumulation curve can be smoothed by repeatedly randomizing sample order and averaging the curves, the estimate curve can be the average of many randomized sample orders.

A well-behaved estimator will level off, even as sample size is increasing. An objective of biodiversity research is to identify richness estimators that rise and level off sooner than sample species richness. If such estimators can be found, community species richness might be estimated with less sampling effort.

There are three general methods of estimating species richness from sample data: extrapolating species-accumulation curves, fitting parametric models of relative abundance, and using nonparametric estimators (Bunge and Fitzpatrick 1993; Colwell and Coddington 1994; Gaston 1996). In the earlier section on measuring the rate of species capture in an inventory, fitting equations to species-accumulation curves was discussed. If the equation is asymptotic, the asymptote of the fitted curve can be used as an estimate of the species richness of the community. An equation commonly used to estimate species richness is the Michaelis-Menten (M-M) equation (Clench 1979; Soberón and Llorente 1993; Colwell and Coddington 1994; Chazdon et al. 1998):

$$S(n) = \frac{S_{max} n}{B + n}$$

where $S(n)$ is the observed number of species, n is the number of samples, and S_{max} and B are fitted constants. When the smoothed species-accumulation curve of the Berlese data are fitted to this equation, $S_{max} = 141$ species (Fig. 13.9). Notice that the fitted M-M curve tends to deviate from the observed curve by overshooting at the beginning and undershooting at the end. This is a common observation (Silva and Coddington 1996), because many species-accumulation curves seem to better fit non-asymptotic curves

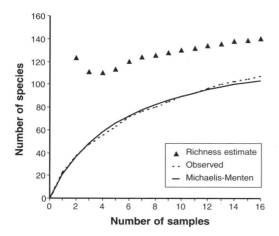

Figure 13.9. Michaelis-Menten estimates of species richness based on the Berlese data. The richness estimate is the asymptote parameter of the fitted M-M equation (see text). The pooled quadrat plot shows a gradual increase in the estimate with increasing sample size. The plotted M-M curve is based on the full data set of 16 samples. Note that although it fits reasonably well, it is "cupped" relative to the observed curve, overshooting early in the curve and undershooting at the end of the curve. This is commonly observed when M-M curves are fit to species-accumulation curves.

such as the logarithmic (Coddington, pers. comm.). Also notice that the pooled-quadrat plot of the M-M richness estimate is not stable, gradually rising as sample size increases (Fig. 13.9). A difficulty with fitting asymptotic curves is that there are many different asymptotic equations and multiple methods of fitting curves to them. This results in a plethora of different estimated richness values for the same observed species-accumulation curve. Which of the different equations or curve-fitting methods is best is unknown, and it may vary from study to study (Colwell and Coddington 1994). Examples of the use of the M-M equation to estimate species richness include Clench (1979) for Lepidoptera, Coddington et al. (1996) and Silva and Coddington (1996) for spiders, and Chazdon et al. (1998) for rainforest trees.

A common richness estimation procedure is to fit relative abundance data to a lognormal curve (see the earlier section on relative abundance distributions) and then estimate the area under the "hidden" portion of the curve (Fig. 13.7). The problems of fitting a continuous distribution to discrete data (witness the drastically different distribution shapes in Fig. 13.8 and imagine how different the richness estimates would be) and the lack of a method for calculating confidence intervals for the estimates (Pielou 1975) argue against its use in most cases (Colwell and Coddington 1994; Silva and Coddington 1996).

Some nonparametric methods show promise for richness estimation. These methods have been developed for the general problem of taking a sample of classifiable objects and estimating the true number of classes in the population (Bunge and Fitzpatrick 1993; Colwell and Coddington 1994). In ecology, such methods have been most frequently applied to estimating population size from mark-recapture data. Estimating richness is essentially the same problem, with the abundance of a species in a sample equivalent to the number of captures of an individual in a mark-recapture study.

A commonly used nonparametric estimator is the first-order jackknife (Burnham and Overton 1978, 1979; Heltshe and Forrester 1983; Colwell and Coddington 1994). The estimate of species richness (S^*_{jack}) is based on the number of uniques (L, species occurring in one sample):

$$S^*_{jack} = S_{obs} + L\left(\frac{n-1}{n}\right)$$

where S_{obs} is the observed number of species and n is the number of samples. Belshaw and Bolton (1994a) use S^*_{jack} to estimate the species richness of litter-soil ants in Ghana.

Chao and colleagues developed a set of nonparametric methods for estimating the number of classes in a sampling universe (Chao 1984, 1987; Chao and Lee 1992; Chao et al. 1993;

Lee and Chao 1994). Colwell and Coddington (1994) and Chazdon et al. (1998) evaluated these methods when applied to the problem of estimating species richness from biological data sets. Two estimators that show considerable promise are Chao2 and the incidence-based coverage estimator (ICE). Both rely on incidence (presence-absence) data. Chao2 is the simpler to calculate:

$$S^*_{\text{chao2}} = S_{\text{obs}} + \frac{L^2}{2M}$$

where L is the number of uniques and M is the number of duplicates. Calculating ICE is more complicated (Lee and Chao 1994; Chazdon et al. 1998), but it is one of the estimators provided in Colwell's EstimateS program (Colwell 1997). Chazdon et al. (1998) found Chao2 and ICE to perform similarly, although ICE was less sensitive to spatial patchiness.

When applied to the Berlese data, the pooled quadrat plot for S^*_{chao2} shows the estimate steadily increasing with sample size (Fig. 13.10). This indicates that this fauna is still far under-sampled and that attempting a richness estimate at all is probably premature without additional sampling. It cannot be emphasized enough that obtaining reliable estimates of species richness from diverse communities is difficult, requiring intensive sampling effort and very large sample sizes.

Are There Patterns of Association among Samples?

When comparing different habitats, seasons, potential conservation units, and so on, one wants to know how different the communities are in species composition. For example, we may know from richness estimation that one community is depauperate relative to another. We may wish to determine whether the depau-

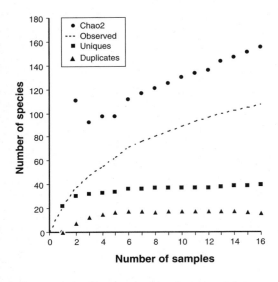

Figure 13.10. Chao2 estimates of species richness based on the Berlese data. The pooled quadrat plot shows a steady increase in the estimate with increasing sample size, which suggests that this data set is inadequate to estimate richness. The number of uniques is not declining (the presence of declining uniques is another indication of an inventory nearing completion). The number of duplicates is showing a slight tendency to decline.

perate community contains a subset of the species in the richer community or a distinct set of species.

Numerous measures of similarity and difference exist in the ecological literature (Pielou 1984; Ludwig and Reynolds 1988). A frequently used index of similarity is Jaccard's index, which is the number of species shared by two species lists (the intersection of the lists) divided by the total number of species in both lists (the union of the lists). Thus when lists have entirely distinct faunas, with no overlap, Jaccard's index equals 0, and when the two lists are identical, Jaccard's index equals 1. A measure of dissimilarity is the number of species unique to one or the other of two lists (complement of the intersection) divided by the total number of species in both lists (the union). This measure, called the Marczewski-Steinhaus distance in the literature of statistical ecology (Orlóci 1978; Pielou 1984), equals 1 – Jaccard's index. Questions in ecology and conservation biology often stress differences rather than similarities. Are two communities different? Is community *a* more different from *b* than it is from *c?* There is greater conservation value in two very different communities than two very similar ones. Because of this tendency to stress differences, Colwell and Coddington (1994) have argued that dissimilarity indexes are to be preferred over similarity indexes for reporting comparisons of communities. They propose *complementarity* as a replacement name for the Marczewski-Steinhaus distance. Complementarity thus becomes a positive measure of the dissimilarity between two species lists, and it varies from 0 to 1.

Sample complementarity (or other measures of dissimilarity or similarity) can be visually examined for gross patterns of association among samples. A matrix of complementarity values for a sample set may reveal patterns, such that within-habitat complementarities tend to be lower than between-habitat complementarities. For example, when complementarities for all pairs of Berlese samples are calculated, there is no apparent difference between within-habitat and between-habitat values (Table 13.4). Roth et al. (1994) used such a matrix approach (with a similarity index) to conclude that samples were more similar within land management categories than between them. Nonindependence of complementarity values in a matrix such as Table 13.4 (individual samples contribute to multiple complementarity values) makes statistical comparison of values in different blocks problematic.

When samples occur along a spatial or temporal gradient (rather than in habitat blocks, as in the earlier examples), complementarities can be plotted as a function of the distance between the sample pairs (for an example, see Belshaw and Bolton 1994a, using Morisita's index). Again statistical analysis of the resultant plot is

Table 13.4 Complementarity of Paired Berlese Samples[a]

	1	2	3	4	5	6	7	8	9	10	11	12	13	14	15
2	0.75														
3	0.83	0.80													
4	0.89	0.83	0.67												
5	0.83	0.83	0.84	0.77											
6	0.79	0.68	0.62	0.74	0.85										
7	0.70	0.72	0.69	0.82	0.78	0.76									
8	0.87	0.92	0.77	0.84	0.85	0.83	0.82								
9	0.78	0.79	0.82	0.80	0.87	0.77	0.76	0.87							
10	0.78	0.85	0.80	0.80	0.91	0.73	0.78	0.83	0.84						
11	0.89	0.79	0.80	0.82	0.90	0.71	0.80	0.82	0.74	0.84					
12	0.80	0.83	0.85	0.78	0.85	0.75	0.79	0.91	0.75	0.87	0.77				
13	0.85	0.82	0.86	0.86	0.94	0.79	0.71	0.91	0.70	0.89	0.76	0.70			
14	0.89	0.89	0.68	0.71	0.91	0.72	0.76	0.81	0.82	0.84	0.74	0.82	0.81		
15	0.81	0.90	0.79	0.79	0.90	0.78	0.83	0.73	0.85	0.81	0.78	0.79	0.85	0.79	
16	0.79	0.76	0.87	0.81	0.84	0.71	0.82	0.91	0.82	0.89	0.79	0.71	0.79	0.78	0.70

[a]Mean complementarity: within-forest type, 0.79 (standard deviation 0.06); between-forest type, 0.82 (standard deviation 0.06).

dubious owing to the nonindependence of points, but visual inspection of the data may nonetheless reveal gross patterns (e.g., more distant samples having higher complementarity).

When species lists are complete for two communities, complementarity is a straightforward parameter describing the difference between them. When complementarity is between samples, however, complications arise (Koch 1987; Cobabe and Allmon 1994; Colwell and Coddington 1994). A major difficulty in assessing community differences is distinguishing "sampling zeroes" from "structural zeroes." Imagine two urns, each of which contains balls of various colors. A sample from the first urn contains red balls; a sample from the second does not. We cannot tell whether the absence of a red ball in the second sample is due to red balls being present in the second urn but not captured in our sample, or red balls being absent in the second urn. In a species-by-sample matrix generated by a sampling program, an entry of "0" or "absent" for a species may be due to undersampling (the species was present in the community, but missed by the sampling), or the species may be truly absent. Gaston (1994) refers to the former as sampling zeroes and the latter as structural zeroes. The inability to distinguish them is a fundamental problem in biodiversity inventory. Teasing out the effect of undersampling is a subject of current investigation (Chen et al. 1995; Colwell 1997).

A major realm for the analysis of patterns of association among samples is ordination and classification (Gauch 1982; Pielou 1984; Ludwig and Reynolds 1988; Kent and Coker 1992; Jongman et al. 1995; see also Mike Palmer's Web site for ordination: http://www.okstate.edu/artsci/botany/ordinate). Pielou introduces her treatment of ordination with a geometric model. Imagine a hyperspace with as many dimensions as there are species in a species-by-sample data matrix. Each dimension is one species. Each sample can then be plotted as a point in the hyperspace, with the coordinates being the abundance values for each species. In the case of presence-absence data, the values are all 0 or 1. The complete data set is represented as a cloud of points, one point for each sample, in this S-dimensional space (S is the number of species). The objective of ordination is to project this cloud of points, which we cannot visualize, onto a two- or three-dimensional subspace, which we can visualize, in such a way that patterns in the data are revealed. Such patterns might be clusters of points, such that samples from particular habitats group together.

Examples of studies of ant communities that use ordination and classification techniques include Andersen (1991d), Andersen and Yen (1992), and Oliver and Beattie (1996b). Large numbers of rare species may cause problems in ordination, and Pielou (1984) suggests that it might be best to exclude them prior to analysis.

Conclusion

Ants are a dominant element of most terrestrial environments and thus are frequent subjects of ecological sampling. Methods of analysis of ecological data are constantly evolving, a function of changing research questions and improved analytical tools. The species-by-sample matrix is likely to remain the basic data structure, with research questions dictating how the samples are taken and new analytical tools influencing the analysis. Measuring true relative abundances in nature will always be problematic because of clumped spatial distributions and biased sampling methods. However, many questions can be asked of ant communities that do not require precise knowledge of relative abundance. We can ask inventory questions regarding degree of completeness and rate of approach to completeness. We can ask whether sample diversity and composition are related to environmental variables, such as habitat, season, year, land management regime, elevation, or diversity of other taxa. We can estimate community species richness given certain assumptions about community boundaries. Cognizance of the relationships between sampling regime, data structure, and analysis options will not only improve the quality of individual projects involving ant communities but also make more likely synthetic analyses that examine results from many separate studies. The global importance of ants in terrestrial ecosystems and their potential value in environmental monitoring justify an emphasis on quantitative sampling and cross-study comparability.

ACKNOWLEDGMENTS

I thank Rob Colwell and Jonathan Coddington for many discussions of inventory methodology and for help with the manuscript of this chapter. I thank the ALAS staff—Danilo Brenes, Carolina Godoy, Nelci Oconitrillo, Maylin Paniagua, and Ronald Vargas—for helping provide the Berlese data set. This work has been supported by National Science Foundation (NSF) grants BSR-9025024, DEB-9401069, DEB-9706976, and by the Office of Forestry, Environment, and Natural Resources, Bureau of Science and Technology, U.S. Agency for International Development, under NSF grant BSR-9025024; by Apple Computer; and by ACI-US.

Chapter 14

The ALL Protocol

A Standard Protocol for the Collection of Ground-Dwelling Ants

Donat Agosti and Leeanne E. Alonso

There is no single best method to sample the ground-dwelling ant fauna. The objectives of each study will determine the appropriate methods and sampling intensity. For example, a wide variety of methods and a greater effort are needed to obtain a thorough inventory of the ants of an area and to collect as many species as possible. In contrast, a rapid assessment of the ant fauna using a few standardized methods and less sampling effort would allow for comparisons between different habitats and would establish a baseline for a longer-term monitoring program.

However, the use of standardized methods that can be reliably repeated in different habitats, at different times of the year, and by different researchers is beneficial. Using the same basic methodology, individual studies can be analyzed in relation to others and can thus be

put into a larger, global context. Here we present a standard protocol for the collection of ground-dwelling ants, the Ants of the Leaf Litter (ALL) Protocol. We expect that this protocol will stimulate further research on ant diversity and that it will be used by a variety of researchers in a diverse array of sites.

The ALL Protocol starts with a minimal configuration, utilizing two of the ant collecting methods that have been proven to sample the largest component of the ground- and leaf litter–inhabiting ant fauna: the mini-Winkler extractor and pitfall traps (Chapter 9). This method is rapid; sampling can be completed in a total of 3 days per site if desired. The sample size, 20 1-m^2 samples of leaf litter and 20 pitfall traps, has been found to be sufficient to sample at least 70% of the ant fauna (Chapter 10).

However, we suggest that researchers start with 50 samples during the first survey to practice the techniques and to determine the actual number of samples needed to collect the desired percentage of ant species (Chapter 10). Depending on the study objectives, other complementary methods can be added to the standard protocol in order to sample a wider range of ant species (Chapters 1 and 9).

The ALL Protocol

Basic Setup
   200-m transect (at least one)
   20 sampling points at 10-m intervals
   48-hour time period
   1–2 people (2 people recommended)
Methods Employed at Each Sampling Point
   Standardized, Repeatable Techniques
      Collect leaf litter, 1 m^2
         Sift litter
         Extract ants from litter using mini-Winkler
      Place 1 pitfall trap
   Optional Techniques to Collect More Species
      Inspect dead wood
      Scrape soil (15 × 15-cm area at 1-cm layers
         down to 10 cm)
      Direct collecting by hand

Overview of the ALL Protocol

The most important points in implementing the ALL Protocol are outlined in this section, along with references to chapters of this book that contain more information. See Appendixes 1 and 3 for complete lists of equipment needed for the sampling methods and specimen processing.

Transects

Before choosing a particular transect, it is worthwhile to walk through the area to get an impression of the overall environmental variation. Chapters 1 and 9 provide guidance on transect placement.

Ecological Data

In addition to the standard collection information (Chapter 11), ecological data must be recorded. For each transect, the following minimal set of parameters should be described: name of collector, date, choice of transect, locality, habitat, season, soil type, temperature, and microhabitat. See Chapter 9 for a complete list of relevant ecological information and explanations of these parameters.

Labeling Field Samples

It is of the utmost importance to label all samples adequately. Most of the labeling can be done prior to the commencement of field work. Vials used for collecting ants by hand or from logs should preferably be prelabeled as well. See Chapters 9 and 11 for more details.

Pitfall Traps

Pitfall traps should be placed 1 m from the transect line on the opposite side of the transect from where the leaf litter samples were taken. Any plastic drinking cup with smooth sides can be used, but it is best to use cups with openings of the same diameter consistently, in order to standardize the samples. Twenty cups are needed. See Chapter 9 for more information on how to set and collect pitfall traps.

Leaf Litter Samples and Mini-Winkler Extraction

See Chapter 9 for complete instructions on how to collect the 1-m^2 leaf litter samples and extract ants using mini-Winkler sacks. The ALL Protocol requires at least 20 mini-Winkler extractors and one sifter.

Sorting Samples in the Laboratory

Ant specimens and other invertebrates can be separated from debris using the salt water extraction method (Chapter 11). After separation the samples should be washed with ethanol.

Identifying Morphospecies

Ants from each sample should be separated from other invertebrates and housed in a separate vial.

Procedures for sorting and identifying specimens to morphospecies are given in Chapter 11.

Labeling Samples and Specimens

All samples must be labeled immediately with proper labels (Chapter 11).

Time Requirements

A minimum of 3 days is needed to use the standard ALL Protocol. The leaf litter collections should be run through the mini-Winkler extractor for a 48-hour period. Pitfall traps should also be left out for 48 hours. Both the mini-Winkler extractor and the pitfalls can be left running for a longer time if desired, but samples should be collected from both at 48 hours in order to obtain the standard sample. More time may produce additional species, but the benefits should be weighed against the advantages of running more transects instead.

For inventories, it is recommended that more than one transect be run and the species-accumulation curves be plotted by sample and transect (Chapter 13). This approach will provide a review of the fraction of the ant fauna sampled and will help determine if additional transects are needed. Additional sampling methods—such as the inspection of dead wood, soil scraping, and direct sampling—may be added in order to maximize the diversity of ants sampled (Chapter 9).

Timetable

An estimate is given in this section for the amount of time required for one person to carry out the standard ALL Protocol. It is recommended that two people carry out the protocol together, to provide assistance with leaf litter gathering, sifting, and other tasks. The estimated total time needed to sample, process, and identify ant specimens from one transect is 161 working hours for a single professional.

Field Work

All times are in hours.

	One Person	Two People
DAY ONE		
Early morning		
1. Mark transect	1.5	1.0
2. Dig in pitfall traps	1.5	1.0
3. Collect Winkler samples	5.0	3.0
Afternoon		
1. Fill in Winkler apparatus	3.0	2.0
Late afternoon/early evening		
1. Direct collecting	1.0	1.0
Total	12.0 hours	8.0 hours
DAY THREE		
Morning		
1. Collect one log	1.0	1.0
2. Direct collecting	1.0	1.0
3. Scrape soil	1.0	1.0
Afternoon		
1. Analyze soil samples	2.0	1.0
2. Collect pitfall traps	2.0	1.5
3. Collect Winkler samples	2.0	1.5
4. Check all labeling	0.5	0.5
Total	9.5 hours	7.5 hours

Laboratory Work, Identification, and Analyses

Mounting, labeling, and identifying specimens from Winkler samples	60
Mounting, labeling, and identifying ant specimens from pitfall traps	60
Mounting, labeling, and identifying ant specimens from other samples	10
Entering and analyzing data	10
Total	140 hours

Data on ant diversity collected using this protocol can be compiled and compared, thereby providing the context needed to begin looking at truly global ant diversity patterns. We encourage researchers who use the ALL Protocol to provide their data to the social insects Web site (http://research.amnh.org/entomology/social_insects/) for inclusion in a global database on ant diversity.

Chapter 15

Applying the ALL Protocol

Selected Case Studies

Brian L. Fisher, Annette K. F. Malsch, Raghavendra Gadagkar,
Jacques H. C. Delabie, Heraldo L. Vasconcelos, and Jonathan D. Majer

Although the ALL Protocol put forward in this book was only recently developed and is published for the first time here, it is derived from the experiences of many myrmecologists and from numerous studies of ant diversity over the years. Before the development of this protocol, studies of ant diversity utilized a wide variety of methods, as described in Chapter 9. The ALL Protocol is the result of an extensive evaluation of these methods in different countries and under a variety of conditions. The studies by Delabie et al. (Chapter 10) provided the strongest data for choosing methods for the ALL Protocol, but a number of other key studies were also influential in its development.

Brief descriptions of these key studies, from diverse parts of the world (Madagascar, Malaysia, India, and Brazil), are provided here to illustrate how the two principal collecting methods of the ALL Protocol (leaf litter extraction using the Winkler technique and pitfall traps) compare, and how they have been used successfully in a variety of biodiversity studies, particularly to measure ant diversity and to detect habitat change. Full descriptions of these studies are presented in Agosti et al. (2000).

Madagascar

A series of studies in Madagascar by Fisher and colleagues (Fisher 1996a, 1996b, 1998, 1999a, 1999b; Fisher and Razafimandimby 1997; Fisher et al. 1998) provides the best insight into the usefulness of the ALL Protocol and the comparative value of its two principal methods, Winkler extraction and pitfall traps. In one particularly informative study, Fisher surveyed ground-dwelling ant diversity at elevation zones separated by 400 m at four rainforest localities: the Reserve Naturelle Integrale d'Andohahela, the Reserve Naturelle Integrale d'Andringitra, the Reserve Speciale d'Anjanaharibe-Sud, and the Western Masoala Peninsula in eastern Madagascar.

At each site, 50 pitfall traps were used and 50 leaf litter (Winkler) samples were taken, in parallel lines 10 m apart along a 250-m transect. Pitfall traps were placed and leaf litter samples gathered every 5 m along the transect. Pitfall traps consisted of test tubes (18 mm internal diameter by 150 mm long), partly filled to a depth of about 50 mm with soapy water and a 5% ethylene glycol solution, inserted into polyvinyl chloride (PVC) sleeves, and buried with the rim flush with the soil surface. Traps were left in place for 4 days. Ants were extracted from samples of leaf litter using Winkler extractors over a 48-hour period (Chapter 9; Fisher 1998).

The observed and predicted number of species sampled by the Winkler, pitfall, and combined methods for 10-, 20-, 30-, 40-, and

Table 15.1 Observed Number of Ant Species Evaluated at Different Sample Sizes for Winkler Sacks, Pitfall Traps, and Both Methods for Each 800-m Zone Site in Madagascar[a]

Methods	Observed Species Richness after:					Estimated Species Richness[b]		
	10 Samples	20 Samples	30 Samples	40 Samples	All (50) Samples	ICE	Jack-knife	M-M
800 m Andohahela								
Winkler	39.6 (59.3)	49.5 (74.1)	55.3 (82.8)	59.5 (89.1)	63 (94.3)	80.0	79.7	66.8
Pitfall	14.8 (43.6)	20.5 (60.3)	24.3 (71.5)	27.4 (80.6)	30 (88.3)	48.4	43.7	34.0
Both methods	44.5 (59.6)	55.5 (74.4)	62.1 (83.2)	66.9 (89.7)	71 (95.2)	90.3	90.6	74.6
785 m Andringitra								
Winkler	52.2 (66.8)	63.3 (81.0)	68.8 (88.1)	72.8 (93.1)	76 (97.2)	87.5	90.7	78.2
Pitfall	10.2 (45.6)	14 (62.4)	16.4 (72.9)	17.6 (78.6)	19 (84.7)	23.4	24.9	22.4
Both methods	53.8 (68.6)	64.1 (81.8)	69.8 (89.0)	74.0 (94.4)	77 (98.2)	88.2	91.7	78.4
825 m Andringitra								
Winkler	43.4 (67.6)	51.7 (80.6)	56.4 (88.0)	60.5 (94.4)	64 (99.8)	78.8	79.7	64.1
Pitfall	8.7 (39.0)	12.2 (54.9)	14.9 (67.0)	17.0 (76.7)	19 (85.6)	32.1	27.8	22.2
Both methods	44.9 (66.5)	53.3 (79.1)	59.4 (88.1)	63.5 (94.1)	67 (99.3)	82.6	83.7	67.4
825 m Masoala								
Winkler	62.2 (57.1)	79.8 (73.3)	91.1 (83.7)	99.3 (91.3)	106 (97.4)	139.84	136.38	108.81
Pitfall	8.9 (38.2)	12.7 (54.5)	15.8 (67.7)	18.27 (78.6)	20 (86.0)	33.05	27.84	23.25
Both methods	62.5 (55.4)	81.2 (71.9)	93.4 (82.7)	102.4 (90.6)	109 (96.5)	141.7	139.4	113.0
875 m Anjanaharibe-Sud								
Winkler	54.4 (58.4)	69.1 (74.1)	78.9 (84.7)	86.7 (93.1)	92 (98.8)	117.17	114.5	93.2
Pitfall	8.5 (37.8)	11.7 (52.1)	14.6 (65.0)	17.4 (77.2)	20 (88.8)	89.2	32.7	22.51
Both methods	56.2 (57.6)	71.9 (73.7)	82.7 (84.7)	91.1 (93.3)	97 (99.3)	126.7	122.5	97.6

[a]Number of species represents the mean of 100 randomizations of sample pooling order.

[b]ICE, incidence-based coverage estimator; jackknife, first-order jackknife estimator; M-M, Michaelis-Menten asymptote (the percentage of the M-M asymptote is given in parentheses in the first five columns).

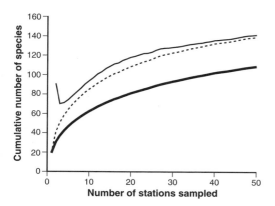

Figure 15.1. Assessment of leaf litter ant sampling technique at 825 m on the Masoala Peninsula, Madagascar. The lower species-accumulation curve (thick line) plots the observed number of species as a function of the number of stations sampled. The upper curves display the nonparametric first-order jackknife (dashed line) and the incidence-based coverage estimator (ICE; solid line), estimated total species richness based on successively larger numbers of samples from the data set. Curves are plotted from the means of 100 randomizations of sample accumulation order.

50-sample sizes are shown in Table 15.1. Within the area of the survey, the Winkler technique collected the majority of ants foraging and living in the leaf litter. Most species collected by pitfall traps were also sampled by Winkler extraction, indicating that whereas pitfall samples in the same area would most likely add additional species, these species would probably have already been obtained by the Winkler method. Although this may hold true for most rainforest sites, it may not apply to all habitats. For example, Fisher and Razafimandimby (1997) found that in dry forest habitats, which contain more areas of open or bare ground, pitfall traps may collect a greater number of unique species.

Species-accumulation curves for the 825-m site on the Masoala Peninsula, the most species-rich site (Fig. 15.1), indicate that within the area

of the survey the techniques employed collected the majority of ants foraging and living in the leaf litter in the area encompassed by the 250-m transect, and that with increased sampling effort using the same methods in the same area, only marginal increases in species richness would be attained. Although additional collecting methods, or a survey in a different area or season at the same elevation, would most likely collect additional species, these results show that the ALL Protocol provides sufficient sampling for statistical estimation and comparison of species richness, and for comparison of faunal similarity and species turnover.

Malaysia

Studies of ant diversity in the Pasoh Forest Reserve of West Malaysia by Malsch provide interesting data on the effects of plot size sampled in ant diversity studies. Situated in Negeri Sembilan, West Malaysia, about 140 km southeast of Kuala Lumpur (2 59′N, 102 19′E), the Pasoh Forest Reserve is a typical example of a Southeast Asian ever-wet lowland rainforest, with primary lowland dipterocarp forest situated between 75 and 150 m above sea level.

A total of nine leaf litter plots (each 25 m²) were investigated. Each plot comprised a 5 × 5-m² area with an additional 3 × 3-m² area nested in the middle of the plot. Each of the two nested areas (16 m² and 9 m²) was sampled separately, and the sum of the two equaled a 25-m² area. This approach enabled the comparison of all nested areas within plots. Ants from the leaf litter were extracted by the Winkler method after 24 hours and then again after an additional 24 hours (Chapter 9).

The nested sampling area design revealed that the size of the leaf litter sample (plot size) can influence the number of ant species collected. The number of species collected per square meter for each 9- and 25-m² plot is shown in Table 15.2. On average, one more species was

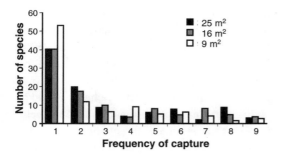

Figure 15.2. Frequency of capture for species in nine 9-m², 16-m², and 25-m² plots in Malaysia.

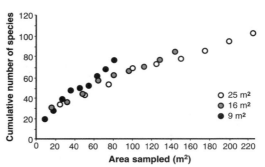

Figure 15.3. Species-accumulation curves for the three plot sizes of 9 m², 16 m², and 25 m² in Malaysia.

found per m² on the 9 m² plots compared to the 25 m² plots. The similarity in ant species composition was also affected by plot size, with Sorensen s similarity values of the 25 m² plots ranging from 37.5% to 63.8%, and those of the 9 m² plots from 28.5% to 66.7%. The mean values are 52.7% for the 25 m² plots and 43.0 % for 9 m² plots. These differences are highly significant ($P < 0.001$, Mann-Whitney U-Test). The higher species turnover in the 9 m² plots resulted from more single captures and a smaller number of repeated captures (Fig. 15.2).

The species-accumulation curves for each plot size reveal that none of the plot sizes samples all ants of the area since none of the curves

levels off (Fig. 15.3). In addition, each plot size produces a different estimation of the overall ant species richness. This is an important consideration when comparing the results from studies using different plot sizes; it is not possible to compare them directly.

These results emphasize the importance of using a consistent, standard plot size across studies in order to make comparisons. The 1-m² litter plot of the ALL Protocol provides this standardization.

Western Ghats, India

Ant diversity was investigated using a variety of sampling techniques in the state of Karnataka, Western Ghats, India (Gadagkar et al. 1990, 1993). A total of 36 1-ha plots from 12 habitat types were sampled in sites representing elevations from sea level to 600 m in forested habitats, in three monoculture plantations, and in a forest that was regularly harvested to produce leaf manure. At each of these sites, sampling was carried out in three 1-ha plots.

Five sampling methods for ground-dwelling ants were employed at each site: vegetation sweeps, pitfall traps, light traps, scented traps, and direct (hand) collecting. Light traps use a luminescent light source to attract insects that

Table 15.2 Number of Ant Species per Square Meter for 9- and 25-m² Plot Samples in Malaysia

Plot	9 m²	25 m²
P1	2.22	1.36
P2	2.44	1.16
P3	2.22	1.44
P4	2.22	1.64
P5	1.55	0.92
P6	2.00	1.24
P7	3.11	1.52
P8	3.22	1.76
P9	3.22	1.68
Mean	2.47	1.41

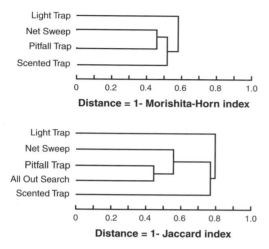

Figure 15.4. Dendrograms comparing different sampling methods by ant species trapped in India. Data pooled from 36 plots for each sampling method.

are active at night. Although light traps are typically used to sample flying insects, they can occasionally be useful for attracting flying alate ants and some nocturnal ant species. Scented traps are essentially a combination of two standard ant sampling techniques (pitfall traps and baits); in this study they consisted of 2.5-liter plastic jars that were baited with unrefined sugarcane and hung at about 1 m from the ground on wooden pegs. Intensive hand collecting was performed in each 1-ha plot to collect representatives of as many species of ants as possible. Two persons made the search for 1 hour between 1400 and 1500 in every case.

In addition to providing the first estimates of ant diversity and abundance for any forest locality of India, the results of this study provide informative comparisons of five different methods of ant sampling. The combination of the four trapping methods used was somewhat more successful than hand collecting, yielding 120 species from 31 genera while hand collecting yielded 101 species from 27 genera. More significant is the fact that the traps and hand col-

lecting yielded different species; while 78 species were obtained by both methods, the traps yielded 42 unique species and hand collecting yielded 20 unique species. It appears, therefore, that in spite of the efficacy of the traps, a combination of trapping and hand collecting may be desirable if a more complete list of ant species at a site is desired.

Of the four trapping methods used, pitfall traps sampled the most species, followed by vegetation sweeps, scented traps, and light traps in that order. The fact that pitfall traps and vegetation sweeps were more successful is not surprising; indeed the fact that scented traps and light traps yielded as many ants as they did is surprising. Not only did the scented traps and light traps yield more ants than expected, they yielded an ant fauna rather different from that obtained by the other methods (Fig. 15.4).

Had this study included leaf litter extraction as in the ALL Protocol, many more ant species would likely have been collected. However, the combination of several sampling methods used in this study, including hand collecting, illustrates that different techniques usually collect different components of the ant fauna. Therefore, if a more thorough inventory of ground-dwelling ants is desired, it is recommended that a few additional methods be used along with the ALL Protocol.

Brazil

Two studies in Brazil, one in the highly fragmented Atlantic rainforests of Bahia and the other in the Brazilian Amazon, reveal the utility of the ALL Protocol as a means of detecting habitat disturbance.

Atlantic Forests

In Bahia, ten 110-m transects were established from the center of a botanical reserve of secondary rainforest in the Centre for Cocoa Research, Itabuna, Bahia (Majer et al. 1997).

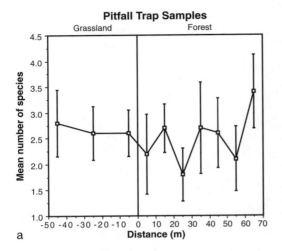

a

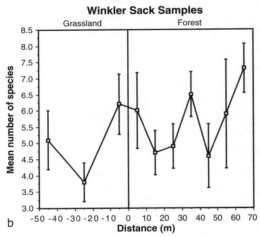

b

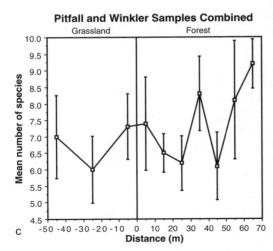

c

Transects were marked at 10-m interval points and were sited so that seven points extended into the reserve, one occurred in the middle of the planted edge 5 m outside the forest, and two were in an adjacent field. The ALL Protocol was followed, with leaf litter samples collected along the transect and extracted using Winkler sacks (for only 24 hours) and pitfall traps set out along the transect and left out for 48 hours.

Figures 15.5a–c illustrate the variation in the mean number of ant species collected by pitfall traps, by Winkler sacks, and by both methods along the transects. The mean number caught by pitfall traps ranged from 1.8 to 3.4, and there was little apparent trend in numbers along the transects, although the maximum richness was encountered at the point farthest into the forest (Fig. 15.5a). The mean number of species extracted by the Winkler sacks ranged from 3.8 to 7.3. It is noteworthy that the highest density was once again reached at the point farthest into the forest and that the lowest density was encountered at the point 25 m into the field (Fig. 15.5b).

The results of this survey reveal an abrupt differentiation in forest and field ant community composition. Five assemblages of ant species were distinguished along the transect. The largest grouping contained species that were ubiquitous along the transects or that were ubiquitous except at points outside the forest. The second group comprised ants that showed a tendency to occur around the outer forest margin, while the third and fourth groups contained ants that were generally found in deeper forest. The fifth group contained 12 species that were found

Figure 15.5. Mean number of ant species (and standard error) sampled by (a) pitfall traps, (b) Winkler sacks, and (c) both methods combined along ten transects extending from the field into the rainforest in Bahia, Brazil. The vertical line indicates the position of the fence around the forest reserve.

only in the field or planted edge. These ant assemblages can be used to monitor ant communities in these different land uses and to detect further changes, even if fairly subtle, in habitats and their microclimates.

Amazon

In the Brazilian Amazon, ground-dwelling ants were collected in three 1-ha forest fragments, in three 10-ha fragments, in two 100-ha fragments, and in one continuous forest area. In each of these nine fragments a 1-ha plot was delimited and, within this, a total of 36 sampling points, distributed at intervals of 20 m, were established. Three methods of ant sampling were used: litter extraction, pitfall traps, and soil samples.

Of the three methods, litter sampling was the most efficient in terms of the number of species collected. The mean number of species collected per plot was significantly greater in the litter than in the pitfall traps and soil, whereas the number of species collected in the pitfall traps was greater than that in the soil (ANOVA, $F_{2,16}$ = 29.87, $P < 0.001$; Table 15.3). Although the number of species recorded per fragment was greater in the litter samples than in the pitfall traps, the total number of species recorded by each of these two methods in all nine forest plots studied was quite similar, and both yielded greater numbers than collections from the soil samples (Table 15.3). Litter sampling was also the best method to predict overall ant species richness (number of species collected using the three methods combined) in each of the study plots.

The number of species that were unique to each method ranged from 20 to 43 species, a number that usually represented more than 20% of all species collected by that method (Table 15.3). This observation indicates that these methods are complementary. Their use in combination, therefore, better characterized the ant fauna of the fragments. Species of Cerapachy-

Table 15.3 Number of Ant Species Collected Using Three Different Sampling Methods in Forest Fragments near Manaus, Brazil

Subfamily	Litter Samples	Pitfall Traps	Soil Samples
Myrmicinae	96	82	53
Ponerinae	33	36	33
Formicinae	11	12	12
Dolichoderinae	3	5	0
Ecitoninae	1	5	1
Cerapachyinae	2	0	5
Pseudomyrmecinae	1	2	0
Leptanilloidinae	0	0	2
Total (unique)	147 (43)	142 (39)	106 (20)
Mean ± SD	54.2 ± 11.6	45.7 ± 7.9	30.3 ± 5.6

inae, for instance, were recorded mostly in the soil samples, whereas those in the Ecitoninae were mostly recorded in the pitfall traps. On the other hand, many Myrmicinae were only recorded in the litter samples (Table 15.3).

No consistent changes in species diversity were found in response to variations in fragment area. It must be stressed, however, that these results reflect a lack of relationship between the *density* of ant species (number per unit area) and forest area, not in overall species *number* and forest area, as the latter relationship is clearly positive and significant. Within two of the three sites studied, the density of ant species increased as forest area increased, whereas in the third site the opposite trend was found. Differences in the history of fragment isolation (resulting in different matrix habitats) may have accounted, at least in part, for these conflicting results (Vasconcelos and Delabie 2000).

Ordination of the study plots according to their similarities in species composition indicated that forest fragmentation does affect the composition of the ground-dwelling ant community. A "site effect" on species composition was also detected, indicating some degree of heterogeneity in species distribution among the three sites studied, even though these sites were only 10–25 km apart.

The results of this study strongly suggest that forest fragmentation affects the structure of ground-dwelling ant communities. The diversity and composition of the ant community would thus be useful to include in a monitoring program of forest fragments to follow and predict future changes.

Conclusion

These five case studies, plus those conducted by Delabie et al. (Chapter 10), provided the comparative basis for selection of methods for the ALL Protocol. They also illustrate the application of the ALL Protocol to address a wide range of research and applied conservation questions in a variety of locations. We hope that these studies will inspire and guide the use of the ALL Protocol, and the inclusion of ground-dwelling ants, in biodiversity studies across the globe.

ACKNOWLEDGMENTS

The Malagasy project was funded in part by grants from the World Wide Fund for Nature, Madagascar; the National Geographic Society (5152-93); and the National Science Foundation (INT 9319515). Research in India was supported by grants from the Ministry of Environment and Forests, Government of India. The Amazon study was supported by the Biological Dynamics of Forest Fragments Project (a project of the Smithsonian Institution and the Brazilian Institute for Amazon Research) and the Brazilian Council for Scientific and Technological Development.

List and Sources of Materials for Ant Sampling Methods

Materials

STUDY AREA SETUP
Measuring tape (50 m or longer)
Random number table
Surveyor flags
Flagging tape
Compass

GENERAL ANT SAMPLING MATERIALS
Forceps (featherweight and watchmaker's #5)
Aspirator and aspirator vials
Ethanol (ethyl alcohol) (70–95%; preferably 95%)
Specimen storage vials (glass or plastic with polypropylene stoppers or leakproof caps; preferably 2 ml polyethylene tubes with silicon O-ring screw caps)
Resealable plastic bags
Field notebooks and pencils
Preprinted or blank sturdy paper specimen tags
Aluminum tags
Indelible ink marker

Trowel or shovel
Gloves

Method-Specific Ant Sampling Materials

BAITS
Bait materials (e.g., canned tuna)
Bait platform materials

PITFALL TRAPPING
Cups for pitfall traps
Scoop for pitfall traps
Shovel for digging trap holes
Killing agent (e.g., propylene glycol)
Tea strainer

QUADRAT SAMPLING, INTENSIVE SAMPLING, AND COLONY SAMPLING
Prefabricated quadrat of polyvinyl chloride (PVC) tubing
White sorting or sifting tray (with or without mesh)

LITTER SAMPLING
Berlese funnels and collecting jars
Support for Berlese funnels
Litter sifter
Winkler sacks
Ground cloth
Large plastic litter sample bags
Machete or large knife

ENVIRONMENTAL MEASUREMENTS
Thermometers for soil and air
Humidity-measuring device
Densiometer
Marked pole for measuring vertical vegetation
 profiles
Wire (flag) for measuring litter depth
Meter stick or tape to measure cover
 types

Specialty Suppliers
Australian Entomology Supplies
Box 250
Bangalow, NSW 2479
Australia
Telephone/fax: 61-66-847188
Entomological equipment: forceps, vials, Berlese
funnels, aspirators

Bioquip Products
17803 LaSalle Avenue
Gardena, CA 90248-3602
USA
Telephone: 1-310-324-0620
Fax: 1-310-324-7931
e-mail: bioquip@aol.com
Entomological equipment: forceps, vials, Berlese
funnels, aspirators, unit trays, insect drawers and
cabinets

A. Daigger & Company, Inc.
199 Carpenter Avenue
Wheeling, IL 60090
USA
Telephone: 1-800-621-7193
Fax: 1-800-320-7200

Gas-collecting bulbs (polyethylene valves), specimen containers (for pitfall traps), sample bags

Fisher Scientific
P.O. Box 3029
Malvern, PA 19355
USA
Telephone: 1-800-766-7000
Vials, including 2 ml polyethylene microcentri-
fuge tubes with silicon O-ring screw caps; nal-
gene bottles

Forestry Suppliers, Inc.
P.O. Box 8397
Jackson, MS 39284-8397
USA
Telephone: 1-800-647-5368, 601-354-3565
 (international)
Fax: 1-800-543-4203, 601-355-5126
 (international)
Materials for study area setup: flags, flagging
tape, measuring tapes, tools, densitometers, note-
books, tags, compasses, wind meters, global
positioning system units

Marizete Pereira dos Santos
Rua do Coqueiro no 60
Bairro Conquista
Cidade Ilhéus
Bahia-Brasil
CEP 45 660 000
Brazil
Telephone: 550-73-231-5888
E-mail: pires@maxnet.com.br
Winkler sacks, litter sifters

Omega Engineering, Inc.
P.O. Box 4047
Stamford, CT 06907-0047
USA
Telephone: 1-800-826-6342
Fax: 1-800-848-4271
Digital handheld thermometers, thermocouples,
humidity meters, thermohygrometers

PGC Scientifics
P.O. Box 7277
Gaithersburg, MD 20898-7277
USA
Telephone: 1-800-424-3300
Fax: 1-800-662-1112
Disposable polypropylene sample containers (for use as pitfall traps), twirl (sample) bags

Sante Traps
1118 Slashes Road
Lexington, KY 40502
USA

Telephone: 859-268-9534
E-mail: santetraps@aol.com
Winkler sacks, litter sifters, Malaise traps

Sarstedt, Inc.
P.O. Box 468
Newton, NJ 28658-0468
USA
Telephone: 1-800-257-5101
Vials, including polyethylene specimen freezer tubes with silicon O-ring screw caps (#72.694.105)

Ant Survey Data Sheet

Observer:_____ Date:_____ Time:_____

Sample code:_____ GPS coordinates:_____

Locality:_____

Habitat type:_____ Elevation:_____ Slope:_____ Aspect:_____

Sample type:_____ Sampling duration:_____

Quadrat size/trap size/litter volume/bait type: _____

Nest type:_____

Temp (air):_____ Temp (soil):_____ Relative humidity (R_H):_____ Wind:_____

Insolation:_____

Percentage Ground Cover

Bare:_____ Litter:_____ Stone:_____ Plant:_____ Other:_____

Soil description:_____ Litter depth:_____

Foliage Height Profile (cm):

Point 1: 0–25:_____ 25–50:_____ 50–100:_____ 100–150:_____ 150–200:_____

Point 2: 0–25:_____ 25–50:_____ 50–100:_____ 100–150:_____ 150–200:_____

Point 3: 0–25:_____ 25–50:_____ 50–100:_____ 100–150:_____ 150–200:_____

Point 4: 0–25:_____ 25–50:_____ 50–100:_____ 100–150:_____ 150–200:_____

Dominant taxa in Foliage Height Profile (FHP): _____

_____

Percentage Canopy Cover

*Point 1:*_____ *Point 2:*_____ *Point 3:*_____ *Point 4:*_____

Dominant canopy taxa:_____

Notes:

List of Materials for Ant Specimen Processing

General Tools

Indelible ink pens (light-, water-, and alcohol-proof)

Ethanol (ethyl alcohol) (70–95%, pure with no denaturants; preferably 95%)

Forceps: two pairs featherweight forceps; two pairs watchmaker's #5 forceps

Grindstone (for sharpening forceps)

Personal computer with text processing and spreadsheet or database software

Laser printer

Labeling Materials

Thin cardstock of neutral pH

Scissors

Fine-point indelible ink pens or markers (point size 0.05, 0.1, or 0.2)

Forceps (watchmaker's #5)

Laser or dot matrix printer (if possible)

Tools for Specific Operations

SALT WATER EXTRACTION

Sodium chloride or table salt (NaCl) or saturated salt solution

Wash bottles (any plastic container that will enable one to squirt a steady stream of liquid)

Heating element

Pot

Jars

Graduated cylinder

Strainer

Funnel

Metal or cloth mesh

Fine paint brush
Spoon
Forceps (featherweight)

MANUAL SORTING OF ANTS FROM DEBRIS
Ethanol (70–95%, preferably 95% if available)
Squeeze bottles
Forceps (featherweight)
Forceps (watchmaker's #5)
Fine paint brushes
Petri dishes of various sizes
Vials (2–5 ml) (for storage of ants; glass with
polyethylene stopper or 2 ml polyethylene
tubes with silicon O-ring screw top cap)
Locality labels
Scissors
Stereoscopic microscope with light source

SORTING TO MORPHOSPECIES
Stereoscopic microscope with light source
Petri dishes of various sizes
Forceps (featherweight)
Forceps (watchmaker's #5)
Ethanol (>70%; does not need to be new;
may be reused from former samples)
Squeeze bottle
Fine brush

MOUNTING OF SPECIMENS
Cardboard or bristol board (heavy, acid-free
stock for points: business card weight)
Insect pins (size 2, size 3)
Point puncher
Glues (white glues, wood glues, all water
soluble)
Fine stick (to apply glue to the points)
Pinning block (a wooden or aluminum block
with pin-sized holes of different depths, to
prevent bending of labels and points when
affixing to pins)
Cork or foam pads (4 × 5 cm) (to mount ants
under the microscope)
Index cards (for white background on which
to arrange specimens before mounting)

Tissues or paper towels (to dry specimens)
Forceps (watchmaker's #5): two pairs

Specimen Storage
Insect cabinets
Drawers (ideally with glass lids and unit
trays; minimally, boxes with soft bottoms
and tightly closing tops)
Unit trays
Labels for unit trays or drawers
Cabinets with strong shelves (for ethanol
collection)
Ethanol (70–95%, preferably 95% if
available)
Vials (2–5 ml for specimens; preferably
glass with polyethylene stopper or 2 ml
polyethylene tubes with silicon O-ring
screw top cap)
Larger containers (e.g., bale-top mason jars)
(to house smaller vials in ethanol)

Shipping of Specimens
Cardboard or wooden boxes of various sizes
(shipping boxes should allow a 12-cm
space for Styrofoam on each side of
enclosed specimen boxes)
Cling film (plastic wrap)
Loose Styrofoam (to pad the contents of the
boxes)
Insect pins
Forceps (watchmaker's #5)
Shipping labels
Strong tape
Customs declaration forms

Identification of Specimens
Unit trays
Tools to mount specimens under microscope
in various positions (or L-shaped holder)
Notepad or personal computer with text pro-
cessing software
Camera lucida attachment for microscope
(useful for drawing details of specimens)
Scissors (to cut labels)

Glossary

Ted R. Schultz and Leeanne E. Alonso

This glossary is included to enable readers to use this manual without necessarily referring to additional references. It relies heavily on a number of works, all of which are recommended, including Jaeger (1955), Lincoln and Boxshall (1987), Torre-Bueno (1989), Hölldobler and Wilson (1990), and Bolton (1994).

acidopore The orifice of the formic acid–projecting system peculiar to the ant subfamily Formicinae.

Aculeata The group of apocritan Hymenoptera, which includes the ants, in which the ovipositor is modified into a sting.

adventive A nonnative species present in a given area because of accidental introduction (e.g., human transport).

Afrotropical Region Sub-Saharan Africa south of the Sahara Desert and the southern half of the Saudi Arabian peninsula, but variously including or excluding Madagascar and nearby islands, which are sometimes referred to separately as the Malagasy Region.

alate In ants, a winged male or winged female (gyne).

alitrunk (mesosoma) In apocritan Hymenoptera, the middle body region from which arise the legs and wings (when present), posterior to the head and anterior to abdominal segment 2 (the petiole in ants). It is formed from the fusion of the thorax and the first abdominal segment.

ALL protocol Ants of the Leaf Litter protocol, the standardized method for sampling ground-dwelling ants described and recommended in this volume. The protocol employs Winkler litter extraction as the primary tool, pitfall trapping as a secondary tool, and other subsidiary methods depending on conditions. See particularly Chapters 9 and 14.

antennal segments The separate sclerotized units into which the antennae are subdivided, connected

223

to each other by flexible membranes. In ants and most other aculeate Hymenoptera, they primitively number 12 in the females and 13 in the males.

anterad Toward the anterior; on an insect body, directed toward the head.

anthropogenic Caused by humans.

Apocrita A suborder of Hymenoptera, including all Hymenoptera except suborder Symphyta, in which segment 1 of the abdomen has become fused with the thorax to form the propodeum and in which the larvae are apodous. Ants are members of the Apocrita.

apodeme A chitinous ingrowth of the arthropod exoskeleton to which muscles are attached.

arbicolous Nesting and/or foraging in trees.

aspirator A suction device for picking up insects.

autapomorphy In phylogenetic systematics, a derived character or character state.

basal At or pertaining to the base or point of attachment nearest the main body of an organism.

basidiomycete A member of a taxonomic subdivision of fungi (the Basidiomycotina) that includes those fungi that produce basidiocarps, or true mushrooms.

Berlese funnel A device for collecting small litter- or soil-dwelling arthropods, consisting of an electric lamp mounted above a funnel containing a piece of screen, hardware cloth, or other mesh. Litter is placed over the mesh and, driven downward by the heating and drying agency of the lamp, arthropods fall into the funnel and thence into a collecting jar filled with alcohol or other killing agent (see Fig. 9.6, page 137).

biocontrol Control of pestiferous organisms through the use of their natural enemies (e.g., predators, parasites, fungal diseases)

biodiversity "The variety of life forms, the ecological roles they perform, and the genetic diversity they contain" (Wilcox 1984:640); the number of species or higher taxa in a given region.

biogeography The study of the geographical distributions of organisms and their habitats, and of the historical and biological factors that produced them.

bioindicator In ecology, an aspect of the environment, usually a species or group or species, of use in monitoring biodiversity, ecological status, or other biological attributes of a particular area.

biomass The mass (including or excluding water weight, as specified) of a circumscribed biological entity or collection of entities (e.g., of a single ant, of all ants in a given location, or of all organisms in a given locality).

bivuoac In army ants, the mass of workers that serve as a protective refuge for the queen and brood.

carina An elevated ridge or keel on the insect integument.

carton In myrmecology, a cardboard-like construction material manufactured by some ants using bits of wood, wood pulp, dried plant matter, and soil, generally used to form protective enclosures around their nests. The resulting structures are referred to as "carton nests."

caste In social insects, any set of individuals in a given colony that is both morphologically distinct and specialized in behavior (morphological castes); more broadly, any set of individuals of a particular morphological type or age group, or both, that performs specialized labor in the colony.

clade A monophyletic group.

cladistic analysis Phylogenetic analysis in which monophyletic groups (clades) are identified based on synapomorphies (shared, derived characters or character states assumed to have been present in a shared common ancestor).

cladogram A branching diagram, most commonly interpreted as a phylogenetic tree, constructed using cladistic analysis.

clypeus That part of the insect head below the frons to which the labrum is attached anteriorly; in most ants, the portion of the "face" (dorsum) of the head capsule that borders the mouth parts, bounded by the antennal sockets and tentorial pits above, the genae (cheeks) on the sides, and the anterior edge of the head capsule below.

coevolution The interdependent evolution of two or more species having an obvious ecological relationship, usually restricted to cases in which interactions are mutually beneficial (mutualisms), but occasionally used more loosely to refer to symbiotic evolution in general.

commensalism A symbiosis in which one partner benefits and the other is neither harmed nor benefited.

complementarity In ecology, the relationship of two habitats that have similar species richnesses but very few species in common.

conspecific Of or pertaining to the same species (opposite of heterospecific).

cotype An imprecise term not recognized by the International Code of Zoological Nomenclature, formerly used to refer to a paratype or syntype.

covariate Of or pertaining to the quality in which two or more quantities vary in a way that preserves a mathematical relationship.

coxa The basal segment of the arthropod leg.

curation The art and science of preserving biological specimens and organizing, caring for, and maintaining collections of such specimens.

cuticle a secretion of the epidermis covering the entire body of an arthropod as well as lining ectodermal invaginations, such as the proctodaeum, stomodaeum, and tracheae.

denticle A small tooth.

dichthadiigyne queen (dichthadiiform ergatogyne, dichthadiiform queen) In army ants, a member of an aberrant reproductive caste characterized by a wingless alitrunk, a huge gaster, and an expanded postpetiole.

disarticulation The separation of one sclerotized component of the arthropod skeleton from a neighboring component to which it was previously attached.

distal At or pertaining to the free end of a morphological structure, farthest away from the main body of the organism.

domatia (myrmecodomatia) Specialized structures, such as inflated stems or hollow thorns, used by ant plants for the housing of ant colonies.

dulosis *See* **slavemaking**

ecological succession The chronological distribution of organisms within an area.

elaiosome An ant-attractive nutritive attachment on seeds manufactured by some plants to encourage dispersal.

endemism The quality of being native to and exclusively restricted to a particular geographical region.

energetics The study of energy transformation within a community or system.

epigaeic Living, or at least foraging, above the surface of the ground (opposite of hypogaeic).

epinotum *See* **propodeum.**

ergatogyne *See* **ergatoid.**

ergatoid (ergatogyne) In ants, any form intermediate between the worker and the queen.

eusociality (true sociality, higher sociality) The condition in which the following three traits are present: cooperation in caring for the young; reproductive division of labor, with more or less sterile individuals working on behalf of individuals engaged in reproduction; and overlap of at least two generations of life stages capable of contributing to colony labor. All ants are eusocial.

Fluon A liquid form of Teflon that, when painted onto vertical surfaces and allowed to dry, forms an effective climbing barrier to most insects, including ants.

Formicidae The family of Hymenoptera that comprises the ants, characterized by the presence of the metapleural gland (secondarily absent in some groups), petiole, and eusociality.

formicosis A disease of the lungs brought on by excessive inhalation of formic acid vapors (produced by ants in the subfamily Formicinae), usually as a result of ant collecting using an aspirator.

foundress In ants, the newly fecundated gyne (queen) that begins the colony life cycle.

frass Solid larval insect excrement.

frons In insects, a sclerite of the head immediately posterior to the clypeus.

fungivorous Feeding on fungi.

furcula In the aculeate hymenopteran sting, a small, forked apodemal sclerite positioned dorsobasally, to which important muscles of the sting attach.

gaster (metasoma) The posterior region of the body in apocritan Hymenoptera; in ants, the portion posterior to the petiole (i.e., true abdominal segments 3–10) or, if a postpetiole is present, the portion posterior to the postpetiole (i.e., true abdominal segments 4–10).

gena The insect "cheek," the area of the head below the eye.

granivorous Feeding on grain, i.e., on the seeds of grasses.

gular teeth *See* (more correctly) **hypostomal teeth.**

gyne (queen) In ants, the female reproductive caste.

habitus Overall general form or appearance.

head capsule The fused sclerites of the arthropod head, which form a hardened, compact case, the cranium.

heterospecific Of or pertaining to a different species (opposite of conspecific).

Holarctic Region The region containing both the Palearctic and Nearctic Regions.

holotype In taxonomy, a single specimen designated as the name-bearing type of a species or subspecies when it was established, or the single specimen on which the taxon was based when no type was specified.

homonym In species-level taxonomy, each of two or more available names established for different nominal taxa having the same spelling or spellings deemed to be the same by the International Code of Zoological Nomenclature. Homonymy is the problematic situation in which two or more species in the same genus have the same name.

homoplasy The apparent independent evolution (by parallelism or convergence) of characteristics that are indistinguishable.

humisol A type of soil that is rich in organic material.

hypogaeic Living primarily below the surface of the ground, or at least beneath cover such as leaf litter, stones, and dead bark (opposite of epigaeic).

hypostoma The anteroventral region of the head; the area of cuticle immediately behind the buccal cavity and forming its posterior margin.

hypostomal teeth (gular teeth) In ants, one or more pairs of triangular or rounded teeth that project forward from the anterior margin of the hypostoma.

inflorescence The arrangement and sequence of development of flowers on a flowering shoot.

infrabuccal pocket In ants, a cavity on the floor of the buccal chamber in which, in most species, indigestible material accumulates and is compacted for later disposal; in ants of the subfamily Pseudomyrmecinae infrabuccal pocket contents are fed to larvae, and in fungus-growers the infrabuccal pocket is used by gynes to transport fungus-garden mycelium.

insolation Exposure to solar radiation

instar In insects, the stage between moults in a nymph or larva.

integument The outer layer of an arthropod, including the basement membrane, epidermis, and cuticle.

interspecific competition Simultaneous demand for limited resources between members of different species.

intraspecific competition Simultaneous demand for limited resources between members of the same species.

karyology The branch of cytology dealing with the study of nuclei, especially the structure of chromosomes.

labrum The "upper lip" of the insect mouth, arising anterior to the clypeus.

latosol A type of soil, occurring primarily in tropical regions, in which plant detritus decays rapidly, leaching silica from the soil in the process known as laterization.

lectotype In taxonomy, one of a series of syntypes that, subsequent to the publication of the original description, is selected and designated through publication to serve the same function as the holotype specimen for the species.

lineage A group of organisms descended from a common ancestor. *See* **clade.**

mandibles The first pair of jaws in insects. In ants, the organs for cutting and otherwise processing food, for biting enemies, for carrying brood, for nest construction, and for generally manipulating the environment, usually stout and jawlike, but variously modified and elongate in some species.

mesocoxa The coxa of the middle leg (i.e., the leg arising from the mesothorax).

mesonotum The dorsal part of the mesothorax (thorax segment 2).

mesosoma *See* **alitrunk.**

mesothorax The middle member of the three main subdivisions of the insect thorax, from which the anterior pair of wings and the middle legs arise.

metacoxa The coxa of the hind leg (i.e., the leg arising from the metathorax).

metanotum The dorsal part of the metathorax (thorax segment 3).

metapleural gland An antibiotic-producing exocrine gland peculiar to ants located at the posteroventral corner of the metapleuron.

metapleuron The lateral region of the metathorax.

metasoma *See* **gaster.**

metathorax The posterior member of the three main subdivisions of the insect thorax, from which the posterior pair of wings and the rear legs arise.

midden A refuse heap.

monophyletic Describing a group consisting of an ancestral species and all of its descendants.

morphocline One of a graded series of states within in a morphological character.

morphospecies A temporary grouping created to distinguish morphologically distinct clusters of specimens from one another prior to rigorous identification (where possible) with nominal species.

mutualism A symbiosis in which both parties benefit.

mycorrhizal fungi Fungi that grow in obligate association with the roots of plants.

myrmecodomatia *See* **domatia.**

myrmecologist A student of myrmecology.

myrmecology The study of ants (family Formicidae).

myrmecophyte A higher plant that lives in obligatory, mutualistic association with ants.

Nearctic Region The northern regions of the New World, including North America and the Central Mexican Plateau.

Neotropical Region The tropical region of the New World (i.e., of Central and South America).

ocellus In insects, a simple eye distinct from the paired compound eyes; in ants, an eye occurring in a group of three on the vertex, present in males and gynes but frequently absent in workers.

oviposition The act of depositing eggs.

Palearctic Region The northern regions of the Old World, including Europe, Africa north of the Sahara, and Asia as far south as the southern edge of the Yangtse-kiang watershed and the Himalayas.

Paleotropical Region The entire tropical region of the Old World, including the Afrotropical, Malagasy, Oriental, and Indo-Australian tropical regions.

palp One of the paired appendages of the maxillary and labial mouth part segments; in ants, the maximum number of maxillary palpal segments is six, the minimum number (rarely seen) is zero; the maximum number of labial palps is four, the minimum number is one.

pantropical Of or pertaining to the tropical regions of the entire world.

paraphyletic Of or pertaining to a taxonomic group that does not include all the descendants of a common ancestor.

parasitism A symbiosis in which one partner benefits at the expense of the other.

paratype In taxonomy, each specimen of a type series other than the holotype; one of the series of specimens examined during the formulation of the original description of the species.

parthenogenesis The development of an individual from a female gamete without fertilization by a male gamete.

patchiness Heterogeneity within an environment with respect to ecological conditions of interest (e.g., those that might favor one species over another or that might subdivide populations with regard to gene flow).

petiole In ants, the second abdominal segment (i.e., the segment immediately posterior to the alitrunk), which is constricted both anteriorly and posteriorly.

phylogram A phylogenetic tree; often used to refer specifically to a phylogenetic tree in which relative branch lengths are specified.

pilosity A covering of hair.

pitfall trap A steep-sided container sunk into the ground so that the opening is even with the surface, often containing a small amount of liquid preservative; used to trap ground-dwelling animals, which fall in and cannot escape.

plicae Folds, wrinkles, or pleats; in the ant proventriculus, the relatively narrow, longitudinal strips of thin, flexible cuticle that connect the broad, sclerotized, cuticular plates of the proventricular bulb.

polydomy The condition in which a single ant colony simultaneously has more than one nest.

polymorphism In social insects, the condition of having more than one caste within the same sex; in ants, the condition of having workers of distinctly different proportions (e.g., minima and maxima workers or soldiers).

polyphyletic Of or pertaining to a taxonomic group that contains members derived from two or more ancestral sources (i.e., that are not part of an immediate line of descent).

posterad Toward the posterior; on an insect body, directed toward the rear of the abdomen.

postpetiole The modified form of the third abdominal segment (i.e., the segment immediately posterior to the petiole) present in some ant groups, in which this segment is constricted posteriorly to form what is essentially a second petiole.

predation The consumption of one animal by another.

presclerite In ants, the distinctly differentiated anterior section of an abdominal sclerite, separated from the remainder of the sclerite by a ridge, constriction, or both.

presternite In ants, a presclerite derived from a sternite.

pretergite In ants, a presclerite derived from a tergite.

proctodeum The insect hindgut.

promesonotum The fused pronotum and mesonotum.

pronotum The dorsal part of the prothorax (thorax segment 1).

propodeum (epinotum) In apocritan Hymenoptera, the first abdominal segment, which has become fused with the thorax to form the alitrunk.

prothorax The anterior member of the three main subdivisions of the insect thorax, from which the front legs arise.

proventriculus In insects generally, the valve separating the crop (anteriorly) and the midgut (posteriorly); in ants, the proventriculus regulates whether food is consumed by the individual (allowed to pass into the midgut) or whether it is retained in the "social stomach" that consists of the combined crops of all colony members.

queen *See* **gyne**.

relictual Of or pertaining to persistent remnants of formerly widespread species or higher taxa currently restricted to certain isolated areas or habitats.

remediation In ecology, the act of returning a disturbed habitat to its natural state.

replete An individual ant worker that functions as a living reservoir, having a crop so distended with liquid food that the abdominal segments are pulled apart and the intersegmental membranes stretched tight.

ruderal Pertaining to or living among rubbish or debris, or inhabiting disturbed sites.

scape The basal antennal segment.

sclerite Any plate of the arthropod body wall bounded by membrane or by sutures.

sclerotization Hardening of cuticle to form the arthropod exoskeleton, as compared with the more flexible, nonsclerotized, membranous cuticular areas.

slavemaking (slavery, dulosis) The condition in which workers of a parasitic (slavemaking) ant species raid the nest of another species, capture brood (usually pupae), and rear them as enslaved nestmates.

slavery *See* **slavemaking**

speciation The process by which novel species arise.

species richness The absolute number of species in an ecological assemblage or community.

speciose Of or pertaining to a clade containing a relatively large number of species.

sternite *See* **sternum.**

sternum (sternite) The ventral (lower) sclerite of a segment.

stomodeum The insect foregut.

subpetiolar process An anteroventral projection on the petiole or its peduncle.

sulcus A groove with a purely functional (rather than developmental) origin.

suprageneric In taxonomy, of or pertaining to taxonomic ranks above the genus level (e.g., subtribe, tribe, subfamily, family).

suture On the insect integument, a groove marking the line of fusion of two developmentally distinct cuticular plates.

symbiont A member of a symbiosis.

symbiosis The living together of two organisms.

synonym In taxonomy, each of two or more scientific names of the same rank used to denote the same taxon.

syntype In taxonomy, each specimen of a type series from which neither a holotype nor a lectotype has been designated.

systematics The classification of living organisms into hierarchical groups emphasizing their phylogenetic relationships.

taxon A defined and named unit consisting of a group of related organisms (e.g., species, genus, tribe, subfamily, family).

tentorial pits The external depressions in the exoskeleton of the head at which corresponding to the roots of the tentorial arms.

tentorium The internal skeleton of the insect head. The points at which the tentorium is confluent with the exoskeleton are marked by the tentorial pits.

tergite *See* **tergum.**

tergosternal fusion A condition of the ant abdominal segments in which the tergite and sternite are continuously fused rather than connected by membrane, so that they are incapable of independent movement relative to each other, occurring in some or all of abdominal segments 2 (petiole) to 4.

tergum (tergite) The dorsal (upper) sclerite of a segment.

termitarium A nest, natural or artificial, or a colony of termites. Frequently used to refer specifically to nest structures constructed by termites.

termitotherous Hunting and preying upon termites.

thermophilic Preferring warm temperatures.

thermoregulation The physiological processes, behavioral processes, or both by which an organism adjusts its body temperature to a level different from the ambient temperature.

thorax The second major subdivision of the insect body, bearing the legs and wings. The thorax is posterior to the head and anterior to the abdomen, and consists of three subdivisions, the prothorax, mesothorax, and metathorax.

trachea A spirally reinforced, elastic air tube that is the principal component of the insect respiratory system. A system of tracheae connects the outside atmosphere with the internal tissues and organs.

tribe The taxonomic rank above genus and below family (i.e., a group of genera).

trophallaxis The exchange of alimentary liquid among colony members and guest organisms.

trophic Pertaining to nutrition.

trophic eggs A nonviable egg laid by an ant queen to serve as food for other members of the colony, usually her offspring.

vertex The top of the insect head between the eyes and posterior to the frons.

Winkler sack (Winkler bag, Winkler eclector, Winkler extractor) A device for collecting small litter- or soil-dwelling arthropods, consisting of one or more bags constructed from cloth mesh. The mesh bags are filled with litter and suspended within an outer cloth enclosure, which includes a funnel that catches insects escaping through the mesh and directs them into a collecting receptacle, usually filled with alcohol (see Figs. 9.4 and 9.5, pages 134 and 135).

xeric Having very little moisture; tolerating or adapted to dry conditions.

zoogeography The branch of biogeography dealing specifically with the distribution of animals.

Literature Cited

Abensperg-Traun, M., and D. Steven. 1995. The effects of pitfall trap diameter on ant species richness (Hymenoptera: Formicidae) and species composition of the catch in a semi-arid eucalypt woodland. Australian Journal of Ecology 20:282–287.

———. 1997. Ant- and termite-feeding in Australian mammals and lizards: A comparison. Australian Journal of Ecology 22:9–17.

Abensperg-Traun, M., G. W. Arnold, D. E. Steven, G. T. Smith, L. Atkins, J. J. Viveen, and M. Gutter. 1996. Biodiversity indicators in semi-arid, agricultural Western Australia. Pacific Conservation Biology 2:375–389.

Adams, E. S. 1994. Territory defense by the ant *Azteca trigona:* Maintenance of an arboreal ant mosaic. Oecologia 97:202–208.

Adams, E., and W. Tschinkel. 1995. Density-dependent competition in fire ants: Effects on colony survivorship and size variation. Journal of Animal Ecology 64:315–324.

Adis, J. 1979. Problems of interpreting arthropod sampling with pitfall traps. Zoologisher Anzeiger 202:177–184

Adis J., Y. D. Lubin, and G. G. Montgomery. 1984. Arthropods from the canopy of inundated and terra firme forests near Manaus, Brazil, with critical considerations on the pyrethrum-fogging technique. Studies on Neotropical Fauna and Environment 4:223–236.

Adis, J., J. W. de Morais, and H. Guimarães de Mesquita. 1987. Vertical distribution and abundance of arthropods in the soil of a Neotropical secondary forest during the rainy season. Studies on Neotropical Fauna and Environment 22:189–197.

Agosti, D. 1990. Review and reclassification of *Cataglyphis* (Hymenoptera, Formicidae). Journal of Natural History 24:1457–1505.

———. 1991. Revision of the oriental ant genus *Cladomyrma,* with an outline of the higher classification of the Formicinae (Hymenoptera:

Formicidae). Systematic Entomology 16:293–310.

———. 1992. Revision of the ant genus *Myrmoteras* of the Malay Archipelago (Hymenoptera, Formicidae). Revue Suisse de Zoologie 99:405–429.

———. 1994a. The phylogeny of the ant tribe Formicini (Hymenoptera: Formicidae) with the description of a new genus. Systematic Entomology 19:93–117.

———. 1994b. A revision of the South American species of the ant genus *Probolomyrmex* (Hymenoptera: Formicidae). Journal of the New York Entomological Society 102:429–434.

Agosti, D., J. Majer, L. Alonso, and T. R. Schultz (eds.). 2000. Sampling Ground-dwelling Ants: Case Studies from the Worlds' Rain Forests. Curtin University of Technology, Perth, Western Australia.

Agosti, D., M. Maryati, and C. Y. C. Arthur. 1994. Has the diversity of tropical ant fauna been underestimated? An indication from leaf litter studies in a West Malaysian lowland rain forest. Tropical Biodiversity 2:270–275.

Agosti, D., J. Moog, and U. Maschwitz. 1999. Revision of the Oriental plant-ant genus *Cladomyrma*. American Museum Novitates 3283, December 8, 1999.

Alexander, R. D. 1974. The evolution of social behavior. Annual Review of Ecology and Systematics 5:325–383.

Allen, C., R. Lutz, and S. Demarais. 1995. Red imported fire ant impacts on northern bobwhite populations. Ecological Applications 5:632–638.

Allen, G. E., and W. F. Buren. 1974. Microsporidan and fungal diseases of *Solenopsis invicta* Buren in Brazil. Journal of the New York Entomological Society 82:125–130.

Allen, G. E., and A. Silveira-Guido. 1974. Occurrence of microsporidia in *Solenopsis richteri* and *Solenopsis* sp. in Uruguay and Argentina. Florida Entomologist 57:327–329.

Allen, M. F., J. A. MacMahon, and D. C. Andersen. 1984. Reestablishment of Endogonaceae on Mount St. Helens: Survival of residuals. Mycologia 76(6):1031–1038.

Andersen, A. N. 1983. A brief survey of ants of Glenaladale National Park, with particular reference to seed-harvesting. Victorian Naturalist 100:233–237.

———. 1986a. Diversity, seasonality and community organization of ants at adjacent heath and woodland sites in southeastern Australia. Australian Journal of Zoology 34:53–64.

———. 1986b. Patterns of ant community organization in mesic southeastern Australia. Australian Journal of Ecology 11:87–99.

———. 1988. Immediate and longer-term effects of fire on seed predation by ants in sclerophyllous vegetation of southeastern Australia. Australian Journal of Ecology 13:285–293.

———. 1990. The use of ant communities to evaluate change in Australian terrestrial ecosystems: A review and a recipe. Proceedings of the Ecological Society of Australia 16:347–357.

———. 1991a. Parallels between ants and plants: Implications for community ecology. Pp. 539–558. *In* C. R. Huxley and D. F. Cutler (eds.), Ant-Plant Interactions. Oxford University Press, Oxford.

———. 1991b. Seed-harvesting by ants in Australia. Pp. 493–503. *In* C. R. Huxley and D. F. Cutler (eds.), Ant-Plant Interactions. Oxford University Press, Oxford.

———. 1991c. Sampling communities of ground-foraging ants: Pitfall catches compared with quadrat counts in an Australian tropical savanna. Australian Journal of Ecology 16:273–279.

Andersen, A. N. 1991d. Responses of ground-foraging ant communities to three experimental fire regimes in a savanna forest of tropical Australia. Biotropica 23:575–585.

———. 1992. The rainforest ant fauna of the northern Kimberley region of Western Australia (Hymenoptera: Formicidae). Journal of the Australian Entomological Society 31:187–192.

———. 1995. A classification of Australian ant communities, based on functional groups which parallel plant life-forms in relation to stress and disturbance. Journal of Biogeography 20:15–29.

———. 1996. Fire ecology and management. Pp. 179–195. *In* C. M. Finlayson and I. Von Oertzen (eds.), Landscape and Vegetation Ecology of the Kakadu Region, Northern Australia. Kluwer Academic Publishers, Amsterdam.

———. 1997a. Functional groups and patterns of organization in North American ant communities: A comparison with Australia. Journal of Biogeography 24:433–460.

———. 1997b. Using ants as bioindicators: Multiscale issues in ant community ecology. Conservation Ecology [online] 1, Article 8.

Andersen, A. N., and R. E. Clay. 1996. The ant fauna of Danggali Conservation park in semi-arid

South Australia: A comparison with Wyperfield (Vic.) and Cape Arid (W.A.) National Parks. Australian Journal of Entomology 35:289–295.

Andersen, A. N., and J. D. Majer. 1991. The structure and biogeography of rainforest ant communities in the Kimberely region of northwestern Australia. Pp. 333–346. *In* N. L. McKenzie, R. B. Johnston, and P. J. Kendrick (eds.), Kimberley Rainforests of Australia. Surrey Beatty and Sons, Chipping Norton, NSW.

Andersen, A. N., and M. E. McKaige. 1987. Ant communities at Rotamah Island, Victoria, with particular reference to disturbance and *Rhytidoponera tasmaniensis*. Proceedings of the Royal Society of Victoria 99:141–146.

Andersen, A. N., and A. D. Patel. 1994. Meat ants as dominant members of Australian ant communities: An experimental test of their influence on the foraging success and forager abundance of other species. Oecologia 98:15–24.

Andersen, A. N., and H. Reichel. 1994. The ant (Hymenoptera: Formicidae) fauna of Holmes Jungle, a rainforest patch in the seasonal tropics of Australia's Northern Territory. Journal of the Australian Entomological Society 33:153–158.

Andersen, A. N., and A. V. Spain. 1996. The ant fauna of the Bowen Basin, in the semi-arid tropics of central Queensland (Hymenoptera: Formicidae). Australian Journal of Entomology 35:213–221.

Andersen, A. N., and A. L. Yen. 1985. Immediate effects of fire on ants in the semi-arid mallee region of north-western Victoria. Australian Journal of Ecology 10:25–30.

———. 1992. Canopy ant communities in the semi-arid Mallee region of North-western Victoria. Australian Journal of Zoology 40:205–214.

Andersen, A. N., M. S. Blum, and T. M. Jones. 1991. Venom alkaloids in *Monomorium* "*rothsteini*" Forel repel other ants: Is this the secret to success by *Monomorium* in Australian ant communities? Oecologia 88:157–160.

Andersen, A. N., S. Morrison, and L. Belbin. 1996. The Role of Ants in Minesite Restoration in the Kakadu Region of Australia's Northern Territory, with Particular Reference to Their Use as Bioindicators. Final Report to the Environmental Research Institute of the Supervising Scientist, Australia.

Andrade, M. L. de. 1998. Fossil and extant species of *Cylindromyrmex* (Hymenoptera: Formicidae). Revue Suisse de Zoologie 105(3):581–664.

Andrade, M. L. de and C. Baroni Urbani. 1999. Diversity and adaption in the ant genus *Cephalotes*, past and present (Hymenoptera, Formicidae*)*. Stuttgarter Beiträge zur Naturkunde, Serie B 271:1–889.

Andrewartha, H., and L. Birch. 1954. The distribution and abundance of animals. University of Chicago Press, Chicago.

Arnett, R. H., Jr. 1985. American Insects. Van Nostrand Reinhold, New York.

Arnett, R. H., Jr., and M. E. Arnett. 1990. The Naturalists Directory and Almanac (International): An Index to Contemporary Naturalists of the World and Their Special Interests, 45th ed. Sandhill Crane Press, Gainesville, Florida.

Arnett, R. H., Jr., and G. A. Samuelson. 1986. The Insect and Spider Collections of the World. E. J. Brill/Flora and Fauna Publications, Gainesville, Florida.

Arnett, R. H., Jr., G. A. Samuelson, and G. M. Nishida. 1993. The Insect and Spider Collections of the World, 2nd ed. Sandhill Crane Press, Gainesville, Florida.

Arnol'di, K. V. 1930. Studien über die Systematik der Ameisen. IV. *Aulacopone*, eine neue Ponerinengattung (Formicidae) in Russland. Zoologischer Anzeiger 89:139–144.

———. 1970. Review of the ant genus *Myrmica* (Hymenoptera, Formicidae) in the European part of the USSR. Zoologicheskii Zhurnal 49:1829–1844. [In Russian.]

———. 1975. A review of the species of the genus *Stenamma* (Hymenoptera, Formicidae) of the USSR and description of new species. Zoologicheskii Zhurnal 54:1819–1829. [In Russian.]

———. 1976a. Review of the genus *Aphaenogaster* (Hymenoptera, Formicidae) in the USSR. Zoologicheskii Zhurnal 55:1019–1026. [In Russian.]

———. 1976b. Ants of the genus *Myrmica* Latr. from Central Asia and the southern Kazakstan. Zoologicheskii Zhurnal 55:547–558. [In Russian.]

———. 1977. Review of the harvester ants of the genus *Messor* (Hymenoptera, Formicidae) in the fauna of the USSR. Zoologicheskii Zhurnal 56:1637–1648. [In Russian.]

Atsatt, P. R. 1981. Lycaenid butterflies and ants: Selection for enemy-free space. American Naturalist 118:538–654.

Autuori, M. 1942. Contribuição para o conhecimento da saúva (*Atta* spp. Hymenoptera: Formicidae).

III. Excavação de um saúveiro (*Atta sexdens rubropilosa* Forel, 1908). Archivos do Instituto de Biolóico, São Paulo 13:137–148.

Bailey, I. W. 1920. Some relations between ants and fungi. Ecology 1:174–189.

———. 1922a. Notes on neotropical ant-plants. I. *Cecropia angulata,* sp. nov. Botanical Gazette 74:369–391.

———. 1922b. The anatomy of certain plants from the Belgian Congo, with special reference to Myrmecophytism. Bulletin of the American Museum of Natural History 45:585–622, plates 26–45.

Balazy, S., A. Lenoir, and J. Wisniewski. 1986. *Aegeritella roussillonensis* n. sp. (Hyphomycetales, Blastosporae), une espèce nouvelle de champignon epizoique sur les fourmis *Cataglyphis cursor* (Fonscolombe) (Hymenoptera, Formicidae) en France. Cryptogamie, Mycologie 7:37–45.

Banschbach, V. S., and J. M. Herbers. 1996. Complex colony structure in social insects. I. Ecological determinants and genetic consequences. Evolution 50:285–297.

Baroni Urbani, C. 1968. Über die eigenartige Morphologie der männlichen Genitalien des Genus *Diplorhoptrum* Mayr und die taxonomischen Schlussfolgerungen. Zeitschrift für Morphologie der Tiere 63:63–74.

Baroni Urbani, C. 1969. Gli *Strongylognathus* del gruppo *huberi* nell'Europa occidentale: Saggio di una revisione basata sulla casta operaia (Hymenoptera Formicidae). Bolletino de la Società Entomolgica Italiana 99–101:132–168.

———. 1975a. Primi reperti del genere *Calyptomyrmex* Emery nel subcontinente Indiano. Entomologica Basiliensis 1:395–411.

———. 1975b. Contributo alla conoscenza dei generi *Belonopelta* Mayr e *Leiopelta* gen. n. (Hymenoptera: Formicidae). Mitteilungen der Schweizerischen Entomologischen Gesellschaft 48:295–310.

———. 1977. Ergebnisse der Bhutan-Expedition 1972 des Naturhistorischen Museums in Basel. Hymenoptera: Fam. Formicidae Genus *Mayriella.* Entomologica Basiliensis 2:411–414.

———. 1978a. Contributo alla conoscenza del genere *Amblyopone* Erichson (Hymenoptera: Formicidae). Mitteilungen der Schweizerischen Entomologischen Gesellschaft 51:39–51.

———. 1978b. Materiali per una revisione dei *Leptothorax* neotropicali appartenenti al sotto-

genere *Macromischa* Roger, n. comb. (Hymenoptera: Formicidae). Entomologica Basiliensis 3:395–618.

Baroni Urbani, C., and M. L. De Andrade. 1993. *Perissomyrmex monticola* n. sp., from Bhutan: The first natural record for a presumed Neotropical genus with a discussion on its taxonomic status. Tropical Zoology 6:89–95.

Baroni Urbani, C., B. Bolton, and P. S. Ward. 1992. The internal phylogeny of ants (Hymenoptera: Formicidae). Systematic Entomology 17:301–329.

Barrer, P. M., and J. M. Cherrett. 1972. Some factors affecting the site and pattern of leaf-cutting activity in the ant *Atta cephalotes* L. Journal of Entomology 47:15–27.

Basset, Y., N. D. Springate, H. P. Aberlenc, and G. Delvare. 1997. A review of methods for sampling arthropods in tree canopies. Pp. 27–52. *In* N. Stork, J. Adis, and R. K. Didham (eds.), Canopy Arthropods. Chapman and Hall, London.

Beattie, A. J. 1985. The Evolutionary Ecology of Ant-Plant Mutualisms. Cambridge University Press, New York.

Beattie, A., and I. Oliver. 1994. Taxonomic minimalism. Trends in Evolution and Ecology 9:488–490.

Beattie, A. J., C. Turnbull, R. B. Knox, and E. G. Williams. 1984. Ant inhibition of pollen function: A possible reason why ant pollination is rare. American Journal of Botany 71:421–426.

Beattie, A. J., C. Turnbull, T. Hough, S. Jobson, and R. B. Knox. 1985. The vulnerability of pollen and fungal spores to ant secretions: Evidence and some evolutionary implications. American Journal of Botany 72:606–614.

Beattie, A. J., C. L. Turnbull, T. Hough, and R. B. Knox. 1986. Antibiotic production: A possible function for the metapleural glands of ants (Hymenoptera: Formicidae). Annals of the Entomological Society of America 79:448–450.

Beccaloni, G. W., and K. J. Gaston. 1995. Predicting the richness of Neotropical forest butterflies: Ithomiinae (Lepidoptera: Nymphalidae) as indicators. Biological Conservation 71:77–86.

Belshaw, R., and B. Bolton. 1993. The effect of forest disturbance on leaf litter ant fauna in Ghana. Biodiversity and Conservation 2:656–666.

———. 1994a. A survey of the leaf litter ant fauna in Ghana, West Africa (Hymenoptera: Formicidae). Journal of Hymenoptera Research 3:5–16.

————. 1994b. A new myrmicine ant genus from cocoa leaf litter in Ghana (Hymenoptera: Formicidae). Journal of Natural History 28:631–634.

Belt, T. 1874. The Naturalist in Nicaragua. John Murray, London.

Benson, W. W., and A. Y. Harada. 1988. Local diversity of tropical and temperate ant faunas. Acta Amazonica 18:275–289.

Bequaert, J. 1922. Ants in their diverse relations to the plant world. Bulletin of the American Museum of Natural History 45:333–584, plates 26–29.

Bernard, F. 1954. Fourmis moissonneuses nouvelles ou peu connues des montagnes d'Algérie et révision des *Messor* du groupe *structor* (Latr.). Bulletin de la Société d'Histoire Naturel de l'Afrique du Nord 45:354-365.

————. 1956. Révision des *Leptothorax* (Hyménoptères Formicidae) d'Europe occidentale, basée sur la biométrie et les genitalia mâles. Bulletin de la Société Zoologique de France 81:151–165.

————. 1979. *Messor carthaginensis* n. sp., de Tunis, et révision des *Messor* du groupe *barbara* (Hym. Formicidae). Bulletin de la Société Entomologique de France 84:265–269.

Bernstein, R. A. 1979. Schedules of foraging activity in species of ants. Journal of Animal Ecology 48:921–930.

Bernstein, R. A., and M. Gobbel. 1979. Partitioning of space in communities of ants. Journal of Animal Ecology 48:931–942.

Beshers, S., and J. Traniello. 1994. The adaptiveness of worker demography in the attine ant *Trachymyrmex septentrionalis*. Ecology 75:763–775.

Bestelmeyer, B. T. 1997. Stress tolerance in some Chacoan dolichoderine ants: Implications for community organization and distribution. Journal of Arid Environments 35:297–310.

Bestelmeyer, B. T., and J. A. Wiens. 1996. The effects of land use on the structure of ground-foraging ant communities in the Argentine chaco. Ecological Applications 6:1225–1240.

Besuchet, C., D. H. Burckhardt, and I. Löbl. 1987. The "Winkler/Moczarski" eclector as an efficient extractor for fungus and litter coleoptera. The Coleopterists' Bulletin 41:392–394.

Billen, J. P. J. 1990. Phylogenetic aspects of exocrine gland development in the Formicidae. Pp. 317–318. *In* G. K. Veeresh, B. Mallik, and C. A. Viraktamath (eds.), Social Insects and the Environment. Proceedings of the 11th International Congress of IUSSI, 1990. Oxford and IBH, New Delhi.

Bingham, C. T. 1903. The fauna of British India, including Ceylon and Burma. *In* Hymenoptera, Vol. II. Ants and Cuckoo-Wasps. Taylor and Francis, London.

Black, R. W., II. 1987. The biology of leaf nesting ants in a tropical wet forest. Biotropica 19:319–325.

Blackburn, T., P. Harvey, and M. Pagel. 1990. Species number, population density and body size relationships in natural communities. Journal of Animal Ecology 59:335–345.

Bolton, B. 1972. Two new species of the ant genus *Epitritus* from Ghana, with a key to the world species (Hym., Formicidae). Entomologists' Monthly Magazine 107:205–208.

————. 1973a. A remarkable new arboreal ant genus (Hym. Formicidae) from West Africa. Entomologists' Monthly Magazine 108:234–237.

————. 1973b. The ant genus *Polyrhachis* F. Smith in the Ethiopian region (Hymenoptera: Formicidae). Bulletin of the British Museum (Natural History). Entomology 28:283–369.

————. 1973c. New synonymy and a new name in the ant genus *Polyrhachis* F. Smith (Hym., Formicidae). Entomologists' Monthly Magazine 109:172–180.

————. 1974a. A revision of the Palaeotropical arboreal ant genus *Cataulacus* F. Smith (Hymenoptera: Formicidae). Bulletin of the British Museum (Natural History). Entomology 30:1–105.

————. 1974b. A revision of the ponerine ant genus *Plectroctena* F. Smith (Hymenoptera: Formicidae). Bulletin of the British Museum (Natural History). Entomology 30:309–338.

————. 1975a. A revision of the ant genus *Leptogenys* Roger (Hymenoptera: Formicidae) in the Ethiopian region with a review of the Malagasy species. Bulletin of the British Museum (Natural History). Entomology 31:235–305.

————. 1975b. A revision of the African ponerine ant genus *Psalidomyrmex* André (Hymenoptera: Formicidae). Bulletin of the British Museum (Natural History). Entomology 32:1–16.

————. 1975c. The *sexspinosa*-group of the ant genus *Polyrhachis* F. Smith (Hym. Formicidae). Journal of Entomology Series B 44:1–14.

————. 1976. The ant tribe Tetramoriini (Hymenoptera: Formicidae). Constituent genera, review of smaller genera and revision of *Triglyphothrix*

Forel. Bulletin of the British Museum (Natural History). Entomology 34:281–379.

———. 1977. The ant tribe Tetramoriini (Hymenoptera: Formicidae). The genus *Tetramorium* Mayr in the Oriental and Indo-Australian regions, and in Australia. Bulletin of the British Museum (Natural History). Entomology 36:67–151.

———. 1979. The ant tribe Tetramoriini (Hymenoptera: Formicidae). The genus *Tetramorium* Mayr in the Malagasy region and in the New World. Bulletin of the British Museum (Natural History). Entomology 38:129–181.

———. 1980. The ant tribe Tetramoriini (Hymenoptera: Formicidae). The genus *Tetramorium* Mayr in the Ethiopian zoogeographical region. Bulletin of the British Museum (Natural History). Entomology 40:193–384.

———. 1981a. A revision of the ant genera *Meranoplus* F. Smith, *Dicroaspis* Emery and *Calyptomyrmex* Emery (Hymenoptera: Formicidae) in the Ethiopian zoogeographical region. Bulletin of the British Museum (Natural History). Entomology 42:43–81.

———. 1981b. A revision of six minor genera of Myrmicinae (Hymenoptera: Formicidae) in the Ethiopian zoogeographical region. Bulletin of the British Museum (Natural History). Entomology 43:245–307.

———. 1982. Afrotropical species of the myrmicine ant genera *Cardiocondyla, Leptothorax, Melissotarsus, Messor* and *Cataulacus* (Formicidae). Bulletin of the British Museum (Natural History). Entomology 45:307–370.

———. 1983. The Afrotropical dacetine ants (Formicidae). Bulletin of the British Museum (Natural History). Entomology 46:267–416.

———. 1984. Diagnosis and relationships of the myrmicine ant genus *Ishakidris* gen. n. (Hymenoptera: Formicidae). Systematic Entomology 9:373–382.

———. 1986. A taxonomic and biological review of the tetramoriine ant genus *Rhoptromyrmex* (Hymenoptera: Formicidae). Systematic Entomology 11:1–17.

———. 1987. A review of the *Solenopsis* genus-group and revision of Afrotropical *Monomorium* Mayr (Hymenoptera: Formicidae). Bulletin of the British Museum (Natural History). Entomology 54:263–452.

———. 1988a. *Secostruma,* a new subterranean tetramoriine ant genus (Hymenoptera: Formicidae). Systematic Entomology 13:263–270.

———. 1988b. A review of *Paratopula* Wheeler, a forgotten genus of myrmicine ants (Hym., Formicidae). Entomologists' Monthly Magazine 124: 125–143.

———. 1990a. Abdominal characters and status of the cerapachyine ants (Hymenoptera, Formicidae). Journal of Natural History 24:53–68.

———. 1990b. [Untitled. A key to the living subfamilies of ants, based on the worker caste.] Pp. 33–34. *In* B. Hölldobler and E. O. Wilson, The Ants. Harvard University Press, Cambridge, Massachusetts.

———. 1990c. Army ants reassessed: The phylogeny and classification of the doryline section (Hymenoptera, Formicidae). Journal of Natural History 24:1339–1364.

———. 1990d. The higher classification of the ant subfamily Leptanillinae (Hymenoptera: Formicidae). Systematic Entomology. 15:267–282.

———. 1991. New myrmicine genera from the Oriental Region (Hymenoptera: Formicidae). Systematic Entomology 16:1–13.

———. 1992. A review of the ant genus *Recurvidris* (Hym.: Formicidae), a new name for *Trigonogaster* Forel. Psyche (Cambridge) 99:35–48.

———. 1994. Identification Guide to the Ant Genera of the World. Harvard University Press, Cambridge, Massachusetts.

———. 1995a. A taxonomic and zoogeographical census of the extant ant taxa (Hymenoptera: Formicidae). Journal of Natural History 29:1037–1056.

———. 1995b. A New General Catalogue of the Ants of the World. Harvard University Press, Cambridge, Massachusetts.

———. 1999. Ant genera of the tribe Dacetonini (Hymenoptera: Formicidae). Journal of Natural History 33:1639–1689.

Bolton, B., and R. Belshaw. 1993. Taxonomy and biology of the supposedly lestobiotic ant genus *Paedalgus* (Hymenoptera: Formicidae). Systematic Entomology 18:181–189.

Bolton, B., and A. C. Marsh 1989. The Afrotropical thermophilic ant genus *Ocymyrmex* (Hymenoptera: Formicidae). Journal of Natural History 23:1267–1308.

Bond, W., and P. Slingsby. 1984. Collapse of an ant-plant mutualism: The Argentine ant (*Iridomyrmex humilis*) and myrmecochorous proteaceae. Ecology 65:1031–1037.

Bonham, C. D. 1989. Measurements for Terrestrial Vegetation. John Wiley and Sons, New York.

Borgmeier, T. 1955. Die Wanderameisen der neotropischen Region. Studia Entomologica 3: 1–720.

————. 1959. Revision der Gattung *Atta* Fabricius (Hymenoptera, Formicidae). Studia Entomologica (n.s.) 2:321–390.

————. 1963. Revision of the North American phorid flies. Part I. The Phorinae, Aenigmatiinae, and Metopininae, except Megaselia (Diptera: Phoridae). Studia Entomologica (n.s.) 6:1–256.

Borror, D., C. Triplehorn, and N. Johnson. 1989. An Introduction to the Study of Insects. W. B. Saunders, Philadelphia.

Brandão, C. R. F. 1989. *Belonopelta minima,* a new species from Brazil. Revista Brasileira de Entomologia 33:135–138.

————. 1990. Systematic revision of the neotropical ant genus *Megalomyrmex* Forel (Hymenoptera: Formicidae: Myrmicinae) with the description of thirteen new species. Arquivos de Zoologia (Museu de Zoologia da Universidade de São Paulo) 31:411–481.

————. 1991. Adendos ao Catálogo Abreviado das Formigas da Região Neotropical (Hymenoptera: Formicidae). Revista Brasileira de Entomologia 35(2):319–412.

Brandão, C. R. F., and J. E. Lattke. 1990. Description of a new Ecuadorian *Gnamptogenys* species (Hymenoptera: Formicidae), with a discussion on the status of the *alfaria* group. Journal of the New York Entomological Society 98:489–494.

Brandão, C. R. F., and R. V. S. Paiva. 1994. The Galapagos ant fauna and the attributes of colonizing ant species. Pp. 1–10. *In* D. F. Williams (ed.), Exotic Ants: Biology, Impact, and Control of Introduced Species. Westview Press, Boulder, Colorado.

Brandão, C. R. F., and P. E. Vanzolini. 1985. Notes on incubatory inquilinism between squamata (Reptilia) and the neotropical fungus-growing ant genus *Acromyrmex* (Hymenoptera: Formicidae). Papeis Avulsos de Zoologia (São Paulo) 36:31–36.

Brandão, C. R. F., J. L . M. Diniz, and E. M. Tomotake. 1991. *Thaumatomyrmex* strip millipedes for prey: A novel predatory behaviour in ants, and the first case of sympatry in the genus (Hymenoptera: Formicidae). Insectes Sociaux 38:335–344.

Brandão, C. R. F., J. L. M. Diniz, D. Agosti, and J. H. Delabie. 1999. Revision of the Neotropical ant subfamily Leptanilloidinae. Systematic Entomology 24:17–36.

Brian, M. V. 1964. Ant distribution in a southern English heath. Journal of Animal Ecology 33:451–461.

Brian, M. V., and A. D. Brian. 1951. Insolation and ant populations in the west of Scotland. Transactions of the Royal Entomological Society of London 102:303–330.

Brian, M. V., M. D. Mountford, A. Abbott, and S. Vincent. 1976. The changes in ant species distribution during ten years' post-fire regeneration of a heath. Journal of Animal Ecology 45:115–133.

Brothers, D. J. 1975. Phylogeny and classification of the aculeate Hymenoptera, with special reference to Mutillidae. University of Kansas Science Bulletin 50:483–648.

Brothers, D. J., and J. M. Carpenter. 1993. Phylogeny of Aculeata: Chrysidoidea and Vespoidea (Hymenoptera). Journal of Hymenoptera Research 2:227–304.

Brown, B. V. 1993. Taxonomy and preliminary phylogeny of the parasitic genus *Apocephalus,* subgenus *Mesophora* (Diptera: Phoridae). Systematic Entomology 18:191–230.

Brown, B. V., and D. H. Feener Jr. 1991a. Behavior and host location cues of *Apocephalus paraponerae* (Diptera: Phoridae), a parasitoid of the giant tropical ant, *Paraponera clavata* (Hymenoptera: Formicidae). Biotropica 23:182–187.

————. 1991b. Life history parameters and description of the larva of *Apocephalus paraponerae* (Diptera: Phoridae), a parasitoid of the giant tropical ant *Paraponera clavata* (Hymenoptera: Formicidae). Journal of Natural History 25:221–231.

Brown, J. H. 1995. Macroecology. University of Chicago Press, Chicago.

Brown, J. H., T. Valone, and C. Curtin. 1997. Reorganization of an arid ecosystem in response to recent climate change. Proceedings of the National Academy of Sciences 94:9729–9733.

Brown, W. L., Jr. 1945. An unusual behavior pattern observed in a Szechuanese ant. Journal of the West China Border Research Society, Series B 15:185–186.

————. 1948. A preliminary generic revision of the higher Dacetini (Hymenoptera: Formicidae). Transactions of the American Entomological Society 74:101–129.

————. 1949a. Synonymic and other notes on Formicidae (Hymenoptera). Psyche (Cambridge) 56:41–49.

————. 1949b. Revision of the ant tribe Dacetini. III. *Epitritus* Emery and *Quadristruma* new genus (Hymenoptera: Formicidae). Transactions of the American Entomological Society 75:43–51.

———. 1949c. Revision of the ant tribe Dacetini. I. Fauna of Japan, China and Taiwan. Mushi 20: 1–25.

———. 1949d. Revision of the ant tribe Dacetini. IV. Some genera properly excluded from the Dacetini, with the establishment of the Basicerotini new tribe. Transactions of the American Entomological Society 75:83–96.

———. 1950. Revision of the ant tribe Dacetini. II. *Glamyromyrmex* Wheeler and closely related small genera. Transactions of the American Entomological Society 76:27–36.

———. 1952. On the identity of *Adlerzia* Forel (Hymenoptera: Formicidae). Pan-Pacific Entomologist 28:173–177.

———. 1953a. Revisionary studies in the ant tribe Dacetini. American Midland Naturalist 50:1–137.

———. 1953b. Characters and synonymies among the genera of ants. Part II. Breviora 18:1–8.

———. 1953c. Revisionary notes on the ant genus *Myrmecia* of Australia. Bulletin of the Museum of Comparative Zoology, Harvard University 111:1–35.

———. 1954a. (1953) The Indo-Australian species of the ant genus *Strumigenys* Fr. Smith: S. wallacei Emery and relatives. Psyche (Cambridge) 60:85–89.

Brown, W. L., Jr. 1954b. Remarks on the internal phylogeny and subfamily classification of the family Formicidae. Insectes Sociaux 1:22–31.

———. 1955. A revision of the Australian ant genus *Notoncus* Emery, with notes on the other genera of Melophorini. Bulletin of the Museum of Comparative Zoology, Harvard College 113:471–494.

———. 1958. Contributions toward a reclassification of the Formicidae. II. Tribe Ectatommini (Hymenoptera). Bulletin of the Museum of Comparative Zoology, Harvard University 118:173–362.

———. 1959. A revision of the dacetine ant genus *Neostruma*. Breviora 107:1–13.

———. 1960. Contributions toward a reclassification of the Formicidae. III. Tribe Amblyoponini (Hymenoptera). Bulletin of the Museum of Comparative Zoology, Harvard University 122:143–230.

———. 1962. The neotropical species of the ant genus *Strumigenys* Fr. Smith: Synopsis and keys to the species. Psyche (Cambridge) 69:238–267.

———. 1964. The ant genus *Smithistruma*: A first supplement to the World revision (Hymenoptera:

Formicidae). Transactions of the American Entomological Society 89:183–200.

———. 1965. Contributions to a reclassification of the Formicidae. IV. Tribe Typhlomyrmecini (Hymenoptera). Psyche (Cambridge) 72:65–78.

———. 1967. Studies on North American ants. II. *Myrmecina*. Entomological News 78:233–240.

———. 1972. *Asketogenys acubecca,* a new genus and species of dacetine ants from Malaya (Hymenoptera: Formicidae). Psyche (Cambridge) 79:23–26.

———. 1973. A comparison of the Hylean and Congo–West African rain forest ant faunas. Pp. 161–185. *In* B. J. Meggers, E. S. Ayensu, and W. D. Duckworth (eds.), Tropical Forest Ecosystems in Africa and South America: A Comparative Review. Smithsonian Institution Press, Washington, D.C.

———. 1974a. A supplement to the revision of the ant genus *Basiceros* (Hymenoptera: Formicidae). Journal of the New York Entomological Society 82:131–140.

———. 1974b. *Concoctio* genus nov. Pilot Register of Zoology. Card No. 29.

———. 1974c. *Concoctio concenta* species nov. Pilot Register of Zoology. Card No. 30.

———. 1974d. *Dolioponera* genus nov. Pilot Register of Zoology. Card No. 31.

———. 1974e. *Dolioponera fustigera* species nov. Pilot Register of Zoology. Card No. 32.

———. 1975. Contributions toward a reclassification of the Formicidae. V. Ponerinae, tribes Platythyreini, Cerapachyini, Cylindromyrmecini, Acanthostichini, and Aenictogitini. Search Agriculture (Ithaca, N.Y.) 5(1):1–115.

———. 1976a. *Cladarogenys* genus nov. Pilot Register of Zoology. Card No. 33.

———. 1976b. *Cladarogenys lasia* species nov. Pilot Register of Zoology. Card No. 34.

———. 1976c. Contributions toward a reclassification of the Formicidae. Part VI. Ponerinae, tribe Ponerini, subtribe Odontomachiti. Section A. Introduction, subtribal characters. Genus *Odontomachus*. Studia Entomologica 19:67–171.

———. 1977a. An aberrant new genus of myrmicine ant from Madagascar. Psyche (Cambridge) 84:218–224.

———. 1977b. A supplement to the world revision of *Odontomachus* (Hymenoptera: Formicidae). Psyche (Cambridge) 84:281–285.

———. 1978. Contributions toward a reclassification of the Formicidae. Part VI. Ponerinae, tribe

Ponerini, subtribe Odontomachiti. Section B. Genus *Anochetus* and bibliography. Studia Entomologica 20:549–638.

———. 1979. A remarkable new species of *Proceratium,* with dietary and other notes on the genus (Hymenoptera: Formicidae). Psyche (Cambridge) 86:337–346.

———. 1980a. *Protalaridris* genus nov. Pilot Register of Zoology. Card No. 36.

———. 1980b. *Protalaridris armata* species nov. Pilot Register of Zoology. Card No. 37.

———. 1985. *Indomyrma dasypyx,* new genus and species, a myrmicine ant from peninsular India (Hymenoptera: Formicidae). Israel Journal of Entomology 19:37–49.

Brown, W. L., Jr., and R. G. Boisvert. 1979. The dacetine ant genus *Pentastruma.* Psyche (Cambridge) 85:201–207.

Brown, W. L., Jr., and W. W. Kempf. 1960. A world revision of the ant tribe Basicerotini. Studia Entomologica (n.s.) 3:161–250.

———. 1967. *Tatuidris,* a remarkable new genus of Formicidae (Hymenoptera). Psyche (Cambridge) 74:183–190.

———. 1969. A revision of the neotropical dacetine ant genus *Acanthognathus* (Hymenoptera: Formicidae). Psyche (Cambridge) 76:87–109.

Brown, W. L., Jr., W. H. Gotwald Jr., and J. Lévieux. 1970. A new genus of ponerine ants from West Africa (Hymenoptera: Formicidae) with ecological notes. Psyche (Cambridge) 77:259–275.

Brues, C. T. 1925. *Scyphodon,* an anomalous genus of Hymenoptera of doubtful affinities. Treubia 6:93–96.

Buckley, R. C. 1982a. Ant-plant interactions: A world review. Pp. 111–141. *In* R. C. Buckley (ed.), Ant-Plant Interactions in Australia. Dr. W. Junk, The Hague.

———. 1982b. A world bibliography of ant-plant interactions. Pp. 143–162. *In* R. C. Buckley (ed.), Ant-Plant Interactions in Australia. Dr. W. Junk, The Hague.

Bukowski, T. C. 1991. Solifugae in *Atta* foraging columns. P. 70. *In* B. A. Loiselle and C. K. Augspurger (eds.), OTS 91-1: Tropical Biology: An Ecological Approach. Organization for Tropical Studies, Duke University, Durham, North Carolina.

Bunge, J., and M. Fitzpatrick. 1993. Estimating the number of species: A review. Journal of the American Statistical Association 88:364–373.

Bünzli, G. H. 1935. Untersuchungen über coccidophile Ameisen aus den Kaffeefeldern von Surinam. Mitteilungen der Schweizerischen Entomologischen Gesellschaft 16:453–593.

Burbridge, A. H., K. Leicester, S. McDavitt, and J. D. Majer. 1992. Ants as indicators of disturbance at Yanchep National Park, Western Australia. Journal of the Royal Society of Western Australia 75:89–95.

Buren, W. F. 1968a. Some fundamental taxonomic problems in *Formica* (Hymenoptera: Formicidae). Journal of the Georgia Entomological Society 3:25–40.

———. 1968b. A review of the species of *Crematogaster,* sensu stricto, in North America (Hymenoptera, Formicidae). Part II. Descriptions of new species. Journal of the Georgia Entomological Society 3:91–121.

Burnham, K. P., and W. S. Overton. 1978. Estimation of the size of a closed population when capture probabilities vary among animals. Biometrika 65:623–633.

———. 1979. Robust estimation of population size when capture probabilities vary among animals. Ecology 60:927–936.

Buschinger, A. 1981. Biological and systematic relationships of social parasitic Leptothoracini from Europe and North America. Pp. 211–222. *In* P. E. Howse and J.-L. Clement (eds.), Biosystematics of Social Insects. Systematics Association Special Volume No. 19. Academic Press, London.

———. 1989. Evolution, speciation, and inbreeding in the parasitic ant genus *Epimyrma* (Hymenoptera, Formicidae). Journal of Evolutionary Biology 2:265–283.

Buschinger, A., and U. Winter. 1983. *Myrmicinosporidium durum* Hölldobler 1933, Parasit bei Ameisen (Hym., Formicidae) in Frankreich, der Schweiz und Jugoslawien wieder aufgefunden. Zoologischer Anzeiger 210:393–398.

Buschinger, A., J. Heinze, K. Jessen, P. Douwes, and U. Winter. 1987. First European record of a queen ant carrying a mealybug during her mating flight. Naturwissenschaften 74:139–140.

Buschinger, A., W. Ehrhardt, K. Fischer, and J. Ofer. 1988. The slave-making ant genus *Chalepoxenus* (Hymenoptera, Formicidae). I. Review of literature, range, slave species. Zoologische Jahrbücher. Abteilung für Systematik, Ökologie und Geographie der Tiere 115:383–401.

Buschinger, A., R. G. Kleespies, and R. D. Schumann. 1995. A gregarine parasite of *Leptothorax* ants from North America. Insectes Sociaux 42:219–222.

Byrne, M. M. 1994. Ecology of twig-dwelling ants in a wet lowland tropical forest. Biotropica 26:61–72.

Caetano, F. H. 1989. Endosymbiosis of ants with intestinal and salivary gland bacteria. *In* W. Schwemmler and G. Gassner (eds.), Insect Endosymbionts: Morphology, Physiology, Genetics, Evolution. CRC Press, Boca Raton, Florida.

Caetano, F. H., and C. Cruz-Landim. 1985. Presence of microorganisms in the alimentary canal of ants of the tribe Cephalotini (Myrmicinae): Location and relationship with intestinal structures. Naturalia 10:37–47.

Cagniant, H. 1997. The ant genus *Tetramorium* (Hymenoptera: Formicidae) in Morocco. Annales de la Société Entomologique de France 33(1):89–100.

Cagniant, H., and X. Espadaler. 1997a. *Leptothorax, Epimyrma* and *Chalepoxenus* of Morocco (Hymenoptera: Formicidae). Key and catalogue of species. Annales de la Société Entomologique de France 33(3):259–284.

Cagniant, H., and X. Espadaler. 1997b. The ant genus *Messor* in Morocco (Hymenoptera: Formicidae). Annales de la Société Entomologique de France 33(4):419–434.

Camilo, G. R., and S. A. Phillips Jr. 1990. Evolution of ant communities in response to invasion by the fire ant *Solenopsis invicta*. Pp. 190–198. *In* R. K. Vander Meer, K. Jaffe, and A. Cedeno (eds.), Applied Myrmecology: A World Perspective. Westview Press, Boulder, Colorado.

Cammell, M. E., M. J. Way, and M. R. Paiva. 1996. Diversity and structure of ant communities associated with oak, pine, eucalyptus, and arable habitats in Portugal. Insectes Sociaux 43:37–46.

Carpenter, S., T. Frost, J. Kitchell, and T. Kratz. 1993. Species dynamics and global environmental change: A perspective from ecosystem experiments. Pp. 267–279. *In* P. Kareiva, J. Kingsolver, and R. Huey (eds.), Biotic Interactions and Global Change. Sinauer Associates, Sunderland, Massachusetts.

Chao, A. 1984. Non-parametric estimation of the number of classes in a population. Scandanavian Journal of Statistics 11:265–270.

———. 1987. Estimating the population size for capture-recapture data with unequal catchability. Biometrics 43:783–791.

Chao, A., and S. M. Lee. 1992. Estimating the number of classes via sample coverage. Journal of the American Statistical Association 87:210–217.

Chao, A., M. C. Ma, and M. C. K. Yang. 1993. Stopping rules and estimation for recapture debugging with unequal failure rates. Biometrika 80:193–201.

Chapela, I. H., S. A. Rehner, T. R. Schultz, and U. G. Mueller. 1994. Evolutionary history of the symbiosis between the fungus-growing ants and their fungi. Science 266:1691–1694.

Chapman, T. A. 1920. Contributions to the life history of *Lycaena euphemus* Hb. Transactions of the Royal Entomological Society of London 1919:450–465.

Chazdon, R. L., R. K. Colwell, J. S. Denslow, and M. R. Guariguata. 1998. Statistical methods for estimating species richness of woody regeneration in primary and secondary rain forests of northeastern Costa Rica. Pp. 285–309. *In* F. Dallmeier and J. A. Comiskey (eds.), Forest Biodiversity Research, Monitoring and Modeling: Conceptual Background and Old World Case Studies. Parthenon, Paris.

Chen, Y.-C., W.-H. Hwang, A. Chao, and C.-Y. Kuo. 1995. Estimating the number of common species. Analysis of the number of common bird species in Ke-Yar Stream and Chung-Kang Stream. Journal of the Chinese Statistical Association 33:373–393. [In Chinese with English abstract.]

Cherrett, J. M. 1986. History of the leaf-cutting ant problem. Pp. 10–17. *In* C. S. Lofgren and R. K. Vander Meer (eds.), Fire Ants and Leaf Cutting Ants: Biology and Management. Westview Press, Boulder, Colorado

Chew, R. M. 1995. Aspects of the ecology of three species of ants (*Myrmecocystus* spp., *Aphaenogaster* sp.) in desertified grassland in southeastern Arizona, 1958–1993. American Midland Naturalist 134:75–83.

Chew, R. M., and J. De Vita. 1980. Foraging characteristics of a desert ant assemblage: Functional morphology and species separation in Cochise County, Arizona. Journal of Arid Environments 3:75–83.

Christian, K., and S. R. Morton. 1992. Extreme thermophilia in a central Australian ant, *Melophorus bagoti*. Physiological Zoology 65:885–905.

Clark, D. B., C. Guayasamín, O. Pazmiño, C. Donoso, and Y. Páez de Villacís. 1982. The tramp ant *Wasmannia auropunctata:* Autecology and effects on ant diversity and distribution on Santa Cruz Island, Galapagos. Biotropica 14:196–207.

Clark, J. 1930. The Australian ants of the genus *Dolichoderus* (Formicidae). Subgenus *Hypoclinea* Mayr. Australian Zoologist 6:252–268.

———. 1936. A revision of Australian species of *Rhytidoponera* Mayr (Formicidae). Memoirs of the National Museum of Victoria 9:14–89.

———. 1951. The Formicidae of Australia, Vol. 1: Subfamily Myrmeciinae. CSIRO, Melbourne.

Clausen, C. P. 1940a. Entomophagous Insects. McGraw-Hill, New York.

———. 1940b. The immature stages of the Eucharidae. Proceedings of the Entomological Society of Washington 42:161–170.

———. 1940c. The oviposition habits of the Eucharidae (Hymenotpera). Journal of the Washington Academy of Sciences 30:504–516.

———. 1941. The habits of the Eucharidae. Psyche (Cambridge) 48:57–69.

Clements, R. O. 1982. Sampling and extraction techniques for collecting invertebrates from grassland. Entomologists' Monthly Magazine 118:133–142.

Clench, H. K. 1979. How to make regional lists of butterflies: Some thoughts. Journal of the Lepidopterist Society 33:216–231.

Cobabe, E. A., and W. D. Allmon. 1994. Effects of sampling on paleoecologic and taphonomic analyses in high-diversity fossil accumulations: An example from the Eocene Gosport Sand, Alabama. Lethaia 27:167–178.

Coddington, J. A., L. H. Young, and F. A. Coyle. 1996. Estimating spider species richness in a southern Appalachian cove hardwood forest. Journal of Arachnology 24:111–128.

Coenen-Stass D., Schaarschmidt, and I. Lamprecht. 1980. Temperature distribution and calorimetric determination of heat production in the nest of the wood ant, *Formica polyctena* (Hymenoptera: Formicidae). Ecology 61:238–244.

Cole, A. C., Jr. 1940. A guide to the ants of the Great Smoky Mountains National Park, Tennessee. American Midland Naturalist 24:1–88.

———. 1949. Notes on *Gesomyrmex* (Hymenoptera: Formicidae). Entomological News 60:181.

———. 1968. *Pogonomyrmex* Harvester Ants: A Study of the Genus in North America. University of Tennessee Press, Knoxville.

Colwell, R. K. 1997. EstimateS: Statistical estimation of species richness and shared species from samples. Version 5. User's guide and application published at: http://viceroy.eeb.uconn.edu/estimates.

Colwell, R. K., and J. A. Coddington. 1994. Estimating terrestrial biodiversity through extrapolation. Philosophical Transactions of the Royal Society of London, Series B 345:101–118.

Colwell, R. K., and G. C. Hurtt. 1994. Nonbiological gradients in species richness and a spurious Rapaport effect. American Naturalist 144:570–595.

Connell, J. 1978. Diversity in tropical rain forests and coral reefs. Science 199:1302–1310.

Connell, J., and W. Sousa 1983. On the evidence needed to judge ecological stability or persistence. American Naturalist 121:789–824.

Convention on Biological Diversity. 1992. Convention on Biological Diversity. United Nations Conference on Environment and Development, Rio de Janeiro.

Cover, S. P., J. E. Tobin, and E. O. Wilson. 1990. The ant community of a tropical lowland rain forest site in Peruvian Amazonia. Pp. 699–700. *In* G. K. Veeresh, B. Mallik, and C. A. Viraktamath (eds.), Social Insects and the Environment. Proceedings of the 11th International Congress of IUSSI, 1990. Oxford and IBH Publishing, New Delhi.

Cowling, R. M., and J. J. Midgely. 1996. The influence of regional phenomena on an emerging global ecology. Global Ecology and Biogeography Letters 5:63–65.

Cranston, P. S., and J. W. H. Trueman. 1997. "Indicator" taxa in invertebrate biodiversity assessment. Memoirs of the Museum of Victoria 56(2):267–274.

Creighton, W. S. 1950. The ants of North America. Bulletin of the Museum of Comparative Zoology, Harvard College 104:1–585.

———. 1957. A study of the genus *Xenomyrmex* (Hymenoptera, Formicidae). American Museum Novitates 1843:1–14.

Crist, T. O., and J. A. Wiens. 1994. Scale effects of vegetation structure on forager movements and seed harvesting by ants. Oikos 69:37–46.

———. 1996. The distribution of ant colonies in a semiarid landscape: Implications for community and ecosystem processes. Oikos 76:301–311.

Crosland, M. W. J. 1988. Effect of a gregarine parasite on the color of *Myrmecia pilosula* (Hymenoptera: Formicidae). Annals of the Entomological Society of America 81:481–484.

Crowell, K. L. 1968. Rates of competitive exclusion by the Argentine ant in Bermuda. Ecology 49:551–555.

Crozier, R. H. 1990. From population genetics to phylogeny: Uses and limits of mitochondrial DNA. Australian Systematic Botany 3:111–124

Culver, D. S. 1974. Species packing in Caribbean and North temperate ant communities. Ecology 55:974–988.

Cushman, J. H., J. H. Lawton, and B. F. J. Manly. 1993. Latitudinal patterns in European ant assemblages: Variation in species richness and body size. Oecologia 95:30–37.

Darlington, P. J., Jr. 1971. The carabid beetles of New Guinea. Part IV. General considerations; analysis and history of fauna; taxonomic supplement. Bulletin of the Museum of Comparative Zoology, Harvard College 142:129–337.

Davidson, D. W. 1977a. Species diversity and community organization in desert seed-eating ants. Ecology 58:711–724.

———. 1977b. Foraging ecology and community organization in desert seed-eating ants. Ecology 58:725–737.

———. 1978. Size variability in the worker caste of a social insect (Veromessor pergandei Mayr) as a function of the competitive environment. American Naturalist 112:523–532.

———. 1980. Some consequences of diffuse competition in a desert ant community. American Naturalist 116:92–105.

———. 1988. Ecological studies of Neotropical ant gardens. Ecology 69:1138–1152.

———. 1997. The role of resource imbalances in the evolutionary ecology of tropical arboreal ants. Biological Journal of the Linnean Society 61:153–181.

———. 1998. Resource discovery versus resource domination in ants: A functional mechanism for breaking the tradeoff. Ecological Entomology 23:484–490.

Davidson, D. W., and L. Patrell-Kim. 1996. Tropical arboreal ants: Why so abundant? Pp. 127–140. In A. C. Gibson (ed.), Neotropical Biodiversity and Conservation. University of California, Los Angeles Botanical Garden Publication Number 1. University of California, Los Angeles.

Davidson, D. W., J. H. Brown, and R. S. Inouye. 1980. Competition and the structure of granivore communities. BioSciences 30(4):233–238.

Davidson, D. W., J. T. Longino, and R. R. Snelling. 1988. Pruning of host plant neighbors by ants: An experimental approach. Ecology 69:801–808.

Davidson, D. W., R. R. Snelling, and J. T. Longino. 1989. Competition among ants for myrmecophytes and the significance of plant trichomes. Biotropica 21:64–73.

De Kock, A. E., and J. H. Giliomee. 1989. A survey of the Argentine ant, Iridomyrmex humilis (Mayr) (Hymenoptera: Formicidae) in South African fynbos. Journal of the Entomological Society of Southern Africa 52:157–164.

De Vries, P. J., D. Murray, and R. Lande. 1997. Species diversity in vertical, horizontal, and temporal dimensions of a fruit-feeding butterfly community in an Ecuadorian rainforest. Biological Journal of the Linnean Society 62:343–364.

Delabie, J. H. C. 1995. Inquilinismo simultáneo de duas espécies de Centromyrmex (Hymenoptera; Formicidae; Ponerinae) em cupinzeiros de Syntermes sp. (Isoptera; Termitidae; Nasutermitinae). Revista Brasileira de Entomologia 39:605–609.

Delabie, J. H. C., and H. G. Fowler. 1995. Soil and litter cryptic ant assemblages of Bahian cocoa plantations. Pedobiologia 39:423–433.

Delabie, J. H. C., I. C. do Nascimento, and C. dos S. F. Mariano. 2000. Importance de l'agriculture cacaoyère pour le mainien de la biodiversité: Étude comparée de la myrmécofaune de différents milieux du sud–est de Bahia, Brésil (Hymenoptera; Formicidae). In Proceedings for the 12th International Cocoa Research Conference, Lagos, Nigeria.

Delabie, J. H. C., A. B. Casimiro, I. C. do Nascimento, A. L. B. do Souza, M. Furst, A. M. V. da Encarnação, M. R. B. Smith, and I. M. Cazorla. 1994. Stratification de la communauté de fourmis (Hymenoptera:Formicidae) dans une cacaoyère brésilienne et conséquences pour le contrôle naturel des ravageurs du cacaoyer. Pp. 823–831. In Proceedings of the 11th International Cocoa Research Conference, Lagos, Nigeria.

Delabie, J. H. C., I. C. do Nascimento, P. Pacheco, and A. B. Casimiro. 1995. Community structure of house-infesting ants (Hymenoptera: Formicidae) in Southern Bahia, Brazil. Florida Entomologist 78:264–270.

Deslippe, R. J., and R. Savolainen. 1994. Role of food supply in structuring a population of Formica ants. Journal of Animal Ecology 63:756–764.

Deyrup, M., J. Trager, and N. Carlin. 1985. The genus *Odontomachus* in the southeastern United States (Hymenoptera: Formicidae). Entomological News 96:188–195.

Di Castri, F., J. Robertson Vernhes, and T. Younes. 1992. Inventorying and monitoring biodiversity. Biology International. 27:1–27.

Diniz, J. L. M. 1990. Revisão sistemática da tribo Stegomyrmicini, com a descripção de uma nova espécie (Hymenoptera, Formicidae). Revista Brasileira de Biologia 34:277–295.

Diniz, J. L. M., and C. R. F. Brandão. 1993. Biology and myriapod egg predation by the Neotropical myrmicine ant *Stegomyrmex vizottoi* (Hymenoptera: Formicidae). Insectes Sociaux 40:301–311.

Dixon, A. F. G. 1985. Aphid Ecology. Chapman and Hall, New York.

Dlussky, G. M. 1964. The ants of the subgenus *Coptoformica* of the genus Formica (Hymenoptera, Formicidae) of the USSR. Zoologicheskii Zhurnal 43:1026–1040. [In Russian.]

———. 1965. Ants of the genus *Formica* L. of Mongolia and northeast Tibet (Hymenoptera, Formicidae). Annales Zoologici (Warsaw) 23:15–43.

———. 1967. Ants of the Genus *Formica* (Hymenoptera, Formicidae, g. *Formica*). Nauka, Moscow. [In Russian.]

———. 1969. Ants of the genus *Proformica* Ruzs. of the USSR and contiguous countries (Hymenoptera, Formicidae). Zoologicheskii Zhurnal 48:218–232. [In Russian.]

Dlussky, G. M., and B. Pisarski. 1971. Rewizja polskich gatunków mrówek (Hymenoptera: Formicidae) z rodzaju *Formica* L. Fragmenta Faunistica (Warsaw) 16:145–224.

Dlussky, G. M., and A. G. Radchenko. 1994. Ants of the genus *Diplorhoptrum* (Hymenoptera, Formicidae) from the central Palearctic. Zoologicheskii Zhurnal 73(2):102–111. [In Russian.]

Dlussky, G. M., and O. S. Soyunov. 1988. Ants of the genus *Temnothorax* Mayr (Hymenoptera: Formicidae) of the USSR. Izvestiya Akademii Nauk Turkmenskoi SSR, Seriya Biologicheskikh Nauk 1988(4):29–37. [In Russian.]

Donisthorpe, H. 1946. *Ireneopone gibber* (Hym., Formicidae), a new genus and species of myrmicine ant from Mauritius. Entomologists' Monthly Magazine 82:242–243.

Dorow, W. H. O. and R. J. Kohout.1995. A review of the subgenus *Hemioptica* Roger of the genus *Polyrhachis* Fr. Smith with description of a new species (Hymenoptera: Formicidae: Formicinae). Zoologische Mededelingen (Leiden) 69(1–14): 93–104.

DuBois, M. B. 1981. Two new species of inquilinous *Monomorium* from North America (Hymenoptera: Formicidae). University of Kansas Science Bulletin 52:31–37.

———. 1986. A revision of the native New World species of the ant genus *Monomorium* (*minimum* group) (Hymenoptera: Formicidae). University of Kansas Science Bulletin 53:65–119.

———. 1998. A revision of the ant genus *Stenamma* in the Palaearctic and Oriental regions (Hymenoptera: Formicidae: Myrmicinae). Sociobiology 32(2):193–403.

Dumpert, K. 1981. The Social Biology of Ants. Translated by C. Johnson. Pitman, Boston.

———. 1985. *Camponotus* (*Karavaievia*) *texens* sp. n. and *C.* (*K.*) *gombaki* sp. n. from Malaysia in comparison with other *Karavaievia* species (Formicidae: Formicinae). Psyche (Cambridge) 92:557–573.

Dumpert, K., U. Maschwitz, A. Weissflog, K. Rosciszewski, and I. H. Azarae. 1995. Six new weaver ant species from Malaysia: *Camponotus* (*Karavaievia*) *striaticeps, C.* (*K.*) *melanus, C.* (*K.*) *nigripes, C.* (*K.*) *belumensis, C.* (*K.*) *gentingensis,* and *C.* (*K.*) *micragyne.* Malaysian Journal of Science 16A:87–105.

Eisner, T. 1957. A comparative morphological study of the proventriculus of ants (Hymenoptera: Formicidae). Bulletin of the Museum of Comparative Zoology, Harvard College 116:439–490.

Elmes, G. W. 1991. Ant colonies and environmental disturbance. Pp. 15–32. *In* P. S. Meadows and A. Meadows (eds.), Environmental Impact of Burrowing Animals and Animal Burrows. Clarendon Press, Oxford.

Emery, C. 1897. Revisione del genere *Diacamma* Mayr. Rendiconti delle Sessione dell' Accademia delle Scienze dell Istituto di Bologna (n.s.) 1:147–167.

———. 1901. Notes sur les sous-familles des Dorylines et Ponérines (Famille des Formicides). Annales de la Société Entomologique de Belgique 45:32–54.

———. 1910. Hymenoptera. Fam. Formicidae. Subfam. Dorylinae. Genera Insectorum 102:1–34.

———. 1911. Hymenoptera. Fam. Formicidae. Subfam. Ponerinae. Genera Insectorum 118:1–125.

————. 1913 (1912). Hymenoptera. Fam. Formicidae. Subfam. Dolichoderinae. Genera Insectorum 137:1–50.

————. 1920. La distribuzione geografica attuale delle formiche. Tentativo di spiegarne la genesi col soccorso di ipotesi filogenetiche e paleogeografiche. Atti della Reale Accademia dei Lincei. Memorie. Classe di Scienze, Fisiche, Matematiche e Naturali (5)13:357–450.

————. 1921. Hymenoptera. Fam. Formicidae. Subfam. Myrmicinae. [part] Genera Insectorum 174A:1–94 + 7 plates.

————. 1922. Hymenoptera. Fam. Formicidae. Subfam. Myrmicinae. [part] Genera Insectorum 174B:95–206.

————. 1925a. Hymenoptera. Fam. Formicidae. Subfam. Formicinae. Genera Insectorum 183:1–302.

————. 1925b. Revision des espèces paléarctiques du genre *Tapinoma*. Revue Suisse Zoologique 32:45–64.

Entomological Society of Canada. 1978. Collections of Canadian Insects and Certain Related Groups. Entomological Society of Canada, Ottawa.

Erickson, J. M. 1971. The displacement of native ant species by the introduced Argentine ant *Iridomyrmex humilis* Mayr. Psyche (Cambridge) 78:257–266.

Erwin, T. L. 1983. Beetles and other arthropods of tropical forest canopies at Manaus, Brazil, sampled by insecticidal fogging. Pp. 59–79. *In* S. L. Sutton, T. C. Whitmore, and A. C. Chadwick (eds.), Ecology and Management of Tropical Rainforest. Blackwell, Oxford.

————. 1986. The tropical forest canopy: The heart of biotic diversity. Pp. 123–129. *In* E. O. Wilson (ed.), Biodiversity. National Academy Press, Washington, D.C.

————. 1989. Sorting tropical forest canopy samples (an experimental project for networking information). Insect Collection News 2(1):8.

Espadaler, X. 1982. *Myrmicinosporidium* sp., parasite interne des fourmis: Etude au meb de la structure externe. Pp. 239–241. *In* A. deHaro and X. Espadaler (eds.), La Communication chez les Sociétés d'Insectes. Colloque Internationale de l'Union Internationale pour l'Etude des Insectes Sociaux, Section française, Barcelona, 1982. Universidad Autónoma de Barcelona, Bellaterra.

Ettershank, G. 1966. A generic revision of the world Myrmicinae related to *Solenopsis* and *Pheidologeton* (Hymenoptera: Formicidae). Australian Journal of Zoology 14:73–171.

Evans, H. E. 1962. A review of nesting behavior of digger wasps of the genus *Aphilanthops,* with special attention to the mechanics of prey carriage. Behavior 19:239–260.

————. 1977. Prey specificity in *Clypeadon* (Hymenoptera: Sphecidae). Pan-Pacific Entomologist 53:144.

Farquharson, C. O. 1914. The growth of fungi on the shelters built over Coccidae by *Cremastogater*-ants. Transactions of the Entomological Society of London 1914:42–50.

————. 1918. *Harpagomyia* and other Diptera fed by *Cremastogaster* ants in S. Nigeria. Proceedings of the Entomological Society of London 1918:29–39.

Feener, D. H., Jr. 1981. Competition between ant species: Outcome controlled by parasitic flies. Science 214:815–817.

Feener, D. H., Jr., and K. A. G. Moss. 1990. Defense against parasites by hitchhikers in leaf-cutting ants: A quantative assessment. Behavioral Ecology and Sociobiology 26:17–29.

Feener, D. H., Jr., L. F. Jacobs, and J. O. Schmidt. 1996. Specialized parasitoid attracted to a pheromone of ants. Animal Behavior 51:61–66.

Fellers, G. M., and J. H. Fellers. 1982. Scavenging rates of invertebrates in an eastern deciduous forest. American Midland Naturalist 107:389–392.

Fellers, J. H. 1987. Interference and exploitation in a guild of woodland ants. Ecology 68:1466–1478.

————. 1989. Daily and seasonal activity in woodland ants. Oecologia 78:69–76.

Fernández, C. F., and M. L. H. Baena. 1997. Ants of Colombia VII: New species of the genera *Lachnomyrmex* Wheeler and *Megalomyrmex* Forel (Hymenoptera: Formicidae). Caldasia 19(1–2): 109–114.

Fernández, C. F. and E. E. Palacio. 1997. Key to northern South America's *Pogonomyrmex* (Hymenoptera: Formicidae) with description of a new species. Revista de Biologia Tropical 45(4): 1649–1661.

Ferreira, L. V., and G. T. Prance. 1998. Species richness and floristic composition in four hectares in the Jaú National Park in upland forests in Central Amazonia. Biodiversity and Conservation 7:1349–1364.

Fisher, B. L. 1996a. Ant diversity patterns along an elevational gradient in the Réserve Naturelle Intégrale d'Andringitra, Madagascar. Fieldiana: Zoology (n.s.) 85:93–108.

———. 1996b. Origins and affinities of the ant fauna of Madagascar. Pp. 457–465. *In* W. L. Lourenço (ed.), Biogéographie de Madagascar. Editions ORSTOM, Paris.

———. 1997. Biogeography and ecology of the ant fauna of Madagascar (Hymenoptera: Formicidae). Journal of Natural History 31:269–302.

———. 1998. Ant diversity patterns along an elevational gradient in the Réserve Spéciale d'Anjanaharibe-Sud and on the western Masoala Peninsula, Madagascar. Fieldiana: Zoology (n.s.) 90:39–67.

———. 1999a. Improving inventory efficiency: A case study of leaf litter diversity in Madagascar. Ecological Applications 9:714–731.

———. 1999b. Ant diversity patterns along an elevational gradient in the Réserve Naturelle Intégrale d'Andohahela, Madagascar. Fieldiana: Zoology (n.s.) 94:129–147.

Fisher, B. L., and S. Razafimandimby. 1997. Les fourmis (Hymenoptera: Formicidae). Pp. 104–109. *In* O. Langrand and S. M. Goodman (eds.), Inventaire Biologique Forêts de Vohibasia et d'Isoky-Vohimena. Recherches pour le Développement, Série Sciences Biologiques No. 12. Centre d'Information et de Documentation Scientifique et Technique and World Wide Fund for Nature, Antananarivo, Madagascar.

Fisher, B. L., H. Ratsirarson, and S. Razafimandimby. 1998. Les Fourmis (Hymenoptera: Formicidae). Pp. 107–131. *In* J. Ratsirarson and S. M. Goodman (eds.). Inventaire Biologique de la Forêt Littorale de Tampolo (Fenoarivo Atsinaanana). Recherches pour le Développement, Série Sciences Biologiques No. 14. Centre d'Information et de Documentation Scientifique et Technique and World Wide Fund for Nature. Antananarivo, Madagascar.

Fittkau, E. J., and H. Klinge. 1973. On biomass and trophic structure of the Central Amazonian rain forest ecosystem. Biotropica 5:2–14.

Fluker, S. S., and J. W. Beardsley. 1970. Sympatric associations of three ants: *Iridomyrmex humilis, Pheidole megacephala,* and *Anoplolepis longipes* in Hawaii. Annals of the Entomological Society of America 63:1290-1296.

Forel, A. 1901. Fourmis termitophages, lestobiose, *Atta tardigrada,* sous-genres d'Euponera. Annales de la Société Entomologique de Belgique 45:389–398.

Fowler, H. G. 1988. Taxa of the neotropical grass-cutting ants, *Acromyrmex* (Hymenoptera: Formicidae: Attini). Científica (Jaboticabal) 16:281–295.

———. 1995. Biodiversity estimates: Ant communities and the rare ant species (Hymenoptera: Formicidae) in a fauna of a sub-tropical island. Revista de Matemática e Estatística, São Paulo 13:29–38.

Fowler, H. G., J. V. E. Bernardi, and L. F. T. di Romagnano. 1990. Community structure and *Solenopsis invicta* in São Paulo. Pp. 199–207. *In* R. K. Vander Meer, K. Jaffe, and A. Cedeno (eds.), Applied Myrmecology: A World Perspective. Westview Press, Boulder, Colorado.

Fox, M. D., and B. J. Fox. 1982. Evidence for interspecific competition influencing ant species diversity in a regenerating heathland. Pp. 99–110. *In* R. C. Buckley (ed.), Ant-Plant Interactions in Australia. Dr. W. Junk, The Hague.

Francoeur, A. 1973. Révision taxonomique des espèces néarctiques du groupe *fusca,* genre *Formica* (Formicidae, Hymenoptera). Mémoires de la Société Entomologique du Québec 3:1–316.

Francoeur, A.1985. *Formicoxenus quebecensis* Francoeur sp. nov. Pp. 378–379. *In* A. Francoeur, R. Loiselle, and A. Buschinger. Biosystématique de la tribu Leptothoracini (Formicidae, Hymenoptera). 1. Le genre *Formicoxenus* dans la région holarctique. Naturaliste Canadian (Québec) 112:343–403.

Franks, N. R., and W. H. Bossert. 1983. The influence of swarm raiding army ants on the patchiness and diversity of a tropical leaf-litter ant community. Pp. 151–163. *In* S. L. Sutton, T. C. Whitmore, and A. C. Chadwick (eds.), Tropical Rain Forest: Ecology and Management. Blackwell, Oxford.

Friese, C. F., and M. F. Allen. 1988. The interaction of harvester ant activity and VA mycorrhizal fungi. Proceedings of the Royal Society of Edinburgh 94B:176.

———. 1993. The interaction of harvester ants and vesicular-arbuscular mycorrhizal fungi in a patchy semi-arid environment: The effects of mound structure on fungal dispersal and establishment. Functional Ecology 7:13–20.

Gadagkar, R., K. Chandrashekara, and P. Nair 1990. Insect species diversity in tropics: Sampling methods and a case study. Journal of the Bombay Natural History Society 87(3):337–353.

Gadagkar, R., P. Nair, K. Chandrashekara, and D. M. Bhat. 1993. Ant species richness and diversity in some selected localities in Western Ghats, India. Hexapoda 5:79–94.

Gaedike, R., 1995. Colleciones entomologicae (1961–1994). Nova Supplementa. Entomologica (Berlin) 6:1–83.

Gallardo, A. 1929. Note sur les moeurs de la fourmi *Pseudoatta argentina*. Revista de la Sociedad Entomológica Argentina 2:197–202.

Gallardo, J. M. 1951. Sobre um Teiidae (Reptilia, Sauria) poco conocido para la fauna Argentina. Comunicaciones Instituto Nacional de Investigaciones in Ciencia Naturales 2:8.

Gallé, L. 1991. Structure and succession of ant assemblages in a north European sand dune area. Holarctic Ecology 14:31–37.

Gaston, K. 1994. Rarity. Chapman and Hall, New York.

———. 1996. Species richness: Measure and measurement. Pp. 77–113. *In* K. J. Gaston (ed.), Biodiversity: A Biology of Numbers and Difference. Blackwell, Cambridge.

Gauch, H. G., Jr. 1982. Multivariate Analysis in Community Structure. Cambridge University Press, Cambridge.

Gauld, I., and B. Bolton. 1988. The Hymenoptera. Oxford University Press, New York.

Goeldi, E. 1897. Die Fortpflanzungsweise von 13 brasilianischen Reptilien. Zoologische Jahrbücher, Abteilung für Systematik, Geographie und Biologie der Tiere 10:640–674.

Goldstein, E. L. 1975. Island biogeography of ants. Evolution 29:750–762.

Goldstein, P. Z. 1999. Functional ecosystems and biodiversity buzzwords. Conservation Biology 13:247–255.

Goncalves, C. R. 1961. O genero *Acromyrmex* no Brasil (Hym. Formicidae). Studia Entomologica 4:113–180.

Gösswald, K. 1932. Ökologische Studien über die Ameisenfauna des mittleren Maingebietes. Zeitschrift für Wissenschaftliche Zoologie 142:1–156.

Gotelli, N. J. 1993. Ant lion zones: Causes of high-density predator aggregations. Ecology 74:226–237.

———. 1996. Ant community structure: Effects of predatory ant lions. Ecology 77:630–638.

Gotwald, W. H., Jr. 1969. Comparative morphological studies of the ants, with particular reference to the mouthparts (Hymenoptera: Formicidae). Memoirs of the Cornell University Agricultural Experimental Station 408:1–150.

———. 1982. Army ants. Pp. 157–254. *In* H. R. Hermann (ed.), Social Insects, Vol. 4. Academic Press, New York.

———. 1995. Army Ants: The Biology of Social Predation. Cornell University Press, Ithaca, New York.

Gotwald, W. H., Jr., and W. L. Brown Jr. 1966. The ant genus *Simopelta* (Hymenoptera: Formicidae). Psyche (Cambridge) 73:261–277.

Goulet, H., and J. T. Huber (eds.). 1993. Hymenoptera of the World: An Identification Guide to Families. Publication 1894/E. Center for Land and Biological Resources Research, Research Branch, Agriculture Canada, Ottawa, Ontario.

Greenslade, P., and P. J. M. Greenslade. 1971. The use of baits and preservatives in pitfall traps. Journal of the Australian Entomological Society 10:253–260.

Greenslade, P. J. M. 1964. Pitfall trapping as a method for studying populations of Carabidae (Coleoptera). Journal of Animal Ecology 33:301–310.

———. 1971. Interspecific competition and frequency changes among ants in Solomon Islands coconut plantations. Journal of Applied Ecology 8:323–352.

———. 1972. Comparative ecology of four tropical ant species. Insectes Sociaux 19:195–212.

———. 1973. Sampling ants with pitfall traps: Digging-in effects. Insectes Sociaux 20:343–353.

———. 1978. Ants. Pp. 109–113. *In* W. A. Low (ed.), The Physical and Biological Features of Kunoth Paddock in Central Australia. Technical Paper No. 4. CSIRO Division of Land Resources, Canberra.

———. 1979. A Guide to Ants of South Australia. South Australian Museum, Adelaide.

Greenslade, P. J. M., and P. Greenslade. 1977. Some effects of vegetation cover and disturbance on a tropical ant fauna. Insectes Sociaux 24:163–182

Gregg, R. E. 1959 (1958). Key to the species of *Pheidole* (Hymenoptera: Formicidae) in the United States. Journal of the New York Entomological Society 66:7–48.

———. 1963. The Ants of Colorado. University of Colorado Press, Boulder.

Grimaldi, D., Agosti, D., and J. M. Carpenter. 1997. New and rediscovered primitive Ants (Hymenoptera: Formicidae) in Cretaceous Amber from New Jersey, and their phylogenetic Relationships. American Museum Novitates 3208:1–43.

Grime J. P. 1979. Plant Strategies and Vegetation Processes. John Wiley and Sons, Chichester, U.K.

Haines, B. L. 1978. Element and energy flows through colonies of the leaf-cutting ant, *Atta colombica,* in Panama. Biotropica 10:270–277.

———. 1983. Leaf-cutting ants bleed mineral elements out of rainforest in southern Venezuela. Tropical Ecology 24:85–93.

Hamilton, W. D. 1964. The genetical evolution of social behaviour. I. Journal of Theoretical Biology 7:1–16.

Hamilton, W. D. 1972. Altruism and related phenomena, mainly in social insects. Annual Review of Ecology and Systematics 3:193–232.

Handel, S. N., and A. J. Beattie. 1990a. La dispersion des graines par les fourmis. Pour la Science 156:54–61.

———. 1990b. Seed dispersal by ants. Scientific American 263:76–83.

Handel, S. N., S. B. Fisch, and G. E. Schatz. 1981. Ants disperse a majority of herbs in a mesic forest community in New York State. Bulletin of the Torrey Botanical Club 108:430–437.

Harada, A. Y., and A. G. Bandeira. 1994. Estratificacão e densidade de invertebrados em solo arenoso sob floresta primária e plantios arbóreos na Amazonia central durante estacão seca. Acta Amazonica 24:103–118.

Harris, R. A. 1979. A glossary of surface sculpturing. California Department of Food and Agriculture Laboratory Services/ Entomology Occasional Papers in Entomology 28:1–31.

Hashimoto, Y. 1996. Skeletomuscular modifications associated with the formation of an additional petiole on the anterior abdominal segments in aculeate Hymenoptera. Japanese Journal of Entomology 64:340–356.

Hashmi, A. A. 1973. A revision of the Neotropical ant subgenus *Myrmothrix* of genus *Camponotus* (Hymenoptera: Formicidae). Studia Entomologica 16:1–140.

Haskins, C. P., and E. F. Haskins. 1965. *Pheidole megacephala* and *Iridomyrmex humilis* in Bermuda—Equilibrium or slow replacement? Ecology 46:736–740.

———. 1988. Final observations on *Pheidole megacephala* and *Iridomyrmex humilis* in Bermuda. Psyche (Cambridge) 95:177–184.

Hayek, L. C., and M. A. Buzas. 1996. Surveying Natural Populations. Columbia University Press, New York.

Heinze, J., and B. Hölldobler. 1994. Ants in the cold. Memorabilia Zoologica 48:99–108.

Heltshe, J. F., and N. E. Forrester. 1983. Estimating species richness using the jackknife procedure. Biometrics 39:1–11.

Heppner, J. B., and G. Lamas. 1982. Acronyms for world museum collections of insects, with an emphasis on Neotropical Lepidoptera. Bulletin of the Entomological Society of America 28: 305–316.

Heraty, J. M. 1985. A revision of the nearctic Eucharitinae (Hymenoptera: Chalcidoidea: Eucharitidae). Proceedings of the Entomological Society of Ontario 116:61-103.

———. 1986. *Pseudochalcura* (Hymenoptera: Eucharitidae), a New World genus of chalcidoids parasitic on ants. Systematic Entomology 11:183–212.

Heraty, J. M., and D. C. Darling. 1984. Comparative morphology of the planidial larvae of the Eucharitidae and Perilampidae (Hymenoptera: Chalcidoidea). Systematic Entomology 9:309–328.

Herbers, J. M. 1985. Seasonal structuring of a north temperate ant community. Insectes Sociaux 32: 224–240.

———. 1989. Community structure in north temperate ants: Temporal and spatial variation. Oecologia 81:201–211.

———. 1994. Structure of an Australian ant community with comparisons to North American counterparts (Hymenoptera: Formicidae). Sociobiology 24:293–306.

Herbers, J. M., and S. Grieco. 1994. Population structure of *Leptothorax ambiguus,* a facultatively polygynous and polydomous ant species. Journal of Evolutionary Biology 7:581–598.

Heyer, W. R., M. A. Donnelly, R. W. McDiarmid, L.-A. C. Hayek, and M. S. Foster. 1994. Measuring and Monitoring Biological Diversity: Standard Methods for Amphibians. Smithsonian Institution Press, Washington, D.C.

Hinkle, G., J. K. Wetterer, T. R. Schultz, and M. L. Sogin. 1994. Phylogeny of the attine ant fungi based on analysis of small subunit ribosomal RNA gene sequences. Science 266: 1695–1697.

Hinton, H. E. 1951. Myrmecophilous Lycaenidae and other Lepidoptera—A summary. Proceedings and Transactions of the South London Entomological and Natural History Society (1949–1950): 111–175.

Hölldobler, B. 1967. Zur Physiologie der Gast-Wirt-Beziehungen (Myrmecophilie) bei Ameisen. I. Das Gastverhältnis der *Atemeles-* und *Lomechusa*-Larven (Col. Staphylinidae) zu Formica (Hym. Formicidae). Zeitschrift für Vergleichende Physiologie 56:1–121.

———. 1968. Verhaltensphysiologische Untersuchungen zur Myrmecophilie einiger Staphyliniden-larven. *In* W. Herre (ed.), Verhandlungen der Deutschen Zoologischen Gesellschaft (Heidelberg, 1967), Zoologischer Anzeiger, Supplement 31:428–434.

———. 1983. Territorial behavior in the green tree ant (*Oecophylla smaragdina*). Biotropica 15:241–250.

Hölldobler, B., and C. J. Lumsden. 1980. Territorial strategies in ants. Science 210:732–739.

Hölldobler, B., and Wilson, E. O. 1977. The number of queens: an important trait in ant evolution. Naturwissenschaften 64:8–15.

———. 1990. The Ants. Belknap Press, Cambridge, Massachusetts.

Hölldobler, K. 1929. Über eine merkwürdige Parasitenerkrankung von *Solenopsis fugax*. Zeitschrift für Parasitenkunde 2:67–72.

———. 1933. Weitere Mitteilungen über Haplosporidien in Ameisen. Zeitschrift für Parasitenkunde 6:91–100.

Holway, D. A. 1995. Distribution of the Argentine ant (*Linepithema humile*) in Northern California. Conservation Biology 9:1634–1637.

Holway, D. A., A. V. Suarez, and T. J. Case. 1998. Loss of intraspecific aggression in the success of a widespread invasive social insect. Science 282:949–952.

Hood, W. G., and W. R. Tschinkel. 1990. Desiccation resistance in arboreal and terrestrial ants. Physiological Entomology 15:23–35.

Horn, W., and I. Kahle. 1935a. Über Entomologische Sammlungen, Entomologen, und Entomo-Museologie. Entomologische Beihefte, Berlin-Dahlem.

———. 1935b. Supplement to: Über Entomologische Sammlungen, Entomologen, und Entomo-Museologie. Entomologische Beihefte, Berlin-Dahlem.

Horn, W., I. Kahle, G. Friese, and R. Gaedike. 1990. Colleciones Entomologicae. Ein Kompendium über den Verbleib Entomologischer Sammlungen der Welt bis 1960. Edition Akademischer Landwirtschaftswissenschaften, Berlin.

Horvitz, C., and D. Schemske. 1990. Spatiotemporal variation in insect mutualists of a neotropical herb. Ecology 71:1085–1097.

Hudson, K., and A. Nichols (eds.). 1975. The Directory of World Museums. Columbia University Press, New York.

Huggert, L., and L. Masner. 1983. A review of myrmecophilic-symphilic diapriid wasps in the Holarctic realm, with descriptions of new taxa and a key to genera (Hymenoptera: Proctotrupoidea: Diapriidae). Contributions of the American Entomological Institute 20:63–89.

Hughes, R. G. 1986. Theories and models of species abundance. American Naturalist 128:879–899.

Human, K., and D. Gordon. 1996. Exploitative and interference competition between the Argentine ant and native ant species. Oecologia 105:405–412.

Hung, A. C. F. 1967. A revision of the ant genus *Polyrhachis* at the subgeneric level. Transactions of the American Entomological Society 93:395–422.

———. 1970. A revision of ants of the subgenus *Polyrhachis* Fr. Smith (Hymenoptera: Formicidae: Formicinae). Oriental Insects 4:1–36.

Huxley, C. R. 1978. The ant-plants *Myrmecodia* and *Hydnophytum* (Rubiaceae), and the relationships between their morphology, ant occupants, physiology and ecology. New Phytologist 80:231–268.

Huxley, C. R., and D. F. Cutler (eds.). 1991. Ant-Plant Interactions. Oxford University Press, Oxford.

Iwanami, Y., and T. Iwadare. 1978. Inhibiting effects of myrmicacin on pollen growth and pollen tube mitosis. Botanical Gazette 139:42–45.

Jackson, D. A. 1984. Ant distribution patterns in a Cameroonian cocoa plantation: Investigation of the ant mosaic hypothesis. Oecologia 62:318–324.

Jacobson, E. 1909. Ein Moskito als Gast und diebischer Schmarotzer der *Cremastogaster difformis* Smith und eine andere schmarotzende Fliege. Tijdschrift voor Entomologie 52:158–164.

Jaeger, E. C. 1955. A Source–Book of Biological Names and Terms (3rd ed.). Charles C. Thomas, Springfield, Illinois.

Janet, C. 1897. Sur les rapports de l'*Antennophorus uhlmanni* Haller avec le *Lasius mixtus* Nyl. Comptes Rendus de l'Académie des Sciences Paris 124:583–585.

Janos, D. P. 1993. Vesicular-arbuscular mycorrhizae of epiphytes. Mycorrhiza 4:1–4.

Janzen, D. H. 1966. Coevolution of mutualism between ants and acacias in Central America. Evolution 20:249–275.

————. 1967. Interaction of the bull's-horn acacia (*Acacia cornigera* L.) with an ant inhabitant (*Pseudomyrmex ferruginea* F. Smith) in eastern Mexico. University of Kansas Science Bulletin 47:315–558.

Janzen, D. H., M. Ataroff, M. Fariñas, S. Reyes, N. Rincon, A. Soler, P. Soriano, and M. Vera. 1976. Changes in the arthropod community along an elevational transect in the Venezuelan Andes. Biotropica 8:193–203.

Jeanne, R. L. 1979. A latitudinal gradient in rates of ant predation. Ecology 60:1211–1224.

Jebb, M. 1991. Cavity structure and function in the tuberous Rubicaceae. Pp. 374–389. *In* C. R. Huxley and D. F. Cutler (eds.), Ant-Plant Interactions. Oxford University Press, Oxford.

Johnson, R. A. 1992. Soil texture as an influence on the distribution of the desert seed-harvester ants *Pogonomyrmex rugosus* and *Messor pergandei*. Oecologia 89:118–124.

Jolivet, P. 1996. Ants and Plants: An Example of Coevolution (Enlarged Edition). Backhuys, Leiden, Netherlands.

Jongman, R. H. G., C. J. F. ter Braak, and O. F. R. van Tongeren. 1995. Data Analysis in Community and Landscape Ecology. Cambridge University Press, Cambridge.

Jouvenaz, D. P. 1986. Diseases of fire ants: Problems and opportunities. Pp. 327–338. *In* C. S. Lofgren and R. K. Vander Meer (eds.), Fire Ants and Leaf-Cutting Ants: Biology and Management. Westview Press, Boulder, Colorado.

Jouvenaz, D. P., and D. W. Anthony. 1979. *Mattesia geminata* sp. n. (Neogregarinida: Ophrocystidae), a parasite of the tropical fire ant, *Solenopsis geminata* (Fabricius). Journal of Protozoology 26:354–356.

Jouvenaz, D. P., G. E. Allen, W. A. Banks, and D. P Wojcik. 1977. A survey for pathogens of fire ants, *Solenopsis* spp., in the Southeastern United States. Florida Entomologist 60:275–279.

Kane, M. D. 1997. Microbial fermentation in insect guts. Pp. 231–265. *In* R. I. Mackie and B. A. White (eds.), Ecology and Physiology of Gastrointestinal Microbes. Vol. 1. Gastrointestinal Fermentations and Ecosystems. Chapman and Hall, New York.

Karawajew, W. 1906. Weitere Beobachtungen über Arten der Gattung *Antennophorus*. Mémoires de la Société des Naturalistes de Kiew 20:209–230.

Kareiva, P., J. Kingsolver, and R. Huey. 1993. Biotic Interactions and Global Change. Sinauer Associates, Sunderland, Massachusetts.

Kaspari, M. 1993a. Body size and microclimate use in Neotropical granivorous ants. Oecologia 96:500–507.

————. 1993b. Removal of seeds from Neotropical frugivore droppings. Oecologia 95:81–99.

————. 1996a. Litter ant patchiness at the 1 m^2 scale: Disturbance dynamics in three Neotropical forests. Oecologia 107:265–273.

————. 1996b. Testing resource-based models of patchiness in four Neotropical litter ant assemblages. Oikos 76:443–454.

————. 1996c. Worker size and seed size selection by harvester ants in a Neotropical forest. Oecologia 105:397–404.

Kaspari, M., and M. Byrne 1995. Caste allocation in litter *Pheidole*. Behavioral Ecology and Sociobiology 37:255–263.

Kaspari, M., and E. L. Vargo. 1995. Colony size as a buffer against seasonality: Bergmann's rule in social insects. American Naturalist 145:610–632.

Kaspari, M., and M. Weiser. 1999. The size–grain hypothesis and interspecific scaling in ants. Functional Ecology 13:530–538.

Kaspari, M., and S. Yanoviak. In press. Bait use in tropical litter and canopy ants—evidence for differences in nutrient limitation. Biotropica.

Kaspari, M., L. Alonso, and S. O'Donnell. 2000a. Three energy variables predict ant abundance at a geographic scale. Proceedings of the Royal Society B 267:485–490.

Kaspari, M., S. O'Donnell, and J. Kercher. 2000b. Energy, density, and constraints to species richness: Studies of ant assemblages along a productivity gradient. American Naturalist 155:280–293.

Keeler, K. 1993. Fifteen years of colony dynamics in *Pogonomyrex occidentalis,* the western harvester ant, in western Nebraska. Southwestern Naturalist 38:286–289.

Kempf, W. W. 1951. A taxonomic study on the ant tribe Cephalotini (Hymenoptera: Formicidae). Revista Entomologia (Rio de Janeiro) 22:1–244.

————. 1952. A synopsis of the *pinelii*-complex in the genus *Paracryptocerus* (Hym. Formicidae). Studia Entomologica 1:1–30.

————. 1957. Sôbre algumas espécies de *Procryptocerus* com a descrição duma espécie nova (Hymenoptera, Formicidae). Revista Brasileira de Biologia 17:395–404.

―――. 1958a. New studies of the ant tribe Cephalotini (Hym. Formicidae). Studia Entomologica (n.s.) 1:1–168.

―――. 1958b. Estudos sôbre *Pseudomyrmex* II. (Hymenoptera: Formicidae). Studia Entomologica (n.s.) 1:433–462.

―――. 1958c. Sôbre algumas formigas neotrópicais do gênero *Leptothorax* Mayr. Anais da Academia Brasileira de Ciencias 30:91–102.

―――. 1959. Two new species of *Gymnomyrmex* Borgmeier, 1954 from southern Brazil, with remarks on the genus (Hymenoptera, Formicidae). Revista Brasileira de Biologia 19:337–344.

―――. 1960a. *Phalacromyrmex,* a new ant genus from southern Brazil (Hymenoptera, Formicidae). Revista Brasileira de Biologia 20:89–92.

―――. 1960b. A review of the ant genus *Mycetarotes* Emery (Hymenoptera, Formicidae). Revista Brasileira de Biologia 20:277–283.

―――. 1960c. Estudo sôbre *Pseudomyrmex* I. (Hymenoptera: Formicidae). Revista Brasileira de Entomologia 9:5–32.

―――. 1961a. A survey of the ants of the soil fauna in Surinam (Hymenoptera: Formicidae). Studia Entomologica 4:481–524.

―――. 1961b. Estudos sôbre *Pseudomyrmex* III. (Hymenoptera: Formicidae). Studia Entomologica 4:369–408.

―――. 1962. Retoques à classificação das formigas neotropicais do gênero *Heteroponera* Mayr (Hym., Formicidae). Papéis Avulsos de Zoologia (São Paulo) 15:29–47.

―――. 1963. A review of the ant genus *Mycocepurus* Forel, 1893 (Hymenoptera: Formicidae). Studia Entomologica 6:417–432.

―――. 1964. A revision of the Neotropical fungus-growing ants of the genus *Cyphomyrmex* Mayr. Part I: Group of *strigatus* Mayr (Hym., Formicidae). Studia Entomologica 7:1–44.

―――. 1965. A revision of the Neotropical fungus-growing ants of the genus *Cyphomyrmex* Mayr. Part II: Group of *rimosus* (Spinola) (Hym., Formicidae). Studia Entomologica 8:161–200.

―――. 1967a. A synopsis of the Neotropical ants of the genus *Centromyrmex* Mayr (Hymenoptera: Formicidae). Studia Entomologica 9:401–410.

―――. 1967b. Estudos sôbre *Pseudomyrmex.* IV (Hymenoptera: Formicidae). Revista Brasileira de Biologia 12:1–12.

―――. 1967c. Three new South American ants (Hym. Formicidae). Studia Entomologica 10:353–360.

―――. 1967d. A new revisionary note on the genus *Paracryptocerus* Emery (Hym. Formicidae). Studia Entomologica 10:361–368.

―――. 1968. A new species of *Cyphomyrmex* from Colombia, with further remarks on the genus (Hymenoptera, Formicidae). Revista Brasileira de Biologia 28:35–41.

―――. 1971. A preliminary review of the ponerine ant genus *Dinoponera* Roger (Hymenoptera: Formicidae). Studia Entomologica 14:369–394.

―――. 1972. Catálogo Abreviado das Formigas da Região Neotropical (Hymenoptera: Formicidae). Studia Entomologica 15:3–344.

―――. 1973a. A revision of the Neotropical myrmicine ant genus *Hylomyrma* Forel (Hymenoptera: Formicidae). Studia Entomologica 16:225–260.

―――. 1973b. A new *Zacryptocerus* from Brazil, with remarks on the generic classification of the tribe Cephalotini (Hymenoptera: Formicidae). Studia Entomologica 16:449–462.

―――. 1974a. A review of the Neotropical ant genus *Oxyepoecus* Santschi (Hymenoptera: Formicidae). Studia Entomologica 17:471–512.

―――. 1975a. A revision of the Neotropical ponerine ant genus *Thaumatomyrmex* Mayr (Hymenoptera: Formicidae). Studia Entomologica 18:95–126.

―――. 1975b. Miscellaneous studies on neotropical ants. VI. (Hymenoptera, Formicidae). Studia Entomologica 18:341–380.

Kempf, W. W., and K. Lenko. 1968. Novas observações e estudos sôbre *Gigantiops destructor* (Fabricius) (Hymenoptera: Formicidae). Papéis Avulsos de Zoologia (São Paulo) 21:209–230.

Kent, M., and P. Coker. 1992. Vegetation Description and Analysis: A Practical Approach. Belhaven Press, London.

King, J. R., A. N. Andersen, and A. D. Cutter. 1998. Ants as bioindicators of habitat disturbance: Validation of the functional group model for Australia's humid tropics. Biodiversity and Conservation 7:1627–1638.

Kistner, D. H. 1979. Social and evolutionary significance of social insect symbionts. Pp. 339–413. *In* H. R. Hermann (ed.), Social Insects, Vol. 1. Academic Press, New York.

―――. 1982. The social insects' bestiary. Pp. 1–244. *In* H. R. Hermann (ed.), Social Insects, Vol. 3. Academic Press, New York.

Kitching, R. L. 1993. Rainforest canopy arthropods: problems for rapid biodiversity assessment. Pp.

26–30. *In* A. J. Beattie (ed.), Rapid Biodiversity Assessment: Proceedings of the Biodiversity Assessment Workshop. Macquarie University, Sydney, Australia.

Kleinfeldt, S. E. 1978. Ant-gardens: The interaction of *Codonanthe crassifolia* (Gesneriaceae) and *Crematogaster longispina* (Formicidae). Ecology 59:449–456.

———. 1986. Ant-gardens: Mutual exploitation. Pp. 283–294. *In* B. Juniper and T. R. E. Southwood (eds.), Insects and the Plant Surface. Edward Arnold, London.

Koch, C. F. 1987. Prediction of sample size effects on the measured temporal and geographic distribution patterns of species. Paleobiology 13:100–107.

Kohout, R. J. 1988. New nomenclature of the Australian ants of the *Polyrhachis gab* Forel species complex (Hymenoptera: Formicidae: Formicinae). Australian Entomological Magazine 15:49–52.

———. 1989. The Australian ants of the *Polyrhachis relucens* species-group (Hymenoptera: Formicidae: Formicinae). Memoirs of the Queensland Museum 27:509–516.

———. 1990. A review of the *Polyrhachis viehmeyeri* species-group (Hymenoptera: Formicidae: Formicinae). Memoirs of the Queensland Museum 28:499–508.

Kohout, R. J., and R. W. Taylor. 1990. Notes on Australian ants of the genus *Polyrhachis* Fr. Smith, with a synonymic list of the species (Hymenoptera: Formicidae: Formicinae). Memoirs of the Queensland Museum 28:509–522.

Kremen, C. 1992. Assessing the indicator properties of species assemblages for natural area monitoring. Ecological Applications. 2:203–217.

Kremen C., R. K. Colwell, T. L. Erwin, D. D. Murphy, R. F. Noss, and M. A. Sanjayan. 1994. Terrestrial arthropod asemblages: Their use in conservation planning. Conservation Biology 7:796–808.

Kugler, C. 1978. A comparative study of the myrmicine sting apparatus (Hymenoptera: Formicidae). Studia Entomologica 20:413–548.

Kugler, C. 1994. A revision of the ant genus *Rogeria* with description of the sting apparatus (Hymenoptera: Formicidae). Journal of Hymenoptera Research 3:17–89.

Kugler, C., and W. L. Brown Jr. 1982. Revisionary and other studies on the ant genus *Ectatomma*, including the description of two new species. Search Agriculture (Ithaca, N.Y.) 24:1–8.

Kugler, J. 1986. The Leptanillinae (Hymenoptera: Formicidae) of Israel and a description of a new species from India. Israel Journal of Entomolgy 20:45–57.

Kupyanskaya, A. N. 1980. Ants of the genus *Formica* Linnaeus (Hymenoptera, Formicidae) of the Soviet Far East. Pp. 95–108. *In* P. A. Ler (ed.), Taxonomy of Insects of the Far East. Akademiya Nauk SSSR, Vladivostok. [In Russian.]

———. 1986. Ants (Hymenoptera, Formicidae) of the group *Myrmica* lobicornis Nylander from the Far East. Pp. 83–90. *In* P. A. Ler (ed.), Systematics and Ecology of Insects from the Far East. Akademiya Nauk SSSR, Vladivostok. [In Russian.]

Kusnezov, N. 1951a. El género *Pogonomyrmex* Mayr (Hym., Formicidae). Acta Zoologica Lilloana (Tucuman) 11:227–333.

———. 1951b. *Myrmelachista* en la Patagonia (Hymenoptera, Formicidae). Acta Zoologica Lilloana (Tucuman) 11:353–365.

———. 1951c. El género *Pheidole* en la Argentina (Hymenoptera, Formicidae). Acta Zoologica Lilloana (Tucuman) 12:5–88.

———. 1951d. El género *Camponotus* en la Argentina (Hymenoptera, Formicidae). Acta Zoologica Lilloana (Tucuman) 12:183–252.

———. 1951e. El estado real del grupo *Dorymyrmex* Mayr (Hymenoptera, Formicidae). Acta Zoologica Lilloana (Tucuman) 10:427–448.

———. 1955. Zwei neue Ameisengattungen aus Tucuman (Argentinien). Zoologischer Anzeiger 154:268–277.

———. 1957. Numbers of species of ants in faunae of different latitudes. Evolution 11:298–299.

Kutter, H. 1931. *Forelophilus,* eine neue Ameisengattung. Mitteilungen der Schweizerischen Entomologischen Gesellschaft 15:193–195.

———. 1945. Ein neue Ameisengattung. Mitteilungen der Schweizerischen Entomologischen Gesellschaft 19:485–487.

———. 1950. Über eine neue, extrem parasitische Ameise. 1. Mitteilung. Mitteilungen der Schweizerischen Entomologischen Gesellschaft 23:81–94.

———. 1973. Zur Taxonomie der Gattung *Chalepoxenus* (Hymenoptera, Formicidae, Myrmicinae). Mitteilungen der Schweizerischen Entomologischen Gesellschaft 46:269–280.

Laakso, J., and H. Setälä. 1997. Nest mounds of red wood ants (*Formica aquilonia*): Hot spots for litter-dwelling earthworms. Oecologia 111:565–569.

Lagerheim, G. 1900. Über *Lasius fuliginosus* und seine Pilzzucht. Entomologisk Tidskrift 21:17–29.

Lamas, G., R. K. Robbins, and D. J. Harvey. 1991. A preliminary survey of the butterfly fauna of Pakitza, Parque Nacional del Manu, Peru, with an estimate of its species richness. Publicaciones del Museo de Historia Natural, Universidad Nacional Mayor de San Marcos A 40:1–19.

Lambeck, R. J. 1997. Focal species: A multi-species umbrella for nature conservation. Conservation Biology 11:849–856.

Lambshead, J., and H. M. Platt. 1985. Structural patterns of marine benthic assemblages and their relationships with empirical statistical models. Pp. 371–380. *In* P. E. Gibbs (ed.), Proceedings of the 19th European Marine Biology Symposium, Plymouth, 1984. Cambridge University Press, Cambridge.

Landres, P. B., J. Verner, and J. W. Thomas. 1988. Ecological uses of vertebrate indicator species: A critique. Conservation Biology 2:316–328.

Lattke, J. E. 1986. Two new species of neotropical *Anochetus* Mayr (Hymenoptera: Formicidae). Insectes Sociaux 33:352–358.

———. 1990. A new genus of myrmicine ants (Hymenoptera: Formicidae) from Venezuela. Entomologica Scandinavica 21:173–178.

———. 1991. Studies of neotropical *Amblyopone* Erichson (Hymenoptera: Formicidae). Contributions in Science. Los Angeles County Museum 428:1–7.

———. 1995. Revision of the ant genus *Gnamptogenys* in the New World (Hymenoptera: Formicidae). Journal of Hymenoptera Research 4:137–193.

———. 1997. Revisión del género *Apterostigma* Mayr (Hymenoptera: Formicidae). Arquivos de Zoologia (Museu de Zoologia da Universidade de São Paulo) 34:121–221.

Lattke, J., and W. Goitía. 1997. El genero *Strumigenys* (Hymenoptera: Formicidae) en Venezuela. Caldasia 19(3):367–396.

Launer, A. E., and D. D. Murphy. 1994. Umbrella species and the conservation of habitat fragments: A case of a threatened butterfly and a vanishing grassland ecosystem. Biological Conservation 69:145–153.

Lavorel, S., S. McIntyre, J. Landsberg, and T. D. A. Forbes. 1997. Plant functional classifications: From general groups to specific groups based on response to disturbance. Trends in Ecology and Systematics 12:474–478.

Lawton, J. 1994. What do species do in ecosystems? Oikos 71:364–374.

Lawton, J. H., D. E. Bifnell, B. Bolton, G. F. Blowmers, P. Eggleton, P. M. Hammond, M. Hodda, R. D. Holt, T. B. Larsen, N. A. Mawdsley, N. E. Stork, D. S. Srivastava, and A. D. Watt. 1998. Biodiversity inventories, indicator taxa and effects of habitat modification in tropical forest. Nature 391:72–76.

Leary, R., and F. Allendorf. 1989. Fluctuating asymmetry as an indicator of stress: Implications for conservation biology. Trends in Ecology and Evolution 4:214–217.

Lee, S. M., and Chao, A. 1994. Estimating population size via sample coverage for closed capture-recapture models. Biometrics 50:88–97.

Le Masne, G. 1941. *Tubicera lichtwardti* Schmitz (Dipt. Phoridae), hôte de *Plagiolepis pygmaea* Latr. (Hym. Formicidae). Bulletin de la Société Entomologique de France 46:110–111.

Lesica, P. 1993. Using plan community diversity in reserve design for pothole prairie on the Blackfeet Indian Reservation, Montana, USA. Biological Conservation 65:69–75.

Lesica, P., and P. Kannowski. 1998. Ants create hummocks and alter structure and vegetation of a mountain fen. American Midland Naturalist 139:58–68.

Lévieux, J. 1976. Étude de la structure du nid de quelques espèces terricoles de fourmis tropicales. Annales de l'Université d'Abidjan, Serie C: Sciences 12:23–33.

———. 1983. The soil fauna of tropical savannas. IV. The ants. Pp. 525–540. *In* F. Bourlière (ed.), Tropical Savannas: Ecosystems of the World. Elsevier, Amsterdam.

Levings, S. C. 1983. Seasonal, annual and among-site variation in the ground ant community of a deciduous tropical forest: Some causes of patchy species distributions. Ecological Monographs 53:435–455.

Levings, S. C., and Traniello, J. F. A. 1981. Territoriality, nest dispersion, and community structure in ants. Psyche (Cambridge) 88:265–319.

Levy R. 1996. Interspecific colony dispersion and niche relations of three large tropical rain forest ant species. Pp. 331–340. *In* D. S. Edwards, W. E. Booth, and S. C. Choy (eds.), Tropical Rainforest Research. Kluwer Academic Publishers, Dordrecht.

Lieberburg, I., P. M. Kranz, and A. Seip. 1975. Bermudian ants revisited: The status and interaction of *Pheidole megacephala* and *Iridomyrmex humilis*. Ecology 56:473–478.

Lincoln, R. J., and G. A. Boxshall. 1987. The Cambridge Illustrated Dictionary of Natural History. Cambridge Univesity Press, Cambridge.

Linnaeus, C. 1758–1759. Systema naturae per regna tria naturae, secundum classes, ordines, genera, species, cum characteribus, differentiis, synonymis, locis (10th ed.). Homiae, Salvii. 2v.

Littledyke, M., and J. M. Cherrett. 1976. Direct ingestion of plant sap from cut leaves by leaf-cutting ants Atta cephalotes (L.) and Acromyrmex octospinosus (Reich) (Formicidae, Attini). Bulletin of Entomological Research 66:205–217.

Longino, J. T. 1991. *Azteca* ants in Cecropia trees: Taxonomy, colony structure, and behaviour. Pp. 271–288. *In* C. R. Huxley and D. F. Cutler (eds.), Ant-Plant Interactions. Oxford University Press, Oxford.

———. 1994. How to measure arthropod diversity in a tropical rainforest. Biology International 28:3–13.

Longino, J. T., and R. K. Colwell. 1997. Biodiversity assessment using structured inventory: Capturing the ant fauna of lowland tropical rainforest. Ecological Applications 7:1263–1277.

Longino, J. T., and D. A. Hartley. 1994. *Perissomyrmex snyderi* (Hymenoptera: Formicidae) is native to Central America and exhibits worker polymorphism. Psyche (Cambridge) 101:195–202.

Lubin, Y. D. 1984. Changes in the native fauna of the Galápagos Islands following invasion by the little red fire ant, *Wasmannia auropunctata*. Biological Journal of the Linnean Society 21:229–242.

Ludwig, J. A., and J. F. Reynolds. 1988. Statistical Ecology: A Primer on Methods and Computing. John Wiley and Sons, New York.

Luff, M. L. 1975. Some features influencing the efficiency of pitfall traps. Oecologia 19:345–357.

Lyford, W. H., 1963. Importance of ants to brown podzolic soil genesis in New England. Harvard Forest Papers 7.

Lynch, J. F. 1981. Seasonal, successional, and vertical segregation in a Maryland ant community. Oikos 37:183–198.

MacArthur, R., and E. O. Wilson. 1967. The Theory of Island Biogeography. Princeton University Press, Princeton, New Jersey.

Mack, A. L. (ed.). 1998. A biological assessment of the Lakekamu basin, Papua New Guinea. RAP Working Papers No. 9. Conservation International, Washington, D.C.

MacKay, W. P. 1993. A review of the New World ants of the genus *Dolichoderus* (Hymenoptera: Formicidae). Sociobiology 22:1–148.

———. 1996. A revision of the ant genus *Acanthostichus* (Hymenoptera: Formicidae). Sociobiology 27:129–179.

———. 1997. A revision of the Neotropical ants of the genus *Camponotus*, subgenus *Myrmostenus* (Hymenoptera: Formicidae). Proceedings of the Entomological Society of Washington 99(1):194–203.

———. 2000. A review of the New World ants of the subgenus *Myrafant*, genus *Leptothorax* (Hymenoptera: Formicidae). Sociobiology 36(2):263–444.

MacKay, W. P., and E. Mackay.1997. A revision of the Neotropical ants of the *montivagus* species complex, genus *Camponotus*, subgenus *Myrmentoma* (Hymenoptera: Formicidae). Sociobiology 30(3):319–334.

MacKay, W. P., and S. B. Vinson. 1989. A guide to the species identification of the New World ants. Sociobiology 16:3–47.

Magurran, A. E. 1988. Ecological Diversity and Its Measurement. Princeton University Press, Princeton, New Jersey.

Majer, J. D. 1976. The maintenance of the ant mosaic in Ghana cocoa farms. Journal of Applied Ecology 13:123–144.

———. 1980. The influence of ants on broadcast and naturally spread seeds in rehabilitated bauxite mined areas. Reclamation Review 3:3–9.

———. 1983. Ants: Bioindicators of minesite rehabilitation, land-use, and land conservation. Environmental Management 7(4):375–383.

———. 1984. Recolonisation by ants in rehabilitated open-cut mines in Northern Australia. Reclamation and Revegetation Research 2:279–298.

———. 1985. Recolonization by ants of rehabilitated mineral sand mines on North Stradbroke Island, Queensland, with particular reference to seed removal. Australian Journal of Ecology 10:31–48.

————. 1990. The abundance and diversity of arboreal ants in northern Australia. Biotropica 22: 191–199.

————. 1992. Ant recolonization of rehabilitated bauxite mines of Poços de Caldos, Brasil. Journal of Tropical Ecology 8:97–108.

————. 1996. The use of pitfall traps for sampling ants: A critique. Proceedings of the Museum of Victoria 56:323–329.

Majer, J. D., and A. E. de Kock. 1992. Ant recolonization of sand mines near Richards Bay, South Africa: An evaluation of progress with rehabilitation. South African Journal of Science 88: 31–36.

Majer, J. D., and J. H. C. Delabie. 1994. Comparison of the ant communities of annually inundated and terra firme forests at Trombetas in the Brazilian Amazon. Insectes Sociaux 41:343–359.

Majer, J. D., and O. G. Nichols. 1998. Long-term recolonization patterns of ants in rehabilitated bauxite mines, Western Australia. Journal of Applied Ecology 35:161–181.

Majer, J. D., Day, J. E., Kabay, E. D., and Perriman, W. S. 1984. Recolonization by ants in bauxite mines rehabilitated by a number of different methods. Journal of Applied Ecology 21:355–375.

Majer, J. D., J. H. C. Delabie, and N. L. McKenzie. 1997. Ant litter fauna of forest, forest edges and adjacent grasslands in the Atlantic rain forest region of Bahia, Brazil. Insectes Sociaux 44:255–266.

Malicky, H. 1969. Versuch einer Analyse der ökologischen Beziehungen zwischen Lycaeniden (Lepidoptera) und Formiciden (Hymenoptera). Tijdschrift voor Entomologie 112:213–298.

Mann, W. M. 1916. The Stanford Expedition to Brazil, 1911, John C. Branner, Director: The ants of Brazil. Bulletin of the Museum of Comparative Zoology, Harvard College 60:399–490.

————. 1921. The ants of the Fiji Islands. Bulletin of the Museum of Comparative Zoology, Harvard College 64:401–499.

————. 1926. Some new neotropical ants. Psyche (Cambridge) 33:97–107.

Marsh, A. C. 1984. The efficacy of pitfall traps for determining the structure of a desert ant community. Journal of the Entomological Society of South Africa 47:115–120.

————. 1985. Forager abundance and dietary relationships in a Namib Desert ant community. South African Journal of Zoology 20:197–203.

————. 1986. Ant species richness along a climatic gradient in the Namib Desert. Journal of Arid Environments 11:235–241.

Martin, J. 1977. Collecting, Preparing, and Preserving Insects, Mites, and Spiders. The Insects and Arachnids of Canada, Part 1. Agriculture Canada, Ottawa.

Maschwitz, U. 1974. Vergleichende Untersuchungen zur Funktion der Ameisenmetathorakaldruse. Oecologia 16:303–310.

Maschwitz, U., and H. Hänel. 1985. The migrating herdsman Dolichoderus (Diabolus) cuspidatus: An ant with a novel mode of life. Behavioral Ecology and Sociobiology 17:171–184.

Maschwitz, U., and B. Hölldobler. 1970. Der Kartonnestbau bei Lasius fuliginosus Latr. (Hym. Formicidae). Zeitschrift für Vergleichende Physiologie 66:176-189.

Maschwitz, U., and P. Schönegge. 1980. Fliegen als Beute- und Brauträuber bei Ameisen. Insectes Sociaux 27:1–4.

Maschwitz, U., K. Koob, and H. Schildknecht. 1970. Ein Beitrag zur Funktion der Metathoracaldrüse der Ameisen. Journal of Insect Physiology 16:387–404.

Masner, L. 1976. Notes on the ecitophilous diapriid genus Mimopria Holmgren (Hymenoptera: Proctotrupoidea, Diapriidae). Canadian Entomologist 108:123–126.

Masuko, K. 1984. Studies on the predatory biology of oriental Dacetine ants (Hymenoptera: Formicidae). I. Some Japanese species of Strumigenys, Pentastruma, and Epitritus, and a Malaysian Labidogenys, with special reference to hunting tactics in short-mandibulate forms. Insectes Sociaux 31:429–451.

————. 1995 (1994). Specialized predation on oribatid mites by two species of the ant genus Myrmecina (Hymenoptera: Formicidae). Psyche (Cambridge) 101:159–173.

May, R. M. 1975. Patterns of species abundance and diversity. Pp. 81–120. In M. L. Cody and J. M. Diamond (eds.), Ecology and Evolution of Communities. Belknap Press, Cambridge, Massachusetts.

————. 1990. How many species? Philosophical Transactions of the Royal Society of London, Series B 330 (1257):293–304.

Mahyé–Nunes, A. 1995. Sinopse do genero Mycetarotes (Hym., Formicidae), com a descriçao de duas especies novas. Boletin de Entomologia Venezolana 10(2):197–205.

Mayr, E. 1942. Systematics and the Origin of Species. Columbia University Press, New York.

McAreavey, J. 1947. New species of the genera *Prolasius* Forel and *Melophorus* Lubbock (Hymenoptera, Formicidae). Memoirs of the National Museum of Victoria 15:7–27.

———. 1957. Revision of the genus *Stigmacros.* Memoirs of the National Museum of Victoria 21:7–64.

McArthur, A. J., and M. Adams. 1996. A morphological and molecular revision of the *Camponotus nigriceps* group (Hymenoptera: Formicidae) from Australia. Invertebrate Taxonomy 10(1):1–46.

McCoy, E. D. 1990. The distribution of insects along elevational gradients. Oikos 58:313–322.

McGlynn, T. P. 1999a. The biogeography, behavior, and ecology of exotic ants (178 pp.). Ph.D. dissertation, University of Colorado, Boulder, Colorado.

McGlynn, T. P. 1999b. The worldwide transfer of ants: geographic distribution and ecological invasions. Journal of Biogeography 26:535–548.

McGlynn, T. P., and C. D. Kelley. 1999. Distribution of a Costa Rican wet forest velvet worm (Onychophora, Peripatidae). Annals of the Entomological Society of America 92:53–55.

McIlveen, W. D., and H. Cole Jr. 1976. Spore dispersal of Endogonaceae by worms, ants, wasps, and birds. Canadian Journal of Botany 54:1486–1489.

Medel, R. G., and R. A. Vásquez. 1994. Comparative analysis of harvester ant assemblages of Argentinian and Chilean arid zones. Journal of Arid Environments 26:363–371.

Menozzi, C. 1929. Revisione delle formiche del genere *Mystrium* Roger. Zoologischer Anzeiger 82:518–536.

———. 1939. Formiche dell'Himalaya e del Karakorum raccolte dalla Spedizione italiana comandata da S. A. R. il Duca di Spoleto (1929). Atti della Società Italiana di Scienze Naturali e del Museo Civico di Storia Naturale (Milan) 78:285–345.

Michener, C. D. 1944. Comparative external morphology, phylogeny, and a classification of bees. Bulletin of the American Museum of Natural History 2:151–326.

Miehe, H. 1911a. Javanische Studien. II. Untersuchungen über die javanische *Myrmecodia.* Abhandlungen der kaiserischen Sächsischen Gesellschaft der Wissenschaften Mathematisch-physischer Klasse 32:312–361.

———. 1911b. Ueber die javanische *Myrmecodia* und die Beziehung zu ihren Ameisen. Biologisches Zentralblatt 31:733–738.

Moffett, M. W. 1985. Revision of the genus *Myrmoteras* (Hymenoptera: Formicidae). Bulletin of the Museum of Comparative Zoology, Harvard College 151:1–53.

———. 1986. Revision of the myrmicine genus *Acanthomyrmex* (Hymenoptera: Formicidae). Bulletin of the Museum of Comparative Zoology, Harvard College 151:55–89.

Moncalvo, J.-M., F. M. Lutzoni, S. A. Rehner, J. Johnson, and R. Vilgalys. 2000. Phylogenetic relationships of agaric fungi based on nuclear large subunit ribosomal DNA sequences. Systematic Biology 49:278–305.

Morton, S. R., and D. W. Davidson. 1988. Comparative structure of harvester ant communities in arid Australia and North America. Ecological Monographs 58:19–38.

Moser, J. C. 1964. Inquiline roach responds to trail-marking substance of leaf-cutting ants. Science 143:1048–1049.

Mueller, U. G., S. A. Rehner, and T. R. Schultz. 1998. The evolution of agriculture in ants. Science 281:2034–2038.

Munger, J. C. 1992. Reproductive potential of colonies of desert harvester ants (*Pogonomyrmex desertorum*): effects of predation and food. Oecologia 90:276–282.

Nepstad, D. C., P. Jipp, P. Mautinho, G. Negreiros, and S. Vieira. 1995. Forest recovery following pasture abandonment in Amazoni: Canopy seasonality, fire resistance and ants. Pp. 333–349. *In* D. J. Rapport, C. L. Gaudet, and P. Calow (eds.): Evaluating and Monitoring the Health of Large-Scale Ecosystems. NATO ASI Series 128.

New, T. R. 1987. Insect conservation in Australia: Towards rational ecological priorities. Pp. 5–20. *In* J. D. Majer (ed.), The Role of Invertebrates in Conservation and Biological Survey. Western Australian Department of Conservation and Land Management, Perth, Australia.

Noss, R. F. 1990. Indicators for monitoring biodiversity: A hierarchical approach. Conservation Biology 4:355–364.

O'Dowd, D. J. 1982. Pearl bodies as ant food: An ecological role for some leaf emergences of tropical plants. Biotropica 14:40–49.

Ogata, K. 1982. Taxonomic study of the ant genus *Pheidole* Westwood of Japan, with a description of a new species (Hymenoptera, Formicidae). Kontyû 50:189–197.

———. 1990. A new species of the ant genus *Epitritus* Emery from Japan (Hymenoptera, Formicidae). Esakia Special Issue 1:197–199.

Ogata, K., and K. Onoyama.1998. A revision of the ant genus *Smithistruma* Brown of Japan, with descriptions of four new species (Hymenoptera: Formicidae). Entomological Science 1(2):277–287.

Ogata, K., and W. L. Brown Jr. 1991. Ants of the genus *Myrmecia* Fabricius: A preliminary review and key to the named species. Journal of Natural History 25:1623–1673.

Oliveira, P. S. 1988. Ant-mimicry in some Brazilian salticid and clubionid spiders (Araneae: Salticidae, Clubionidae). Biological Journal of the Linnean Society 33:1–15.

Oliveira, P. S., and I. Sazima. 1984. The adaptive bases of ant-mimicry in a Neotropical aphantochilid spider (Araneae: Aphantochilidae). Biological Journal of the Linnean Society 22:145–155.

Oliveira, P. S., M. Galetti, F. Pedroni, and L. P. C. Morellato. 1995. Seed cleaning by *Mycocepurus goeldii* ants (Attini) facilitates germination in *Hymenaea courbaril* (Caesalpiniaceae). Biotropica 27:518–522.

Oliver, I., and A. J. Beattie. 1996a. Invertebrate morphospecies as surrogates for species: A case study. Conservation Biology 10:99–109.

———. 1996b. Designing a cost-effective invertebrate survey: A test of methods for rapid assessment of biodiversity. Ecological Applications 6:594–607.

Oliver I., A. J. Beattie, and A. York. 1998. Spatial fidelity of plant, vertebrate and invertebrate assemblags in multiple–use forest in Eastern Australia. Conservation Biology 12(4):822–835.

Olson, D. M. 1991. A comparison of the efficacy of litter sifting and pitfall traps for sampling leaf litter ants (Hymenoptera: Formicidae) in a tropical wet forest, Costa Rica. Biotropica 23:166–172.

———. 1992. Rates of predation by ants (Hymenoptera: Formicidae) in the canopy, understory, leaf litter, and edge habitats of a lowland rainforest in southwestern Cameroon. Pp. 101–109. *In* F. Hall and O. Pascal (eds.), Biologie d'une Canopie de Forêt Equatoriale II. Fondation Elf, Paris.

———. 1994. The distribution of leaf litter invertebrates along a Neotropical altitudinal gradient. Journal of Tropical Ecology 10:129–150.

Onoyama, K. 1998. Taxonomic notes on the ant genus *Crematogaster* in Japan (Hymenoptera: Formicidae). Entomological Science 1(2):227–232.

Orlóci, L. 1978. Multivariate Analysis in Vegetation Research. Dr. W. Junk, The Hague.

Orr, A. G., and J. K. Charles. 1994. Foraging in the giant forest ant, *Camponotus gigas* (Smith) (Hymenoptera: Formicidae): Evidence for temporal and spatial specialization in foraging activity. Journal of Natural History 28:861–872.

Orr, M. R. 1992. Parasitic flies (Diptera: Phoridae) influence foraging rhythms and caste division of labor in the leaf-cutter ant *Atta cephalotes* (Hymenoptera: Formicidae). Behavioral Ecology and Sociobiology 30:395–402.

Overal, W. L., and A. G. Bandeira. 1985. Nota sobre habitos de *Cylindromyrmex striatus* Mayr, 1870, na Amazonia (Formicidae, Ponerinae). Revista Brasileira de Entomologia 29:521–522.

Paine, R. 1968. A note on trophic complexity and community stability. American Naturalist 102:91–93.

Palmer, A., and C. Strobeck. 1986. Fluctuating asymmetry: Measurement, analysis, patterns. Annual Review of Ecology and Systematics 17:391–421.

Palacio, E. 1997. Ants of Colombia VI. Two new species of *Octostruma* (Hymenoptera: Formicidae: Basicerotini). Caldasia 19(3):409–418.

Palmer, M., R. Ambrose, and N. Poff. 1997. Ecological theory and community restoration ecology. Restoration Ecology 5:291–300.

Passera, L. 1994. Characteristics of tramp species. Pp. 23–43. *In* D. F. Williams (ed.), Exotic Ants: Biology, Impact, and Control of Introduced Species. Westview Press, Boulder, Colorado.

Peakall, R., S. N. Handel, and A. J. Beattie. 1991. The evidence for, and importance of, ant pollination. Pp. 421–429. *In* C. R. Huxley and D. F. Cutler (eds.), Ant-Plant Interactions. Oxford University Press, Oxford.

Pearson, D. L., and F. Cassola. 1992. World-wide species richness patterns of tiger beetles (Coleoptera: Cicindelidae): Indicator taxon for biodiversity and conservation studies. Conservation Biology 6:376–391.

Peeters, C. 1991. The occurrence of sexual reproduction among ant workers. Biological Journal of the Linnean Society 44:141–152.

Perfecto, I., and R. R. Snelling. 1995. Biodiversity and the transformation of a tropical agroecosystem: Ants in coffee plantations. Ecological Applications 5:1084–1097.

Petal, J., H. Jakubczyk, A. Breymeyer, and E. Olechowicz. 1971. Productivity investigation of two types of meadows in the Vistula Valley. X. Role of the ants as predators in a habitat. Ekologia polska 19:213–222.

Petersen, B. 1968. Some novelties in presumed males of Leptanillinae (Hym., Formicidae). Entomologiske Meddelelser 36:577–598.

Pharo, E. J., A. J. Beattie, and D. Binns. 1999. Vascular plant diversity as a surrogate for bryophyte and lichen diversity. Conservation Biology 13:282–292.

Pielou, E. C. 1966. Species diversity and pattern diversity in the study of ecological succession. Journal of Theoretical Biology 10:370–383.

———. 1969. An Introduction to Mathematical Ecology. John Wiley and Sons, New York.

———. 1975. Ecological Diversity. John Wiley and Sons, New York.

———. 1984. The Interpretation of Ecological Data: A Primer on Classification and Ordination. John Wiley and Sons, New York.

Pierce, N. E. 1987. The evolution and biogeography of association between Lycaenid butterflies and ants. Pp. 89–116. *In* P. H. Harvey and L. Partridge (eds.), Oxford Surveys in Evolutionary Biology, Vol. 4. Oxford University Press, New York.

Pierce, N. E., and S. Easteal. 1986. The selective advantage of attendant ants for the larvae of a lycaenid butterfly, *Glaucopsyche lygdamus.* Journal of Animal Ecology 55:451–462.

Pierce, N. E., and P. S. Mead. 1981. Parasitoids as selective agents in the symbiosis between lycaenid butterfly larvae *Glaucopsyche lygdamus* and ants. Outlook 211:1185–1187.

Pisarski, B. 1966. Études sur les fourmis du genre *Strongylognathus* Mayr (Hymenoptera, Formicidae). Annales Zoologici (Warsaw) 23:509–523.

Poggi, R., and C. Conci. 1996. Elenco delle collezioni entomologiche conservate nelle strutture pubbliche italiane. Memorie della Società Entomològica Italiana 75:3–157.

Poole, R. W. 1974. An Introduction to Quantitative Ecology. McGraw-Hill, New York.

Porter, S. D. 1999. FORMIS99: A Master Bibliography of Ant Literature (computer database). Published by the author, Gainesville, Florida. http://cmave.usda.ufl.edu/~formis/

Porter, S. D., and M. A. Bowers. 1981. Emigration of an *Atta* colony. Biotropica 12:232.

Porter, S. D., and C. D. Jorgensen. 1988. Longevity of harvester ant colonies in southern Idaho. Journal of Range Management 41:104–107.

Porter, S. D., and D. A. Savignano. 1990. Invasion of polygyne fire ants decimates native ants and disrupts arthropod community. Ecology 71:2095–2106.

Porter, S. D., H. G. Fowler, and W. P. Mackay. 1992. Fire ant mound densities in the United States and Brazil (Hymenoptera: Formicidae). Journal of Economic Entomology 85:1154–1161.

Porter, S. D., H. G. Fowler, S. Campiolo, and M. A. Pesquero. 1995a. Host specificity of several Pseudacteon (Diptera: Phoridae) parasites of fire ants (Hymenoptera: Formicidae) in South America. Florida Entomologist 78:70–75.

Porter, S. D., M. A. Pesquero, S. Campiolo, and H. G. Fowler. 1995b. Growth and development of *Pseudacteon* phorid fly maggots (Diptera: Phoridae) in the heads of Solenopsis fire ant workers (Hymenoptera: Formicidae). Environmental Entomology 24:475–479.

Porter, S. D., R. K. Vander Meer, M. A. Pesquero, S. Campiolo, and H. G. Fowler. 1995c. *Solenopsis* (Hymenoptera: Formicidae) fire ant reactions to attacks of pseudacteon flies (Diptera: Phoridae) in southeastern Brazil. Annals of the Entomological Society of America 88:570–575.

Prendergast, J. R., R. M. Quinn, J. H. Lawton, B. C. Eversham, and D. W. Gibbons. 1993. Rare species, the coincidence of diversity hotspots and conservation strategies. Nature 365:335–337.

Preston, F. W. 1948. The commonness, and rarity, of species. Ecology 29:254–283.

Prins, A. J. 1982. Review of *Anoplolepis* with reference to male genitalia, and notes on *Acropyga* (Hymenoptera, Formicidae). Annals of the South African Museum 89:215–247.

———. 1983. A new ant genus from southern Africa (Hymenoptera, Formicidae). Annals of the South African Museum 94:1–11.

Quinlan, R. J., and J. M. Cherrett. 1979. The role of substrate preparation in the symbiosis between the leaf cutting ant *Acromyrmex octospinosus* (Reich) and its food fungus. Ecological Entomology 2:161–170.

————. 1979. The role of fungus in the diet of the leaf-cutting ant *Atta cephalotes* (L.). Ecological Entomology 4:151–160.

Quiroz-Robledo, L., and J. Valenzuela-González. 1995. A comparison of ground ant communities in a tropical rainforest and adjacent grassland in Los Tuxtlas, Veracruz, Mexico. Southwestern Entomologist 20:203–213.

Raaijmakers, J. G. W. 1987. Statistical analysis of the Michaelis-Menten equation. Biometrics 43:793–803.

Radchenko, A. G. 1985. Ants of the genus *Strongylognathus* (Hymenoptera: Formicidae) in the European part of the USSR. Zoologicheskii Zhurnal 64:1514–1523. [In Russian.]

————. 1989a. The ants of the genus *Chalepoxenus* (Hymenoptera, Formicidae) of the USSR fauna. Vestnik Zoologii 1989(2):37–41. [In Russian.]

————. 1989b. Ants of the *Plagiolepis* genus of the European part of the USSR. Zoologicheskii Zhurnal 68(9):153–156. [In Russian.]

————. 1991. Ants of the genus *Strongylognathus* (Hymenoptera, Formicidae) of the USSR fauna. Zoologicheskii Zhurnal 70(10):84–90. [In Russian.]

————. 1992. Ants of the genus *Tetramorium* (Hymenoptera, Formicidae) of the USSR fauna. Report 1. Zoologicheskii Zhurnal 71(8):39–49. [In Russian.]

————. 1993. Ants of the subfamily Cerapachyinae from Vietnam. Zhurnal Ukrains'koho Entomolohichnoho Tovarystva 1:43–47.

————. 1994a. Identification table for ants of the genus *Leptothorax* (Hymenoptera, Formicidae) from central and eastern Palearctic. Zoologicheskii Zhurnal 73(7–8):146–158. [In Russian.]

————. 1994b. A review of the ant genus *Leptothorax* (Hymenoptera, Formicidae) of the central and eastern Palearctic. Communication 1. Subdivision into groups. Groups *acervorum* and *bulgaricus*. Vestnik Zoologii 1994(6):22–28. [In Russian.]

————. 1995. Palearctic ants of the genus *Cardiocondyla* (Hymenoptera, Formicidae). Entomologicheskoe Obozrenie 74:447–455.

————. 1996a. A key of the ant genus *Camponotus* (Hymenoptera, Formicidae) in Palaearctic Asia. Zoologicheskii Zhurnal 75(8):1195–1203. [In Russian].

————. 1996b. Ants of the genus *Plagiolepis* Mayr (Hymenoptera, Formicidae) of Central and Southern Palaearctic. Entomologicheskoe Obozrenie 75(1):178–187.

————. 1997. Review of the ants of *scabriceps* group of the genus *Monomorium* Mayr (Hymenoptera, Formicidae). Annales Zoologici (Warsaw) 46(3–4):211–224.

————. 1998. A key to ants of the genus *Cataglyphis* Foerster (Hymenoptera, Formicidae) of Asia. Entomologicheskoe Obozrenie 77(2):502–508, 527. [In Russian].

Radchenko, A. G., and G. R. Arakelian. 1990. Murav'i gruppy *Tetramorium ferox* Ruzsky iz Kryma i Kavkaza. Biologicheskii Zhurnal Armenii 43:371–378.

Radchenko, A. G., and G. W. Elmes. 1998. Taxonomic revision of the *ritae* species–group of the genus *Myrmica* (Hymenoptera, Formicidae). Vestnik Zoologii 32(4):3–27.

Radchenko, A. G., W. Czechowski, and W. Czechowska. 1997. The genus *Myrmica* Latr. (Hymenoptera, Formicidae) in Poland—A survey of species and a key for their identification. Annales Zoologici (Warsaw) 47(3–4): 481–500.

Rahbek, C. 1995. The elevational gradient of species richness: A uniform pattern? Ecography 18:200–205.

————. 1997. The relationship among area, elevation, and regional species richness in Neotropical birds. American Naturalist 149:875–902.

Redford, K. H. 1987. Ants and termites as food: Patterns of mammalian myrmecophagy. Pp. 349–399. *In* H. H. Genoways (ed.), Current Mammalogy, Vol. 1. Plenum Press, New York.

Reichel, H., and A. N. Andersen. 1996. The rainforest ant fauna of Australia's Northern Territory. Australian Journal of Zoology 44:81–95.

Retana, J., X. Cerda, and X. Espadaler. 1991. Arthropod corpses in a temperate grassland: A limited supply? Holarctic Ecology 14:63–67.

Rettenmeyer, C. W. 1962. The diversity of arthropods found with Neotropical army ants and observations on the behavior of representative species. Proceedings of the North Central Branch of the Entomological Society of America 17:14–15.

————. 1963. Behavioral studies of army ants. University of Kansas Science Bulletin 44:281–465.

Rettenmeyer, C. W., R. Chadab Crepet, M. G. Naumann, and L. Morales. 1983. Comparative foraging by Neotropical army ants. Pp. 59–73.

In P. Jaisson (ed.), Social Insects in the Tropics. Université Paris-Nord, Paris.

Rettig, E. 1904. Ameisenpflanzen-Pflanzenameisen. Beihefte zum Botanischen Zentralblatt 17:89–122.

Rickson, F. R. 1971. Glycogen plastids in Muellerian body cells of *Cecropia peltata,* a higher green plant. Science 173:344–347.

Rigato, F. 1994a. *Dacatria templaris* gen. n., sp. n. A new myrmicine ant from the Republic of Korea. Deutsche Entomologische Zeitschrift (Neue Folge) 41:155–162.

———. 1994b. Revision of the myrmicine ant genus *Lophomyrmex,* with a review of its taxonomic position (Hymenoptera: Formicidae). Systematic Entomology 19:47–60.

Rissing, S. W. 1987. Annual cycles in worker size of the seed-harvester ant *Veromessor pergandei* (Hymenoptera: Formicidae). Behavioral Ecology and Sociobiology 20:117–124.

Robertson, H. G. 1990. Unravelling the *Camponotus fulvopilosus* species complex (Hymenoptera: Formicidae). Pp. 327–328. *In* G. K. Veeresh, B. Mallik, and C. A. Viraktamath (eds.), Social Insects and the Environment. Proceedings of the 11th International Congress of IUSSI, 1990. Oxford and IBH, New Delhi.

Robinson, S. R., and J. Terborgh. 1990. Bird communities of the Cocha Cashu Biological Station in Amazonian Peru. Frogs, snakes, and lizards of the INPA-WWF reserves near Manaus, Brazil. Pp. 199–216. *In* A. H. Gentry (ed.), Four Neotropical Rainforests. Yale University Press, New Haven, Connecticut.

Roepke, W. 1930. Ueber einen merkwürdigen Fall von "Myrmekophilie" bei einer Ameise (*Cladomyrma* sp.?) auf Sumatra beobachtet. Miscellanea Zoologica Sumatrana 45:1–3.

Rogerson, C. T. 1970. The hypocrealean fungi (Ascomycetes, Hypocreales). Mycologia 62:865–910.

Rohlfien, K. 1979. Aus der Geschichte der entomologischen Sammlungen des ehemaligen Deutschen Entomologischen Instituts. Beiträge für Entomologie (Berlin) 29:415–438.

Romero, H., and K. Jaffe. 1989. A comparison of methods for sampling ants (Hymenoptera: Formicidae) in savannas. Biotropica 21: 348–352.

Room, P. M. 1971. The relative distribution of ant species in Ghana's cocoa farms. Journal of Animal Ecology 40:735–751.

———. 1975. Diversity and organization of the ground foraging ant fauna of forest, grassland and tree crops in Papua New Guinea. Australian Journal of Ecology 23:71–89.

Rosciszewski, K. 1994. *Rostromyrmex,* a new genus of myrmicine ants from peninsular Malaysia (Hymenoptera: Formicidae). *Entomologica Scandinavica* 25(2):159–168.

———. 1995. Die Ameisenfauna eines tropischen Tieflandregenwaldes in Südostasien: Eine faunistisch und ökologische Bestandsaufnahme. Thesis, Johann Wolfgang Goethe Universität, Frankfurt am Main, Germany.

Rosengren, R., and P. Pamilo. 1983. The evolution of polygyny and polydomy in mound-building *Formica* ants. Acta Entomologica Fennica 42:65–77.

Rosenzweig, M. L. 1995. Species Diversity in Space and Time. Cambridge University Press, Cambridge.

Ross, K. G., and J. C. Trager. 1990. Systematics and population genetics of fire ants (*Solenopsis saevissima* complex) from Argentina. Evolution 44: 2113–2134.

Roth, D. S., I. Perfecto, and B. Rathcke. 1994. The effects of management systems on ground-foraging ant diversity in Costa Rica. Ecological Applications 4:423–436.

Ryti, R. T., and T. J. Case. 1988a. Field experiments on desert ants: Testing for competition between colonies. Ecology 69:1993–2003.

———. 1988b. The regeneration niche of desert ants: Effects of established colonies. Oecologia 75:303–306.

———. 1992. The role of neighborhood competition in the spacing and diversity of ant communities. American Naturalist 139:355–374.

Sabrosky, C. W. 1959. A revision of the genus Pholeomyia in North America (Diptera Milichiidae). Annals of the Entomological Society of America 52:316–331.

Sachtleben, H. 1961. Second supplement to: Über entomologische Sammlungen, Entomologen, und Entomo-Museologie. Beiträge zur Entomologie 11:481–540.

Sampson, D. A., E. A. Rickart, and P. C. Gonzales. 1997. Ant diversity and abundance along an elevational gradient in the Philippines. Biotropica 29:349–363.

Santschi, F. 1923a. Revue des fourmis du genre *Brachymyrmex* Mayr. Anales del Museo Nacional de Historia Natural de Buenos Aires 31:650–678.

———. 1923b. Descriptions de nouveaux Formicides éthiopiens et notes diverses. I. Revue de Zoologie Africaine 11:259–295.

———. 1929. Fourmis du Maroc, d'Algérie et de Tunisie. Bulletin et Annales de la Société Royale Entomologique de Belgique 69:138–165.

———. 1936. Fourmis nouvelles ou intéressantes de la République Argentine. Revista Entomologia (Rio de Janeiro) 6:402–421.

———. 1937. Les sexués du genre *Anillidris* Santschi. Bulletin de la Société Entomologique de France 42:68–70.

Savolainen, R. 1990. Colony success of the submissive ant *Formica fusca* within territories of the dominant *Formica polyctena*. Ecological Entomology 15:79–85.

Savolainen, R., and K. Vepsäläinen. 1988. A competition hierarchy among boreal ants: Impact on resource partitioning and community structure. Oikos 51:135–155.

Schimper, A. F. W. 1888. Die Wechselbeziehungen zwischen Pflanzen und Ameisen im tropischen Amerika. Botanische Mitteilungen aus den Tropen (Jena) 1:1–98.

———. 1898. Pflanzengeographie auf Physiologischer Grundlage. Pp. 149–170. Jena.

Schindler, D. 1990. Experimental perturbations of whole lakes as tests of hypotheses concerning ecosystem structure and function. Oikos 57: 25–41.

Schmid-Hempel, P. 1992. Worker castes and adaptive demography. Journal of Evolutionary Biology 5:1–12.

Schödl, S. 1998. Taxonomic revision of Oriental *Meranoplus* F. Smith, 1853 (Insecta: Hymenoptera: Formicidae: Myrmicinae). Annalen des Naturhistorischen Museums In Wien. Serie B. Botanik und Zoologie 100B:361–394.

Schroth, M., and U. Maschwitz. 1984. Zur Larvalbiologie und Wirtsfindung von *Maculinea teleius* (Lepidoptera: Lycaenidae), eines Parasiten von *Myrmica laevinodis* (Hymenoptera: Formicidae). Entomologia Generalis 9:225–230.

Schulenberg, T. S., and K. Awbrey (eds.). 1997. The Cordillera del Condor of Ecuador and Peru: A biological assessment. RAP Working Papers No. 7. Conservation International, Washington, D.C.

Schultz, T. R. 1998. Phylogeny of the fungus–growing ants (Myrmicinae: Attini): Evidence from DNA sequences (nuclear elongation factor–1 alpha and mitochondrial cytochrome oxidase I) and morphology. P. 429. *In* M. P. Schwarz and K. Hogendoorn (eds.), Social Insects at the Turn of the Millenium: Proceedings of the XIII International Congress of IUSSI, Adelaide, Australia, 29 December 1998–3 January 1999. Flinders University Press, Adelaide, Australia.

Schultz, T. R., and R. Meier. 1995. A phylogenetic analysis of the fungus-growing ants (Hymenoptera: Myrmicinae: Attini) based on morphological characters of the larvae. Systematic Entomology 20:337–370.

Schumacher, A., and W. G. Whitford. 1976. Spatial and temporal variation in Chihuahuan desert ant faunas. Southwestern Naturalist 21:1–8.

Schupp, E. W., and D. H. Feener. 1991. Phylogeny, lifeform, and habitat dependence of ant-defended plants in a Panamanian forest. Pp. 175–197. *In* C. R. Huxley and D. F. Cutler (eds.), Ant-Plant Interactions. Oxford University Press, Oxford.

Seastedt, T. R., and D. A. Crossley Jr. 1984. The influence of arthropods on ecosystems. BioScience 34:157–161.

Seifert, B. 1988a. A revision of the European species of the ant subgenus *Chthonolasius* (Insecta, Hymenoptera, Formicidae). Entomologische Abhandlungen und Berichte aus dem Staatlichen Museum für Tierkunde in Dresden 51:143–180.

———. 1988b. A taxonomic revision of the *Myrmica* species of Europe, Asia Minor, and Caucasia (Hymenoptera, Formicidae). Abhandlungen und Berichte des Naturkundemuseums Görlitz 62(3):1–75.

———. 1990. Supplementation to the revision of European species of the ant subgenus *Chthonolasius* Ruzsky, 1913 (Hymenoptera: Formicidae). Doriana 6(271):1–13.

———. 1992. A taxonomic revision of the Palaearctic members of the ant subgenus *Lasius* s. str. (Hymenoptera: Formicidae). Abhandlungen und Berichte des Naturkundemuseums Görlitz 66(5): 1–67.

Shattuck, S. O. 1987. An analysis of geographic variation in the *Pogonomyrmex occidentalis* complex (Hymenoptera: Formicidae). Psyche (Cambridge) 94:159–179.

———. 1990. Revision of the dolichoderine ant genus *Turneria* (Hymenoptera: Formicidae). Systematic Entomology 15:101–117.

———. 1991. Revision of the dolichoderine ant genus *Axinidris* (Hymenoptera: Formicidae). Systematic Entomology 16:105–120.

———. 1992a. Review of the dolichoderine ant genus *Iridomyrmex* Mayr with descriptions of three new genera (Hymenoptera: Formicidae). Journal of the Australian Entomological Society 31:13–18.

———. 1992b. Generic revision of the ant subfamily Dolichoderinae (Hymenoptera: Formicidae). Sociobiology 21:1–181.

———. 1992c. Higher classification of the ant subfamilies Aneuretinae, Dolichoderinae and Formicinae (Hymenoptera: Formicidae). Systematic Entomology 17:199–206.

———. 1993. Revision of the *Iridomyrmex purpureus* species-group (Hymenoptera: Formicidae). Invertebrate Taxonomy 7:113–149.

———. 1994. Taxonomic Catalog of the Ant Subfamilies Aneuretinae and Dolichoderinae (Hymenoptera: Formicidae). University of California Publications in Entomology 112.

Shattuck, S. O. 1996a. Revision of the *Iridomyrmex discors* species-group (Hymenoptera: Formicidae). Australian Journal of Entomology 35(1):37–42.

Shattuck, S. O. 1996b. The Australian ant genus *Froggattella* Forel (Hymenoptera: Formicidae) revisited. Australian Journal of Entomology 35(1):43–47.

Sheela, S., and T. C. Narendran. 1997. A new genus and a new species of Myrmicinae (Hymenoptera: Formicidae) from India. Journal of Ecobiology 9(2):88–91.

Silva, D., and J. A. Coddington. 1996. Spiders of Pakitza (Madre de Dios, Peru): Species richness and notes on community structure. Pp. 253–311. *In* D. E. Wilson and A. Sandoval (eds.), Manu: The Biodiversity of Southeastern Peru. Smithsonian Institution, Washington, D.C.

Silvestri, F. 1925 (1924). A new myrmecophilous genus of Coccidae (Hemiptera) from India. Records of the Indian Museum 26:311–315.

Skwarra, E. 1934. Ökologie der Lebensgemeinschaften mexikanischer Ameisenpflanzen. Zeitschrift für Morphologie und Oekologie der Tiere 29:306–373.

Smallwood, J. 1982. Nest relocations in ants. Insectes Sociaux 29:138–147.

Smith, J. B. 1906. An Explanation of Terms Used in Entomology. Brooklyn Entomological Society, Brooklyn, New York.

Smith, M. R. 1944. The genus *Lachnomyrmex,* with the description of a second species (Hymenoptera: Formicidae). Proceedings of the

Entomological Society of Washington 46:225–228.

———. 1947a. Ants of the genus *Apsychomyrmex* Wheeler (Hymenoptera: Formicidae). Revista Entomologia (Rio de Janeiro) 17:468–473.

———. 1947b. A new genus and species of ant from Guatemala (Hymenoptera, Formicidae). Journal of the New York Entomological Society 55:281–284.

———. 1953a. A revision of the genus *Romblonella* W. M. Wheeler (Hymenoptera: Formicidae). Proceedings of the Hawaiian Entomological Society 15:75–80.

———. 1953b. A new *Romblonella* from Palau, and the first description of a *Romblonella* male (Hymenoptera, Formicidae). Journal of the New York Entomological Society 61:163–167.

———. 1956a. A key to the workers of *Veromessor* Forel of the United States and the description of a new subspecies (Hymenoptera, Formicidae). Pan-Pacific Entomologist 32:36–38.

———. 1956b. A list of the species of *Romblonella* including two generic transfers (Hymenoptera, Formicidae). Bulletin of the Brooklyn Entomological Society 51:18.

———. 1961. A study of New Guinea ants of the genus *Aphaenogaster* Mayr (Hymenoptera, Formicidae). Acta Hymenopterologica 1:213–238.

———. 1962. A remarkable new *Stenamma* from Costa Rica, with pertinent facts on other Mexican and Central American species (Hymenoptera: Formicidae). Journal of the New York Entomological Society 70:33–38.

Smith, T. M., H. H. Shugart, and F. I. Woodward (eds.). 1997. Plant Functional Types: Their Relevance to Ecosystem Properties and Global Change. Cambridge University Press, Cambridge.

Snelling, R. R. 1973. Studies on California ants. 7. The genus *Stenamma* (Hymenoptera: Formicidae). Contributions in Science (Los Angeles County Museum) 245:1–38.

———. 1975. Descriptions of new Chilean ant taxa (Hymenoptera: Formicidae). Contributions in Science (Los Angeles County Museum) 274:1–19.

———. 1976. A Revision of the Honey Ants, Genus *Myrmecocystus* (Hymenoptera: Formicidae). Los Angeles County Museum of Natural History Bulletin 24.

———. 1979a. Three new species of the Palaeotropical arboreal ant genus *Cataulacus* (Hymenoptera: Formicidae). Contributions in Science (Los Angeles County Museum) 315:1–8.

———. 1979b. *Aphomomyrmex* and a related new genus of arboreal African ants (Hymenoptera: Formicidae). Contributions in Science (Los Angeles County Museum) 316:1–8.

———. 1981. The taxonomy and distribution of some North American *Pogonomyrmex* and descriptions of two new species (Hymenoptera: Formicidae). Bulletin of the Southern California Academy of Sciences 80:97–112.

———. 1982. A revision of the honey ants, genus *Myrmecocystus,* first supplement (Hymenoptera: Formicidae). Bulletin of the Southern California Academy of Sciences 81:69–86.

———. 1988. Taxonomic notes on Nearctic species of *Camponotus,* subgenus *Myrmentoma* (Hymenoptera: Formicidae). Pp. 55–78. *In* J. C. Trager (ed.), Advances in Myrmecology. E. J. Brill, Leiden, Netherlands.

———. 1995a. Systematics of Nearctic ants of the genus *Dorymyrmex* (Hymenoptera: Formicidae). Contributions in Science (Los Angeles County Museum) 454:1–14.

Snelling, R. R., and J. H. Hunt. 1975. The ants of Chile (Hymenoptera: Formicidae). Revista Chilena de Entomologia 9:63–129.

Snelling, R. R., and J. T. Longino. 1992. Revisionary notes on the fungus-growing ants of the genus *Cyphomyrmex, rimosus* group (Hymenoptera: Formicidae: Attini). Pp. 479–494. *In* D. Quintero and A. Aiello (eds.), Insects of Panama and Mesoamerica: Selected Studies. Oxford University Press, Oxford.

Snyder, L. E., and J. M. Herbers. 1991. Polydomy and sexual allocation ratios in the ant *Myrmica punctiventris.* Behavioral Ecology and Sociobiology 28:409–415.

Soberón, M. J., and J. Llorente B. 1993. The use of species accumulation functions for the prediction of species richness. Conservation Biology 7:480–488.

Southwood, T. R. E. 1978. Ecological Methods: With Particular Reference to the Study of Insect Populations. Chapman and Hall, London.

Spellerberg, I. F. 1991. Monitoring Ecological Change. Cambridge University Press, Cambridge.

———. 1992. Evaluation and Assessment for Conservation. Chapman and Hall, London.

Steghaus-Kovac, S., and U. Maschwitz. 1993. Predation on earwigs: A novel diet specialization within the genus *Leptogenys* (Formicidae, Ponerinae). Insectes Sociaux 40:337–340.

Stein, M. B., H. G. Thorvilson, and J. W. Johnson. 1990. Seasonal changes in bait preference by the red imported fire ant, *Solenopsis invicta* (Hymenoptera: Formicidae). Florida Entomologist 73: 117–123.

Stevens, G. C. 1989. The latitudinal gradient in geographical range: How so many species coexist in the tropics. American Naturalist 133:240–256.

———. 1992. The elevational gradient in altitudinal range: An extension of Rapoport's latitudinal rule to altitude. American Naturalist 140:893–911.

Stork, N. E. 1991. The composition of the arthropod fauna of Bornean lowland rain forest trees. Journal of Tropical Ecology 7:161–180.

Stork, N. E., and T. M. Blackburn. 1993. Abundance, body size and biomass of arthropods in tropical forest. Oikos 67:483–489.

Stradling, D. J. 1978. The influence of size on foraging in the ant, *Atta cephalotes,* and the effect of some plant defence mechanisms. Journal of Animal Ecology 47:173–188.

Sudd, J. H. and N. R. Franks. 1987. The Behavioural Ecology of Ants. Blackwell, Glasgow, U.K.

Sugihara, G. 1980. Minimal community structure: An explanation of species abundance patterns. American Naturalist 116:770–787.

Talbot, M. 1943. Population studies of the ant *Prenolepis imparis* Say. Ecology 24:31–44.

———. 1975. A list of the ants of the Edwin George Reserve, Livingston Country, Michigan. Great Lakes Entomologist 8:245–246.

Taylor, L. R. 1978. Bates, Williams, Hutchinson—A variety of diversities. Pp. 1–18. *In* L. A. Mound and N. Waloff (eds.), Diversity of Insect Faunas: 9th Symposium of the Royal Entomological Society. Blackwell, Oxford.

Taylor, L. R., I. P. Woiwod, and J. N. Perry. 1978. The density dependence of spatial behavior and the rarity of randomness. Journal of Animal Ecology 47:383–406.

Taylor, R. W. 1960. Taxonomic notes on the ants *Ponera* leae Forel and *Ponera norfolkensis* (Wheeler) (Hymenoptera: Formicidae). Pacific Science 14:178–180.

———. 1965. A monographic revision of the rare tropicopolitan ant genus *Probolomyrmex* Mayr (Hymenoptera: Formicidae). Transactions of the Royal Entomological Society of London 117: 345–365.

———. 1967. A monographic revision of the ant genus *Ponera* Latreille (Hymenoptera: Formicidae). Pacific Insects Monographs 13:1–112.

———. 1968a. Notes on the Indo-Australian basicerotine ants (Hymenoptera: Formicidae). Australian Journal of Zoology 16:333–348.

———. 1968b. A new Malayan species of the ant genus *Epitritus,* and a related new genus from Singapore (Hymenoptera: Formicidae). Journal of the Australian Entomological Society 7:130–134.

———. 1970a. Characterization of the Australian endemic ant genus *Peronomyrmex* Viehmeyer (Hymenoptera: Formicidae). Journal of the Australian Entomological Society 9:209–211.

———. 1970b. Notes on some Australian and Melanesian basicerotine ants (Hymenoptera: Formicidae). Journal of the Australian Entomological Society 9:49–52.

———. 1973. Ants of the Australian genus *Mesostruma* Brown (Hymenoptera: Formicidae). Journal of the Australian Entomological Society 12:24–38.

———. 1977. New ants of the Australasian genus *Orectognathus,* with a key to the known species (Hymenoptera: Formicidae). Australian Journal of Zoology 25:581–612.

———. 1978a. *Nothomyrmecia macrops:* A living-fossil ant rediscovered. Science 201:979–985.

———. 1978b. A taxonomic guide to the ant genus *Orectognathus* (Hymenoptera: Formicidae). CSIRO Divison of Entomolgy Reports 3:1–11.

———. 1978c. Melanesian ants of the genus *Amblyopone* (Hymenoptera: Formicidae). Australian Journal of Zoology 26:823–839.

———. 1979a. New Australian ants of the genus *Orectognathus,* with summary description of the twenty-nine known species (Hymenoptera: Formicidae). Australian Journal of Zoology 27:773–788.

———. 1979b. Notes on the Russian endemic ant genus *Aulacopone* Arnoldi (Hymenoptera: Formicidae). Psyche (Cambridge) 86:353–361.

———. 1980. Australian and Melanesian ants of the genus *Eurhopalothrix* Brown and Kempf—Notes and new species (Hymenoptera: Formicidae). Journal of the Australian Entomological Society 19:229–239.

———. 1983. Descriptive taxonomy: Past, present and future. Pp. 93–134. *In* E. Highley and R. W. Taylor (eds.), Australian Systematic Entomology: A Bicentenary Perspective. CSIRO, Melbourne.

———. 1985. The ants of the Papuasian genus *Dacetinops* (Hymenoptera: Formicidae: Myrmicinae). Series Entomologica (Hague) 33:41–67.

———. 1989. Australasian ants of the genus *Leptothorax* Mayr (Hymenoptera: Formicidae: Myrmicinae). Memoirs of the Queensland Museum 27:605–610.

———. 1990a. [Untitled. Anomalomyrmini Taylor tribe n., *Anomalomyrma* Taylor gen. n., *Protanilla* Taylor gen. n.] Pp. 278–279. *In* B. Bolton, The higher classification of the ant subfamily Leptanillinae (Hymenoptera: Formicidae). Systematic Entomology 15:267–282.

———. 1990b. New Asian ants of the tribe Basicerotini, with an on-line computer interactive key to the twenty-six known Indo-Australian species (Hymenoptera: Formicidae: Myrmicinae). Invertebrate Taxonomy 4:397–425.

———. 1990c. The nomenclature and distribution of some Australian and New Caledonian ants of the genus *Meranoplus* Fr. Smith (Hymenoptera: Formicidae: Myrmicinae). General and Applied Entomology 22:31–40.

———. 1990d. Notes on the ant genera *Romblonella* and *Willowsiella,* with comments on their affinities, and the first descriptions of Australian species (Hymenoptera: Formicidae: Myrmicinae). Psyche (Cambridge) 97:281–296.

Tennant, L. E., and S. D. Porter. 1991. Comparison of diets of two fire ant species (Hymenoptera: Formicidae): Solid and liquid components. Journal of Entomological Science 26:450–465.

Terayama, M. 1985a. Two new species of the ant genus *Myrmecina* (Insecta; Hymenoptera; Formicidae) from Japan and Taiwan. Edaphologia 32:35–40.

———. 1985b. Two new species of the genus *Acropyga* (Hymenoptera, Formicidae) from Taiwan and Japan. Kontyû 53:284–289.

———. 1987. A new species of *Amblyopone* (Hymenoptera, Formicidae) from Japan. Edaphologia 36:31–33.

Terayama, M. 1996. Taxonomic studies on the Japanese Formicidae, part 2. Seven genera of Ponerinae, Cerapachyinae und Myrmicinae. Nature and Human Activities 1:1–8.

Terayama, M., and K. Ogata. 1988. Two new species of the ant genus *Probolomyrmex* (Hymenoptera, Formicidae) from Japan. Kontyû 56:590–594.

Terayama, M. and K. Onoyama. 1999. The ant genus *Leptothorax* Mayr (Hymenoptera: Formicidae) in Japan. Memoirs of the Myrmecological Society of Japan 1:71–97.

Terayama, M., and S. Yamane. 1989. The army ant genus *Aenictus* (Hymenoptera, Formicidae) from Sumatra, with descriptions of three new species. Japanese Journal of Entomology 57:597–603.

Terayama, M., C.-C. Lin, and W.-J. Wu. 1995. The ant genera *Epitritus* and *Kyidris* from Taiwan (Hymenoptera: Formicidae). Proceedings of the Japanese Society of Systematic Zoology 53:85–89.

———. 1996. The Taiwanese species of the ant genus *Smithistruma* (Hymenoptera, Formicidae). Japanese Journal of Entomology 64(2):327–339.

Terron, G. 1974. Découverte au Cameroun de deux espèces nouvelles du genre *Prionopelta* Mayr (Hym.: Formicidae). Annales de la Faculté des Sciences, Université Fédéral du Cameroun (Yaoundé) 17:105–119.

———. 1981. Deux nouvelles espèces éthiopiennes pour le genre *Proceratium* (Hym.: Formicidae). Annales de la Faculté des Sciences, Université Fédéral du Cameroun (Yaoundé) 28:95–103.

Thaxter, R. 1888. The Entomophthoreae of the United States. Memoirs of the Boston Society of Natural History 4:133–201.

———. 1908. Contribution toward a monograph of the Laboulbeniaceae, pt. II. Memoirs of the American Academy of Arts and Sciences 13: 217–469.

Thompson, C. R., and C. Johnson. 1989. Rediscovered species and revised key to the Florida thief ants (Hymenoptera: Formicidae). Florida Entomologist 72:697–698.

Tillyard, R. J. 1926. The Insects of Australia and New Zealand. Angus and Robertson, Sydney.

Tilman, D. 1996. Biodiversity: Population versus ecosystem stability. Ecology 77:350–363.

Tinaut, A. 1990. *Teleutomyrmex kutteri,* spec. nov.: A new species from Sierra Nevada (Granada, Spain). Spixiana 13:201–208.

Tobin, J. E. 1991. A neotropical rain forest canopy ant community: Some ecological considerations. Pp. 536–538. *In* C. R. Huxley and D. F. Cutler (eds.), Ant-Plant Interactions. Oxford University Press, Oxford.

———. 1994. Ants as primary consumers: Diet and abundance in the Formicidae. Pp 279–307. *In* J. H. Hunt and C. A. Nalepa (eds.), Nourishment and Evolution in Insect Societies. Westview Press, Boulder, Colorado.

———. 1997. Competition and coexistence of ants in a small patch of rainforest canopy in Peruvian Amazonia. Journal of the New York Entomological Society 105:105–112.

Tohmé, G., and H. Tohmé. 1981. Les fourmis du genre *Messor* en Syrie. Position systématique. Description de quelques ailés et de formes nouvelles. Répartition géographique. Ecologia Mediterranea 7(1):139–153.

Topoff, H. 1990. Slave-making ants. American Scientist 78:520–528.

Torre-Bueno, J. R. de la. 1937. A Glossary of Entomology. Brooklyn Entomological Society, Brooklyn, New York.

Torre-Bueno, J. R. de la. 1989. The Torre–Bueno Glossary of Entomology. Compiled by S.W. Nichols, and including Supplement A by G. S. Tulloch. New York Entomological Society, New York.

Trager, J. C. 1984. A revision of the genus *Paratrechina* (Hymenoptera: Formicidae) of the continental United States. Sociobiology 9:49–162.

———. 1991. A revision of the fire ants, *Solenopsis geminata* group (Hymenoptera: Formicidae: Myrmicinae). Journal of the New York Entomological Society 99:141–198.

Trivers, R. L., and H. Hare. 1976. Haplodiploidy and the evolution of the social insects. Science 191:249–263.

Tschinkel, W. R. 1991. Insect sociometry, a field in search of data. Insectes Sociaux 38:77–82.

———. 1992. Brood raiding and the population dynamics of founding and incipient colonies of the fire ant, *Solenopsis invicta.* Ecological Entomology 17:179–188.

———. 1993. Sociometry and sociogenesis of colonies of the fire ant *Solenopsis invicta* during one annual cycle. Ecological Monographs 63: 425–457.

Tschinkel, W. R., and Howard, D. F. 1978. Queen replacement in orphaned colonies of the fire ant, *Solenopsis invicta.* Behavioral Ecology and Sociobiology 3:297–310.

Tulloch, G. S. 1962. Torre-Bueno's Glossary of Entomology, Supplement A. Brooklyn Entomological Society, Brooklyn, New York.

Turk, F. A. 1953. A new genus and species of pseudoscorpion with some notes on its biology. Proceedings of the Zoological Society of London 122:951–954.

Ule, E. 1902. Ameisengärten im Amazonasgebiet. Botanische Jahrbücher für Systematik, Pflanzengeschichte und Pflanzengeographien 30:45–52.

Umphrey, G. J., 1996. Morphometric discrimination among sibling species in the *fulva-rudis-texana* complex of the ant genus *Aphaenogaster* (Hymenoptera: Formicidae). Canadian Journal of Zoology 74(3):528–559.

United Nations Environment Programme. 1995. Global Biodiversity Assessment. Cambridge University Press, Cambridge.

Upton, M. 1991. Methods for Collecting, Preserving and Studying Insects and Allied Forms. Australian Entomological Society Miscellaneous Publications [Brisbane, Australia] 3:1–86.

Valone, T., and J. Brown. 1995. Effects of competition, colonization, and extinction on rodent species diversity. Science 267:880–883.

Vander Meer, R., and L. Alonso. 1998. Pheromone directed behavior in ants. Pp. 159–192. *In* R. Vander Meer, M. Breed, M. Winston, and K. Espelie (eds.), Pheromone Communication in Social Insects. Westview Press, Boulder, Colorado.

Vasconcelos, H. L., and J. H. C. Delabie. 2000. Ground ant communities from central Amazonia forest fragments. Pp. 59–70. *In* D. Agosti, J. Majer, L. Alonso, and T. R. Schultz (eds.), Sampling Ground–Dwelling Ants: Case Studies from the Worlds' Rain Forests. Curtin University School of Environmental Biology Bulletin No. 18. Perth, Australia.

Vaz-Ferreira, R., L. C. de Zolessi, and F. Achával. 1970. Oviposicion y desarrollo de Ofidios y Lacertilios en hormigueros de *Acromyrmex*. Physis 29:431–459.

———. 1973. Oviposición y desarrollo de ofidos y lacertilios en hormigueros de *Acromyrmex*. II. Trabajos del Cinco Congresso Latinoamericano de Zoologia, Montevideo 1:232–244.

Veerhagh, M. 1990. The Formicidae of the rain forest in Panguana, Peru: The most diverse local ant fauna ever recorded. Pp. 217–218. *In* G. K. Veeresh, B. Mallik, and C. A. Viraktamath (eds.), Social Insects and the Environment. Proceedings of the 11th International Congress of IUSSI, 1990. Oxford and IBH, New Delhi.

Vinson, S. B. 1991. Effect of the red imported fire ant (Hymenoptera: Formicidae) on a small plant-decomposing arthropod community. Environmental Entomology 20:98–103.

Von Ihering, H. 1891. Die Wechselbeziehungen zwischen Pflanzen und Ameisen in den Tropen. Ausland 1891:474–477.

Wang, C., and J. Wu. 1991. Taxonomic studies on the genus *Polyrhachis* Mayr of China (Hymenoptera, Formicidae). Forest Research 4:596–601. [In Chinese.]

Wang, C., G. Xiao, and J. Wu. 1989a. Taxonomic studies on the genus *Camponotus* Mayr in China (Hymenoptera, Formicidae) [part]. Forest Research 2:221–228. [In Chinese.]

———. 1989b. Taxonomic studies on the genus *Camponotus* Mayr in China (Hymenoptera, Formicidae). [conclusion]. Forest Research 2:321–328. [In Chinese.]

Wang, M. 1993. Taxonomic study of the ant tribe Odontomachini in China (Hymenoptera: Formicidae). Scientific Treatise on Systematic and Evolutionary Zoology 2:219–230. [In Chinese.]

Wang, M., G. Xiao, and J. Wu. 1988. Taxonomic studies on the genus *Tetramorium* Mayr in China (Hymenoptera, Formicidae). Forest Research 1:264–274. [In Chinese.]

Ward, P. S. 1980. A systematic revision of the *Rhytidoponera impressa* group (Hymenoptera: Formicidae) in Australia and New Guinea. Australian Journal of Zoology 28:475–498.

———. 1984. A revision of the ant genus *Rhytidoponera* (Hymenoptera: Formicidae) in New Caledonia. Australian Journal of Zoology 32:131–175.

———. 1985. The Nearctic species of the genus *Pseudomyrmex* (Hymenoptera: Formicidae). Quaestiones Entomologicae 21:209–246.

———. 1987. Distribution of the introduced Argentine ant (*Iridomyrmex humilis*) in natural habitats of the lower Sacramento Valley and its effects on the indigenous ant fauna. Hilgardia 55:1–16.

———. 1988. Mesic elements in the western Nearctic ant fauna: Taxonomic and biological notes on *Amblyopone, Proceratium,* and *Smithistruma* (Hymenoptera: Formicidae). Journal of the Kansas Entomological Society 61:102–124.

———. 1989. Systematic studies on pseudomyrmecine ants: Revision of the *Pseudomyrmex oculatus* and *P. subtilissimus* species groups, with taxonomic comments on other species. Quaestiones Entomologicae 25:393–468.

———. 1990. The ant subfamily Pseudomyrmecinae (Hymenoptera: Formicidae): Generic revision and relationship to other formicids. Systematic Entomology 15:449–489.

———. 1993. Systematic studies on *Pseudomyrmex* acacia-ants (Hymenoptera: Formicidae: Pseudomyrmecinae). Journal of Hymenoptera Research 2:117–168.

———. 1994. *Adetomyrma,* an enigmatic new ant genus from Madagascar (Hymenoptera: Formicidae), and its implications for ant phylogeny. Systematic Entomology 19:159–175.

———. 1999a. Systematics, biogeography and host plant associations of the *Pseudomyrmex viduus* group (Hymenoptera: Formicidae), *Triplaris-* and *Tachigali*-inhabiting ants. Zoological Journal of the Linnean Society 126:451–540.

———. 1999b. Deceptive similarity in army ants of the genus *Neivamyrmex* (Hymenoptera: Formicidae): Taxonomy, distribution and biology of *N. californicus* (Mayr) and *N. nigrescens* (Cresson). Journal of Hymenoptera Research 8:74–97.

Ward, P. S., B. Bolton, S. O. Shattuck, and W. L. Brown Jr. 1996. A Bibliography of Ant Systematics. University of California Publications in Entomology 116.

Wasmann, E. 1902. Zur Kenntnis der myrmecophilen *Antennophorus* und anderer auf Ameisen und Termiten reitende Acarinen. Zoologischer Anzeiger 25:66–76.

Watkins, J. F., II. 1976. The identification and distribution of New World army ants (Dorylinae: Formicidae). Baylor University Press, Waco, Texas.

———. 1977. The species and subspecies of *Nomamyrmex* (Dorylinae: Formicidae). Journal of the Kansas Entomological Society 50:203–214.

———. 1982. The army ants of Mexico (Hymenoptera: Formicidae: Ecitoninae). Journal of the Kansas Entomological Society 55:197–247.

———. 1985. The identification and distribution of the army ants of the United States of America (Hymenoptera, Formicidae, Ecitoninae). Journal of the Kansas Entomological Society 58:479–502.

Watt, J. C. 1979. Abbreviations for entomological collections. New Zealand Journal of Zoology 6:519–520.

Weber, N. A. 1943. Parabiosis in Neotropical "ant gardens." Ecology 24:400–404.

———. 1944. The neotropical coccid-tending ants of the genus *Acropyga* Roger. Annals of the Entomological Society of America 37:89–122.

———. 1947. A revision of the North American ants of the genus *Myrmica* Latreille with a synopsis of the Palearctic species. I. Annals of the Entomological Society of America 40:437–474.

———. 1948. A revision of the North American ants of the genus *Myrmica* Latreille with a synopsis of the Palearctic species. II. Annals of the Entomological Society of America 41:267–308.

———. 1950a. The African species of the genus *Oligomyrmex* Mayr (Hymenoptera, Formicidae). American Museum Novitates 1442:1–19.

———. 1950b. A revision of the North American ants of the genus *Myrmica* Latreille with a synopsis of the Palearctic species. III. Annals of the Entomological Society of America 43:189–226.

———. 1952. Studies on African Myrmicinae. I (Hymenoptera, Formicidae). American Museum Novitates 1548:1–32.

———. 1972a. The Attines: The fungus-culturing ants. American Scientist 60:448–456.

———. 1972b. Gardening Ants: The Attines. American Philosophical Society, Philadelphia.

Weber, N. A., and J. L. Anderson. 1950. Studies on central African ants of the genus *Pseudolasius* Emery (Hymenoptera, Formicidae). American Museum Novitates 1443:1–7.

Wehner, R., A. C. Marsh, and S. Wehner. 1992. Desert ants on a thermal tightrope. Nature 357:586–587.

Went, F. W., J. Wheeler, and G. C. Wheeler. 1972. Feeding and digestion in some ants (*Veromessor* and *Manica*). BioScience 22:82–88.

Westman, W. 1986. Resilience: Concepts and measures. Pp. 5–19. *In* B. Dell, A. Hopkins, and B. Lamont (eds.), Resilience in Mediterranean Ecosystems. Dr. W. Junk, The Hague.

Wetterer, J. K. 1991. Allometry and the geometry of leaf-cutting in *Atta cephalotes.* Behavioral Ecology and Sociobiology 29:347–351.

Wetterer, J. K., T. R. Schultz, and R. Meier. 1998. Phylogeny of fungus–growing ants (tribe Attini) based on mtDNA sequence and morphology. Molecular Phylogenetics and Evolution 9:42–47.

Wheeler, D., and S. Levings. 1988. The impact of the 1983 El Niño drought on the litter arthropods of Barro Colorado Island, Panama. Pp. 309–326. *In* J. C. Trager (ed.), Advances in Myrmecology. E. J. Brill, New York.

Wheeler, G. C., and E. W. Wheeler. 1930. Two new ants from Java. Psyche (Cambridge) 37:193–201.

Wheeler, G. C., and J. Wheeler. 1986. The Ants of Nevada. Natural History Museum of Los Angeles County, Los Angeles.

Wheeler, J. 1968. Male genitalia and the taxonomy of *Polyergus* (Hymenoptera: Formicidae). Proceedings of the Entomological Society of Washington 70:156–164. [Erratum: Proceedings of the Entomological Society of Washington 70:254.]

Wheeler, W. M. 1905. The North American ants of the genus *Liometopum.* Bulletin of the American Museum of Natural History 21:321–333.

———. 1908. Studies on myrmecophiles. II. *Hetaerius.* Journal of the New York Entomological Society 16:135–143.

———. 1910. Two new myrmecophilous mites of the genus *Antennophorus.* Psyche (Cambridge) 17:1–6.

———. 1913. Observations on the Central American *Acacia* ants. Transactions of the Second International Entomological Congress of Oxford (1912) 2:109–139.

———. 1914. Notes on the habits of *Liomyrmex.* Psyche (Cambridge) 21:75–76.

———. 1918. The Australian ants of the ponerine tribe Cerapachyini. Proceedings of the American Academy of Arts and Sciences 53:215–265.

———. 1922a. Ants of the American Museum Congo expedition. A contribution to the myrmecology of Africa. II. The ants collected by the American Museum Congo Expedition. Bulletin of the American Museum of Natural History 45:39–269.

———. 1922b. Ants of the American Museum Congo expedition. A contribution to the myrmecology of Africa. VII. Keys to the genera and subgenera of ants. Bulletin of the American Museum of Natural History 45:631–710.

———. 1924. Ants of Krakatau and other islands in the Sunda Strait. Treubia 5:239–258.

———. 1925. A new guest-ant and other new Formicidae from Barro Colorado Island, Panama. Biological Bulletin of the Marine Biological Laboratory (Woods Hole) 49:150–181.

———. 1928. The Social Insects: Their Origin and Evolution. Kegan Paul, Trench, Trubner, London.

———. 1934. A second revision of the ants of the genus *Leptomyrmex* Mayr. Bulletin of the Museum of Comparative Zoology, Harvard College 77:69–118.

———. 1935. Ants of the genus *Acropyga* Roger, with description of a new species. Journal of the New York Entomological Society 43:321–329.

———. 1936. Ecological relations of ponerine and other ants to termites. Proceedings of the American Academy of Arts and Sciences 71:159–243.

———. 1942. Studies of Neotropical ant-plants and their ants. Bulletin of the Museum of Comparative Zoology, Harvard University 90:1–262.

Wheeler, W. M. and W. M. Mann. 1942. [Untitled. *Allomerus decemarticulatus* Mayr subsp. *Novemarticulatus* Wheeler & Mann, subsp. nov.] Pp. 188–189. *In* W. M. Wheeler, Studies of Neotropical ant-plants and their ants. Bulletin of the Museum of Comparative Zoology, Harvard College 90:1–262.

Whitcomb, W. H., A. Bhatkar, and J. C. Nickerson. 1973. Predators of *Solenopsis invicta* queens prior to successful colony establishment. Environmental Entomology 2:1101–1103.

Whitford, W. G. 1978. Structure and seasonal activity of Chihuahua desert ant communities. Insectes Sociaux 25:79–88.

Whitford, W. G., and G. Ettershank. 1975. Factors affecting foraging activity in Chihuahuan desert harvester ants. Environmental Entomology 4:689–696.

Wiernasz, D. C., and B. J. Cole. 1995. Spatial distribution of *Pogonomyrmex occidentalis:* Recruitment, mortality, and overdispersion. Journal of Animal Ecology 64:519–527.

Wilcox, B. A. 1984. In situ conservation of genetic resources: Determinants of minimum-area requirements. Pp. 639–647. *In* J. A. McNeeley and K. R. Miller (eds.), National Parks, Conservation, and Development: The Role of Protected Areas in Sustaining Society. Proceedings of the World Congress on National Parks, Bali, Indonesia, 11–22 October 1982. Smithsonian Institution Press, Washington, D.C.

Wilcox, B. A., D. D. Murphy, P. R. Ehrlich, and G. T. Austin. 1986. Insular biogeography of the montane butterfly faunas in the Great Basin: Comparison with birds and mammals. Oecologia 69:188–194.

Willey, R. B., and W. L. Brown Jr. 1983. New species of the ant genus *Myopias* (Hymenoptera: Formicidae: Ponerinae). Psyche (Cambridge) 90:249–285.

Williams, D. F. 1994. Exotic ants: Biology, impact, and control of introduced species. Westview Press, Boulder, Colorado.

Williams, R. N. (ed.). 1978. Worldwide directory of institutions with entomologists, part I: Latin

America. Bulletin of the Entomological Society of America 24:179–193.

Willis, E. O. 1983. A study of ant-following birds of northeastern Brazil. Research Reports of the National Geographic Society 15:745–748.

Willis, E. O., and Y. Oniki. 1978. Birds and army ants. Annual Review of Ecology and Systematics 9:243–263.

Wilson, D. E., F. R. Cole, J. D. Nichols, R. Rudran, and M. S. Foster. 1996. Measuring and Monitoring Biological Diversity. Standard Methods for Mammals. Smithsonian Institution Press, Washington, D.C.

Wilson, E. O. 1953. The ecology of some North American dacetine ants. Annals of the Entomological Society of America 46:479–495.

———. 1955. A monographic revision of the ant genus *Lasius*. Bulletin of the Museum of Comparative Zoology, Harvard College 113:1–201.

———. 1958. Patchy distributions of ant species in New Guinea rain forests. Psyche (Cambridge) 65:26–38.

———. 1959. Some ecological characteristics of ants in New Guinea rain forests. Ecology 40:437–447.

———. 1961. The nature of the taxon cycle in the Melanesian ant fauna. American Naturalist 95:169–193.

———. 1962a. Behavior of *Daceton armigerum* (Latreille), with a classification of self-grooming movements in ants. Bulletin of the Museum of Comparative Zoology, Harvard College 127:401–422.

———. 1962b. The Trinidad cave ant *Erebomyrma* (= *Spelaeomyrmex*) *urichi* (Wheeler), with a comment on cavernicolous ants in general. Psyche (Cambridge) 69:63–72.

———. 1964. The true army ants of the Indo-Australian area (Hymenoptera: Formicidae: Dorylinae). Pacific Insects Monographs 6:427–483.

———. 1971. The Insect Societies. Belknap Press, Cambridge, Massachusetts.

———. 1976. Which are the most prevalent ant genera? Studia Entomologica 19:187–200.

———. 1984. Tropical social parasites in the ant genus *Pheidole*, with an analysis of the anatomical parasitic syndrome (Hymenoptera: Formicidae). Insectes Sociaux 31:316–334.

———. 1985. The principles of caste evolution. Fortschritte der Zoologie 31:307–324.

———. 1987. The arboreal ant fauna of Peruvian Amazon forests: A first assessment. Biotropica 19:245–251.

———. 1989. *Chimaeridris,* a new genus of hook-mandibled myrmicine ants from tropical Asia (Hymenoptera: Formicidae). Insectes Sociaux 36:62–69.

———. 1993. The Diversity of Life. W.W. Norton, New York.

Wilson, E. O., and W. L. Brown Jr. 1953. The subspecies concept and its taxonomic application. Systematic Zoology 2:97–111.

———. 1956. New parasitic ants of the genus *Kyidris,* with notes on ecology and behavior. Insectes Sociaux 3:439–454.

———. 1984. Behavior of the cryptobiotic predaceous ant *Eurhopalothrix heliscata* n. sp. (Hymenoptera: Formicidae: Basicerotini). Insectes Sociaux 31:408–428.

Wilson, E. O., and R. W. Taylor. 1967. An estimate of the potential evolutionary increase in species density in the Polynesian ant fauna. Evolution 21:1–10.

Wilson, E. O., T. Eisner, G. C. Wheeler, and J. Wheeler. 1956. *Aneuretus simoni* Emery, a major link in ant evolution. Bulletin of the Museum of Comparative Zoology, Harvard College 115:81–99.

Wing, M. W. 1968. Taxonomic revision of the Nearctic genus *Acanthomyops* (Hymenoptera: Formicidae). Memoirs of the Cornell University Agricultural Experiment Station 405:1–173.

Wisdom, W., and W. G. Whitford. 1981. Effects of vegetation change on ant communities of arid rangelands. Environmental Entomology 10:893–897.

Wolda, H. 1992. Trends in abundance of tropical forest insects. Oecologia 89:47–52.

Wu, J. 1990. Taxonomic studies on the genus *Formica* L. of China (Hymenoptera: Formicidae). Forest Research 3:1–8. [In Chinese.]

Wu, J., and C. Wang. 1990. A taxonomic study on the genus *Tetraponera* Smith in China (Hymenoptera: Formicidae). Scientia Silvae Sinica 26:515–518. [In Chinese.]

Xu, Z. 1994a. A taxonomic study of the ant genus *Lepisiota* Santschi from Southwestern China (Hymenoptera: Formicidae: Formicinae). Journal of Southwest Forestry College 14(4):231–237.

Xu, Z. 1994b. A taxonomic study of the ant genus *Brachyponera* Emery in Southwestern China (Hymenoptera: Formicidae: Ponerinae). Journal of Southwest Forestry College 14(3):181–185.

Xu, Z. 1995a. Two new species of the ant genus *Prenolepis* from Yunnan China (Hymenoptera: Formicidae). Zoological Research 16(4):337–341.

Xu, Z. 1995b. A taxonomic study of the ant genus *Dolichoderus* Lund in China (Hymenoptera: Formicidae: Dolichoderinae). Journal of Southwest Forestry College 15(1):33–39.

Xu, Z. 1997. A taxonomic study of the ant genus *Pseudolasius* Emery in China (Hymenoptera: Formicidae). Zoological Research 18(1):1–6.

Xu, Z. 1999. Systematic studies on the ant genera of *Carebara, Rhopalomastix* and *Kartidris* in China (Hymenoptera: Formicidae: Myrmicinae). Acta Biologica Plateau Sinica 14:129–136.

Xu, Z., and W. Zhang. 1996. A new species of the genus *Gnamptogenys* (Hymenoptera: Formicidae: Ponerinae) from southwestern China. Entomotaxonomia 18(1):55–58.

Yamauchi, K. 1978. Taxonomical and ecological studies on the ant genus *Lasius* in Japan (Hymenoptera: Formicidae). I. Taxonomy. Science Reports of the Faculty of Education, Gifu University (Natural Sciences) 6:147–181.

Yasumatsu, K., and W. L. Brown Jr. 1951. Revisional notes on *Camponotus herculeanus* Linné and close relatives in Palearctic regions (Hymenoptera: Formicidae). Journal of the Faculty of Agriculture, Kyushu University 10:29–44.

———. 1957. A second look at the ants of the *Camponotus herculeanus* group in eastern Asia. Journal of the Faculty of Agriculture, Kyushu University 11:45–51.

Yasumatsu, K., and Y. Murakami. 1960. A revision of the genus *Stenamma* of Japan (Hymenoptera, Formicidae, Myrmicinae). Esakia 1:27–31.

Young, A. M. 1986. Notes on the distribution and abundance of ground- and arboreal-nesting ants (Hymenoptera: Formicidae) in some Costa Rican cacao habitats. Proceedings of the Entomological Society of Washington 88:550–571.

Zhou, S.-Y., and Z.-M. Zheng. 1999. Taxonomic study of the ant genus *Pheidole* Westwood from Guangxi, with descriptions of three new species. Acta Zootaxonomica Sinica 24(1):83–88.

Zimmerman, B. L., and M. T. Rodriguez. 1990. Frogs, snakes, and lizards of the INPA-WWF reserves near Manaus, Brazil. Pp. 426–454. *In* A. H. Gentry (ed.), Four Neotropical Rainforests. Yale University Press, New Haven, Connecticut.

Contributors

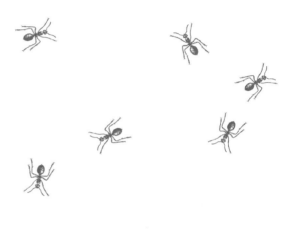

Donat Agosti
Department of Entomology
American Museum of Natural History
Central Park West at 79th Street
New York, NY 10024-5192
USA
Mailing address:
c/o Swiss Embassy
P.O. Box 633
10 Abdel Khalek Saroit
Cairo
Egypt
e-mail: agosti@amnh.org

Leeanne E. Alonso
Rapid Assessment Program
Conservation International
1919 M Street NW, Suite 600
Washington, DC 20036
USA
e-mail: l.alonso@conservation.org

Gary D. Alpert
Museum of Comparative Zoology
Harvard University
Cambridge, MA 02138
USA
e-mail: galpert@oeb.harvard.edu

Alan N. Andersen
CSIRO Wildlife and Ecology
Tropical Ecosystems Research Centre
PMB 44
Winnellie, Darwin, NT 0821
Australia
e-mail: Alan.Andersen@terc.csiro.au

Brandon T. Bestelmeyer
Department of Biology
MSC 3AF Box 30001
New Mexico State University
Las Cruces, NM 80003
USA
e-mail: bbestelm@jornada.nmsu.edu

Dattatray Manjunath Bhat
Centre for Ecological Science
Indian Institute of Science
Bangalore 560 012
India

C. Roberto F. Brandão
Director of Science
Museu de Zoologia
Universidade de Sao Paulo
Caixa Postal 7171
São Paulo, SP 01064-970
Brazil
e-mail: crfbrand@usp.br

William L. Brown Jr. (deceased)
Department of Entomology
Comstock Hall
Cornell University
Ithaca, NY 14853
USA

Sofia Campiolo
Departamento de Ciências Biológicas
Universidade Estadual Santa Cruz
Km 16
Rodovia Ilhéus–Itabuna, BA 45650-000
Brazil
e-mail: campiolo@jacaranda.uescba.com.br

Krishnappa Chandrashekara
Centre for Ecological Science
Indian Institute of Science
Bangalore 560 012
India

Stefan Cover
Museum of Comparative Zoology
Harvard University
Cambridge, MA 02138
USA
e-mail: scover@oeb.harvard.edu

Jacques H. C. Delabie
Laboratorio de Mirmecologia
Centro de Pesquisas do Cacau (CEPEC)
CEPLAC–Ministério da Agricultura e do
    Abastecimento
Caixa Postal 7
Itabuna, BA 45600-000
and Departamento de Ciências Agrárias e
    Ambientais
Universidade Estadual Santa Cruz
Ilhéus, BA 45660-000
Brazil
e-mail: delabie@nuxnet.com.br

Brian L. Fisher
Department of Entomology
California Academy of Sciences
Golden Gate Park
San Francisco, CA 94118
e-mail: bfisher@calacademy.org

Raghavendra Gadagkar
Centre for Ecological Science
Indian Institute of Science
Bangalore 560 012
India
e-mail: ragh@ces.iisc.ernet.in

David Gladstein
Department of Invertebrates
American Museum of Natural History
Central Park West at 79th Street
New York, NY 10024-5192
USA
e-mail: daveg@amnh.org

Ana Yoshi Harada
Departamento de Zoologia
Museu Paraense Emilio Geoldi (MPEG)
Caixa Postal 399
Belem, PA 66040-170
Brazil
ayharada@museu-goeldi.br

Kye S. Hedlund
Department of Computer Science
University of North Carolina
Chapel Hill, NC 27599-3175
USA
e-mail: hedlund@cs.unc.edu

Michael Kaspari
Department of Zoology
University of Oklahoma
Norman, OK 73019-0235
USA
e-mail: mkaspari@ou.edu

John E. Lattke
Museo del Instituto de Zoología Agrícola
Facultad de Agronomía
Universidad Central de Venezuela
Apartado 4579
Maracay 2101-A
Venezuela
e-mail: piquihuye@hotmail.com

John T. Longino
Evergreen State College
Olympia, WA 98505
USA
e-mail: longinoj@evergreen.edu

Jonathan D. Majer
School of Environmental Biology
Curtin University of Technology
P.O. Box U1987
Perth, WA 6845
Australia
e-mail: imajerj@info.curtin.edu.au

Annette K. F. Malsch
AK Ethoökologie
Zoologisches Institut
J. W. Goethe Universität
Siesmeyerstrasse 70
60054 Frankfurt
Germany
e-mail: Malsch@zoology.uni-frankfurt.de

Terrence P. McGlynn
Department of Biology
University of San Diego
5998 Alcalá Park
San Diego, CA 92110
USA
e-mail: mcglynn@acusd.edu

Ivan C. do Nascimento
Departamento de Ciências Agrárias e
Ambientais
Universidade Estadual Santa Cruz
Ilhéus, BA 45660-000
Brazil
e-mail: icardoso@hotmail.com

Padmini Nair
Centre for Ecological Science
Indian Institute of Science
Bangalore 560 012
India

Ted R. Schultz
Department of Entomology, MRC 188
National Museum of Natural History
Smithsonian Institution
Washington, DC 20560
USA
e-mail: schultz@onyx.si.edu

Rogerio Silvestre
Museu de Zoologia
Universidade de Sao Paulo
Caixa Postal 7171
São Paulo, SP 01064-970
Brazil
e-mail: rogestre@usp.br

Heraldo L. Vasconcelos
Department of Ecology
Instituto de Pesquisas da Amazonia,
Coordenação de Pesquisas em Entomologia
Caixa Postal 478
Manaus, AM 69011-970
Brazil
e-mail: heraldo@inpa.gov.br

Philip S. Ward
Department of Entomology
University of California
One Shields Avenue
Davis, CA 95616
USA
e-mail: psward@ucdavis.edu

Edward O. Wilson
Museum of Comparative Zoology
Harvard University
Cambridge, MA 02138
USA

I. W. Wright
School of Environmental Biology
Curtin University of Technology
P.O. Box U1987
Perth, WA 6845
Australia
e-mail: ianw@cs.curtin.edu.au

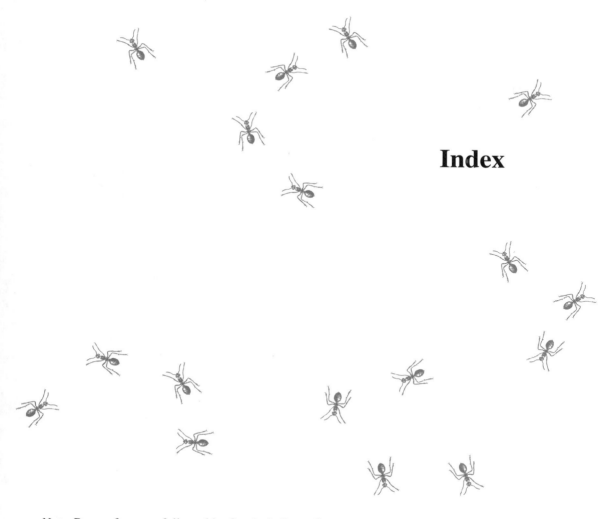

Index

Note: Page references followed by f and t indicate figures and tables, respectively.

abundance class, 195–196, 196f, 197f
abundance curve, 195–196, 196f
Africa, 107, 115, 207–209
ALL. *See* Ants of the Leaf Litter
ant colony
  activity, 97
  alates, 10, 12
  behavior and composition, 10–11
  caste system, 10, 19
  collection methods. *See* collection
    methods
  density, 16, 18
    measurement of, 97, 139
  differences from vertebrate
    societies, 11
  dynamic nature, 90–92
  life cycle
    founding, 11–12
    growth, 12
    reproduction, 12
  queen, 10, 12, 17, 163

  sedentary nature, 90
  size, 12, 16, 19
  temperature and, 11, 12
  workers, 10–11, 163, 163t
ant curation, 155–171
  databases, 170–171
  dissecting, 160
    materials, 160
  labeling, 160–162
    data records, 162
    importance of, 160
    information to include, 124,
      161–162, 161f
    materials, 124, 160–161, 221
    placement, 124, 161
    size, 161
  major ant collections, 167, 168
  mounting, 158–160, 159f
    cleaning, 159–160
    materials, 158–159, 222
    protocol, 158–159

  natural history museums and
    collections, 174–175,
    176t–182t, 183–184
  private collections, 185
  reasons for, 155
  reference collections, 7, 167–168
  sorting, 155–158
    by characters, 162–165
    manually, 156
    materials, 156, 222
    to morphospecies, 6, 157–158,
      162–165
    salt water extraction, 156
    to species, 6, 157, 165–166
    to subfamily and genus, 6, 162
  specimen shipping, 170
    materials, 170, 222
  specimen storage, 168–170, 169f
    materials, 169–170, 222
  taxonomists, 166–167
  type specimens, 167, 184

"ant gardens," 37–38, 42–43
ant niche, 13–15
  food, 14
  nest
    entrances, 13, 139
    habitats, 13–14
    temporal considerations, 14–15
Ants of the Leaf Litter (ALL)
    Conference, xviii–xix
Ants of the Leaf Litter (ALL) proto-
    col, xviii–xix, 204–206
  collection methods, xviii–xix, 204,
    205
  field work
    ecological data, 205
    labeling, 205
    transect lines, 205
  group, 8
  laboratory work
    identifying morphospecies,
      205–206
    labeling samples, 206
    sorting samples, 205
  materials, 215–216, 221–222
  sample size, xviii–xix, 204
  social insects Web site, 6, 166,
    174
  time requirements, 204, 206
  timetable, 206
arboreal species, sampling of, 142
association, measures of, 200–202
Australia, 112, 115

baiting, 126–129
  and ant composition, 126
  bait platform, 127
  and behavioral interactions, 128
  and behavioral studies, 128
  biases, 129, 188
  data output, 128
  evaluation, 128–129
  materials, 126, 127, 215
  methods, 127–128
  nutrient composition, 127, 129
  objectives, 126–127
  single sample (or snapshot studies),
    128
  temporal considerations, 126–127,
    128
Berlese funnel, 136–138
  biases, 138
  data output, 137
  evaluation, 137–138
  funnels
    composition, 136
    preparation of, 136–137, 137f
    size of, 137

  materials, 136
  methods, 136–137
biological diversity. *See also*
    diversity
  ants as indicators of, xviii, 80–88
  definition of, xvii
  measuring, xviii, 80, 192–195
biology, general ant, 3t
biomass, 3t, 4f, 18
Brazil, 211–214, 212f, 213t

castes, 10, 19, 162–163
characters, 70–71, 70f, 164–165
Chihuahuan Desert, 92, 93f
collection methods, 122–144
  bait, 126–129
  Berlese funnel, 136–138
  direct sampling, 141–142
  leaf litter collection, 133–138
  pitfall traps, 129–132, 205
  soil collection, 138, 140
  using combinations of, 123, 144,
    146, 153–154,
    153t
  Winkler sack extraction, 133–136,
    205
colony sampling, 138–139
  data output, 139
  evaluation, 139
  geographical considerations, 138
  materials, 138, 215
  methods, 138–139
  nest entrances, 138, 139
  objectives, 138
  spatial extent available, 138
  transects, dimensions of, 138
community
  definition of, 18
  diversity, 18, 20–22
  form, 18–19
  function, 18, 19–20
  types of, 27, 27f
  variability, 89–90
computer technology
  databases, xviii, 170–171
  EstimateS computer program,
    148
  social insects Web site, 6, 166,
    174
control plots
  importance of, 96
  and species richness, 93
Convention on Biological Diversity,
    xvii

data analysis, 186–203
  of abundance data

  competing models of,
    195–197, 197f
  distributions of, 192–195
  challenges to, 122
  of diversity
    comparisons of, 122, 192–195
    definition of, 192
    measures of, 192–194
  ecological attributes, 186
  list use in, 187
  matrix use in, 186–187, 201–202,
    201f
  organization for, 186–187
  of sample association patterns,
    200–202
    matrix approach for, 201–202,
    201f
    measures of, 200–201
    ordinations, 202
  of species density data, 91–92, 93f
  of species richness data, 197–200
    estimations of, 147–149,
    198–200, 199f
    and sample area, 198
    and sample size, 152–153, 198,
    200, 200f
  taxonomic attributes, 186
data documentation
  local, 125
  regional, 125
  sample, 125–126
  site, 125
data sheet
  coding for, 125
  for field use, 219–220
  for quadrat sampling, 132
deformation, of ants, 97
direct sampling, 141–142
  data output, 141
  evaluation, 141–142
  geographical considerations, 141
  materials, 141
  methods, 141
  nest series, 141
  objectives, 141
  supplemental nature of, 142
  temporal considerations, 141
dissection, 160
diversity
  alpha, 100, 106, 193
  of ants
    reasons for studying, xv–xvi,
    xviii, 2–4
    reasons not typically studied,
    1–2
  beta, 100, 121
  biological diversity

ants as indicators of, xviii,
80–88
definition of, xvii
measuring, xviii, 80, 192–195
data analysis techniques for, 122,
192–195
definition of, 18
general facts, 3t
indexes of, 97–98, 193–194,
194t
taxonomic, 87
dominance diversity indexes, 193

ecological importance of ants, 2, 3t
ecological variation, leaf litter ants
and, 120
ecosystems
inertia of, 93–94
malleability of, 94–95
oscillation of, 95–96
recovery of, 92–96, 93f
resilience of, 94
variability within, 89–90
elaiosomes, 38
environmental stress. *See also*
mortality
ants' response to, 26–27
causes of, 25–26
definition of, 25
environmental stressors. *See also*
mortality
definition of, 92
food availability, effect on ants,
26
foraging surface, effect on ants,
26
and impact on individuals, 26, 97
and impact on inertia, 93–94
and impact on malleability,
94–95
and impact on populations, 26–27,
97
low temperature, effect on ants, 12,
15, 25–26
nest site availability, effects on ants,
16–17, 26
estimators
complementarity, 201–202, 201f
EstimateS computer program,
148
incidence-based coverage estimator,
147–148, 200
jackknife methods, 147–148, 199
nonparametric, 199–200
parametric, 198–199
of species abundance, 147–148
species-accumulation curves,

148–152, 150f, 152f, 188–192,
189f, 190f, 195f
of species richness, 147–149,
198–200, 199f
extraction methods. *See* collection
methods

field notes, 124, 125–126, 219–220
field work, 122–144
approach, 123
arboreal habitats, 142
baiting, 126–129
colony counts, 138–139
cost effectiveness of, 3t
data documentation, 124, 125–
126
direct sampling, 141–142
efficacy of, 23, 143t, 144
environmental variables of,
142–144
herbaceous habitats, 142
intensive sampling, 139–141
litter extraction, 133–138
maps, 126
materials, 124, 126, 215–216
methods, 124–125
monitoring colony density, 16, 18,
97, 139
optimal survey period of, 23
pitfall trapping, 129–132, 205
problems with, 123
quadrat sampling, 132–133
food, 14, 26
foraging
and ant size, 14
and baiting, 126
fossils, 173
functional groups, 27–34, 28t, 29f. *See
also* taxonomy
climate specialists, 28–29
composition of, 31–33, 32f, 33f
cryptic species, 29
distribution of dominance, 30–31,
30f
dominant Dolichoderinae, 28,
30–31, 31t
generalized Myrmicinae, 29, 30–31,
31t
opportunists, 29
reasons for studying, 33
specialist predators, 29
subordinate Camponotini, 28

global ecology, definition of, 25

habitat. *See also* environmental stress;
environmental stressors

changes in, 89
disturbance of
ants' response to, 26–27, 90
definition of, 25
documentation of, 125
habitat groups. *See also* taxonomy
canopy ants, 21
desert ants, 13, 92, 93f
grassland ants, 13–14, 91, 92f
leaf litter ants, 14, 99–121
Neotropical ants, 90–91, 91f, 112,
115
tropical ants, 14, 18–19
hand collecting. *See* direct sampling
haplodiploidy, 10
holotypes, 168
humans, effect on environment,
xvii, 89

incidence-based coverage estimator
(ICE), 147–148, 200. *See also*
estimators
India, 210–211, 211f
indicator taxa
ants as, xviii, 81, 86–88
correlation between other taxa and,
81–86, 82t–84t
criteria of, 80–81
limitations of studying, 85–86
possible use of, 86–87
reasons for studying, 80
using multiple species of, 87–88
intensive sampling, 139–141
biases, 140–141, 188
data output, 140
evaluation, 140–141
geographical considerations,
139
materials, 139–140, 215
methods, 140
nest series and, 139
objective, 139
sample
delineation of, 140
duration of sample collection,
140
size of, 140
subsamples and, 140
interaction, 35–44
commensalism, 36, 37
with fungi, 41–43
with Homopterans, 39
mutualism, 36, 37
nests, guests in, 39–41
parasitism, 17, 18, 36, 40–41
symbiotic, 36
with animals, 39–41

interaction, symbiotic (*continued*)
  between ants, 41
  with plants, 36–38
  trophic, 36
    with animals, 38–39
    with plants, 36
International Rules, 78
International Union for the Study of
    Social Insects (IUSSI), xviii
introduced species, 43–44, 43t
  impact on ecosystems, 12–13, 20,
    22, 44
  impact on rehabilitation, 96
  tramp species, 43
IUSSI (International Union for the
    Study of Social Insects),
    xviii

jackknife methods, 147–148, 199. *See
    also* estimators

keystone species, 20

labeling, 124, 160–162, 221
leaf litter
  collection, 133–138
  population density in, 14
litter techniques, 133–138
  collection rates, 136
  geographical considerations, 106,
    120
  objectives, 133
  sacks
    assembly of, 133, 134f
    duration of litter extraction in,
      136
    size of, 133
    Winkler, 134–136, 134f
  Winkler extraction
    materials, 133, 216
    methods, 133–136, 135f, 205

macroecology, 25
Madagascar, 111–112, 115, 207–209,
    208t, 209f
Malaysia, 115, 209–210, 210f, 210t
malleability, 94–95
maps, geographical information, 126
materials, 215–217, 221–222
  aspirators, 125, 125f
  bait, 126, 127f
  Berlese funnels, 136–137
  calculator, 124
  compass, 124
  data sheets, 219–220
  ethanol solution, 124, 130, 169
  ethylene glycol solution, 130

field notebook, 126
flagging, 124
forceps, 125
global positioning system, 126
litter sifter, 133–134, 134f, 140
maps, 126
meter tapes, 124
paper tags, 124
pitfall traps, 130
plastic twirl bags, 124
PVC, 132
stakes, 124
vials, 124, 169
Winkler sacks, 134–135, 134f
mining sites, recovery of ecosystems
    and, 95–96, 95f
monitoring programs
  ALL protocol, 204–206
  for changes in diversity, 97–98
  for changes in individuals, 97
  for changes in population, 97
  difficulties of, 89–90
  establishing baselines for, 89
  and inertia, 93–94
  long-term studies, 90–92
  and malleability, 94–95
  methodology, 96–97
  and multiple simultaneous control
    sites, 92–93
  and nonlinearities, 96
  and resilience, 94
  types. *See* field work
morphospecies
  characters, 162–165
  differences between castes,
    162–163
  materials used in sorting, 157–
    158
  mounting by, 157–158
mortality. *See also* environmental
    stress; environmental stressors
  of colonies, 22
  and harvesting, 17
  and parasites, 17, 18
  and predators, 17–18
  and regulation of population, 15,
    17–18
  and weather, 17
museums, 174–175, 176t–182t,
    183–184

nests
  availability, 16–17, 26
  entrances, 13, 139
  guests in, 39–41
  habitats, 13–14
  series, 139, 141

New World, 112

observer bias, in quadrat sampling,
    132
Old World, 112
ordering, of numerical data, 186–187
ordination methods, 202

Papua New Guinea, 114
paratypes, 168
passive species inventory, 124
patchiness, 18
perturbation
  recovery from, 92–96, 93f
  studies, 96
phorid flies, 18, 22
pitfall trapping, 129–132, 205
  biases in, 131–132, 188
  data output, 131
  evaluation, 131–132
  geographical considerations, 131,
    132
  killing agents, 130, 131
  materials, 130, 215
  methods, 130–131
  objectives, 129–130
  temporal considerations, 131
  traps, 129f
    composition of, 130
    placement of, 130–131, 149
    size of, 130
polymorphism, 163, 163t
population. *See also* estimators
  definition of, 15
  density, 14, 91–92
  pitfall trapping estimation of, 130,
    131–132
  regulation of, 15–18
  Winkler sack estimation of, 136
predators
  ants as, 14, 38
  effects on ants, 17–18, 27,
    38–39

quadrat sampling, 132–133
  data output, 132
  design, 132
  evaluation, 133
  materials, 132, 215
  methods, 132
  objectives, 132
  quadrat
    composition of, 132
    data collection in, 132
    size of, 132
    temporal considerations, 132,
      133

rainfall
impact on recovery after stress, 94, 94f
impact on resilience after stress, 94, 94f
rapid assessment, 80
resources
and colony density and size, 14, 15, 16
competition for, 15–16
and dominance hierarchies, 22
manipulation of, 16–17
and regulation of population, 15–17
and removal of colonies, 16
and territoriality, 16

salt water extraction
materials, 156, 221–222
reasons for, 156
sample, representativeness of, 145–146, 187–188
sample size. *See also* estimators
determination of, xviii–xix, 153
experiment on, 147–148, 152–153
with Winkler sacks, 133–134, 152, 153t
sampling
and ALL protocol, 204–206
arboreal habitats, 142
baiting, 126–129
challenges to, 123
colony-based studies, 123
colony counts, 138–139
comparing methods of, 143t, 144
considerations in, 4, 23, 123–124, 187–188
design bias, 124, 187–188
direct, 141–142
ease of, 4–5
efficacy analysis, 143t, 144, 191–192, 192f
environmental variables of, 123, 142–144
forager-based studies, 123
geographical scale of, 157
habitats, 123, 191–192
herbaceous habitats, 142
impact of biology on, 123
intensive, 139–141
investigator bias, 124
labeling of, 124
multiple methods in, 123, 144, 146, 153–154, 153t
overview of, 4–5

passive versus active, 124
pitfall trapping, 129–132, 205
quadrat, 132–133
and random distribution, 123
rapid, 80
rate of species accumulation, 148–149, 188–192, 191f
size. *See* sample size
spatial aggregation, 123, 187
species inventory, 123
shipping, of specimens, 170, 222
SI/MAB (Smithsonian Institution/Monitoring and Assessment of Biodiversity), 8
slavemaking, 17, 41
Smithsonian Institution/Monitoring and Assessment of Biodiversity (SI/MAB), 8
species abundance
and baiting techniques, 126
class, 195–197, 196f, 197f
curve, 195–196, 196f
data analysis of, 192–197
estimators of, 147–148
factors in, 12, 195–196, 196f
and litter techniques, 133
and pitfall trapping techniques, 130, 131
and quadrat sampling techniques, 132
reasons for studying, xvii
species composition
and baiting techniques, 126, 129
and colony sampling techniques, 139
and direct sampling techniques, 141
diversity in, 21–22
fluctuations in, 90–92, 96
and intensive sampling techniques, 140
and litter techniques, 133
long-term studies of, 90–92
and pitfall trapping techniques, 130, 131
and quadrat sampling techniques, 132
reorganization of, 98
species density, data analysis techniques for, 91–92, 93f
species distribution, reason for studying, xvii–xviii
species identification. *See also* taxonomy
in the laboratory, 6, 157, 165–166

species richness
altitude and, 109–111, 109t, 110f
and baiting techniques, 126
classifying of data, 197–200
and colony sampling techniques, 139
data analysis of, 147–148, 197–200
definition of, 97–98
and direct sampling techniques, 141
diversity of, 21
estimators of, 147–149, 198–200, 199f
fluctuations in, 90–92, 95
and intensive sampling techniques, 139, 140
latitude and, 21, 109–111, 109t, 110f
and litter techniques, 152–153
measures of association, 200–202
oscillation of, 95–96
and pitfall trapping techniques, 131
and quadrat sampling techniques, 132
recovery of, 92–96
sampling methods for. *See* collection methods; sampling
specimens. *See* ant curation
standard protocols, benefits of, xvii–xix, 204. *See also* Ants of the Leaf Litter protocol
standardization, of techniques, xvii–xix, 204
supplies. *See* materials
symbioses, 36. *See also* interaction, symbiotic
facultative, 36
obligate, 36

Tasmania, 82, 83–84t
taxonomic diversity. *See* diversity
taxonomic groups, correlations between, 85–87
taxonomists, partnerships between, 6–7, 166–167
taxonomy, 45–79. *See also* functional groups; habitat groups; indicator taxa
of Aneuretinae, Dolichoderinae, and Formicinae, 72–73
of Apomyrmini, 74
challenges of, 45, 172–173

taxonomy (*continued*)
  characters, 70–71, 70f
    alitrunk (mesosoma), 71,
      164–165
    gaster, 71, 165
    glands, 70
    head, 164
    petiole and postpetiole, 70–71,
      165
  collections, 174–175, 176t–182t,
    183–184
  electronic resources, xviii, 174

genera, 45, 46t–69t, 70, 173
  of Myrmeciinae,
    Nothomyrmeciinae, and
    Prionomyrmecini, 73
  of Myrmicinae, 73–74
  problems with, 74, 77–79, 173
  of Pseudomyrmecinae, 73
  published sources, 45, 77, 173–
    174
  revisionary, 74–77, 121
    examples of, 75–77
  species, 173

subfamilies, 173
temperature, effects of, 11, 12, 15,
  25–26
terrestrial habitat, descriptions of,
  13–14
time-constrained search, 80
tramp species, 43, 43t
traps, pitfall, 129–132, 205

Winkler
  extraction, 133–136, 135f, 205
  sack, 134–135, 134f